Observing the Moon

The Modern Astronomer's Guide, Second Edition

Written by an experienced and well-known lunar observer, this is a hands-on primer for the aspiring observer of the Moon. Whether you are a novice or are already experienced in practical astronomy, you will find plenty in this book to help you raise your game to the next level and beyond. In this thoroughly updated Second Edition, the author provides extensive practical advice and sophisticated background knowledge of the Moon and of lunar observation. It incorporates the latest developments in lunar imaging techniques, including digital photography, CCD imaging, and webcam observing, and essential advice on collimating all common types of telescope.

Learn what scientists have discovered about our Moon, and what mysteries remain still to be solved. Find out how you can take part in the efforts to solve these mysteries, as well as enjoying the Moon's spectacular magnificence for yourself!

GERALD NORTH graduated in physics and astronomy. A former teacher and college lecturer, he was also a Guest Observer of the Royal Greenwich Observatory. He is now a freelance astronomer and author. He is a long-term member of the British Astronomical Association, and has served in several posts in their Lunar Section. His other observing guides include the acclaimed *Advanced Amateur Astronomy* (Second Edition, Cambridge University Press, 1997) and *Observing Variable Stars, Novae and Supernovae* (with Nick James, Cambridge University Press, 2004).

Observing the Moon

The modern astronomer's guide

SECOND EDITION

GERALD NORTH BSc

CAMBRIDGE
UNIVERSITY PRESS

CAMBRIDGE UNIVERSITY PRESS
Cambridge, New York, Melbourne, Madrid, Cape Town, Singapore, São Paulo

Cambridge University Press
The Edinburgh Building, Cambridge CB2 8RU, UK

Published in the United States of America by Cambridge University Press, New York

www.cambridge.org
Information on this title: www.cambridge.org/9780521874076

First Edition published 2000
Reprinted 2000, 2001, 2002
Second Edition published 2007

Printed in the United Kingdom at the University Press, Cambridge

A catalogue record for this publication is available from the British Library

ISBN 978-0-521-87407-6 hardback

CONTENTS

PREFACE

PREFACE TO THE FIRST EDITION

Interest in the Moon periodically ebbs and flows, like the tides it causes in our oceans. The years leading up to the *Apollo* manned landings marked a particularly high tide. Since then there has been a very deep low tide – but the tide is turning once again. Recently we have had the *Clementine* and *Lunar Prospector* probes and professional studies of the Moon are on the increase. It is not unreasonable to expect that within the next two or three decades people will once again be walking on the eerie lunar surface. When it does happen we will be back to stay this time.

We already know a great deal about our Moon but many mysteries remain. A few of these mysteries might be solved by the modern-day backyard observer. Nonetheless, there are many other motives for the amateur devoting time and energy to study the Moon, or any of the other celestial bodies, through his/her telescope, aside from any wish to do cutting-edge science. I will not waste space listing the other possible motives here. All that really matters is that you, the reader of this book, have an interest in the Moon which you wish to explore. If so, then this is the book for you!

I intend this book to be a 'primer', a guide for the interested amateur astronomer who is yet to become a lunar specialist. Of course I have provided details about practical matters, such as equipment and techniques, but I have also included a limited amount of the history of the study of the Moon and, particularly, of lunar science. Without the science (and to a less important extent, the history) the subject would be sterile and any practical work beyond simple sight-seeing would be pointless.

To 'shoehorn' everything I needed to say into the book-length available has not been easy. The facts of commercial life apply to books as to any other commodity. This book is highly illustrated and was expensive to produce because of this. To keep the cost to you from becoming astronomical in every sense of the word, I have had to keep its length to within very tight limits set by the publisher. Consequently, time and time again I have had to refer you, the reader, to other publications to expand on points that I had not room enough to adequately cover in this book.

However, that shortcoming is also a strength. As I said, this book is a 'primer'. It is certainly not intended to be the definitive history of lunar studies, nor of our scientific understanding of the Moon. I can't really say that it is the last word on practical techniques and equipment for the practising amateur astronomer, either. What I can claim for this book is that it contains enough working knowledge to give any tyro lunar observer a flying start. Beyond that, this book is intended to be a 'springboard' to further studies and practical work. Please do follow up the references I give. Go beyond that and seek further ones on your own. Your knowledge of the Moon and how it has been studied will expand beyond any limits set by the finite size of any one single-volume work.

I hope you like this book and find it interesting. Much more importantly, I hope that you discover for yourself the thrills of examining the Moon's mountains, craters and other surface structures through your telescope's eyepiece. Aside from the awesome spectacle of the views, you will find real fascination in understanding how the Moon got to be as it is.

Gerald North
Bexhill on Sea

PREFACE TO THE SECOND EDITION

The new level of interest in the Moon that I noted in the Preface to the First Edition has been maintained in the years since. Meanwhile much has changed in the arena of practical amateur astronomy. New equipment and techniques have allowed amateurs to make significant advances in the quality of their work and some of the older ways of doing things have fallen by the wayside. The First Edition of this book proved to be popular and it was reprinted a number of times. However things have changed so much since that First Edition was first published it is now time for this new one. Consequently I have re-written much of this book to reflect the amateur astronomer's world of the early twenty-first century. I hope you enjoy reading it – and I hope that you will obtain whatever telescopic equipment you can and turn it to the Moon. Things certainly have moved on in practical astronomy but the Moon remains as beautiful, as thrilling, and as mysterious as ever.

Gerald North
Norfolk

ACKNOWLEDGEMENTS

I am very grateful to the following people for allowing me to reproduce examples of their work in this book: Terry Platt, Gordon Rogers, Tony Pacey, Nigel Longshaw, Andrew Johnson, Roy Bridge, Commander Henry Hatfield, John Gionis, Michael Butcher, Martin Mobberley and Damian Peach. Special thanks are also due to Dr T W Rackham (who sadly has died since the first edition was published) and Manchester University, England, also to Ewen. A. Whitaker and the Lunar and Planetary Laboratory, University of Arizona, USA, and the National Aeronautics and Space Administration (NASA), for allowing me to reproduce many of their excellent photographs. Full acknowledgements are given in the captions accompanying the illustrations within this book.

In addition, Mr John Hill had, for the First Edition, gone to considerable trouble to furnish me with materials and it is very sad for me now to have to record his death along with my thanks. For this edition I have also been given a tremendous amount of help by my many friends of the Breckland Astronomical Society, especially John Gionis, Michael Butcher and Malcolm Dent. In particular Michael Butcher has spent many hours building a photographic-based key map (Figure 7.1), that replaces my hand-drawn version in the First Edition.

Finally I must not forget to thank Dr Simon Mitton and his staff at Cambridge University Press for all their hard work in making the First Edition the success it was. Now I have to thank Vince Higgs along with his staff at the Press again for their sterling work on this Second Edition.

Gerald North
Norfolk

CHAPTER 1

"Magnificent desolation"

No, not a still from a science-fiction movie but a real (*Apollo 17*) astronaut by the "Station 6 Boulder" on the North Massif of the Moon's Taurus–Littrow Valley! The South Massif can be seen on the far side of the valley. The *Apollo 17* mission in December 1972 was the last expedition to the Moon's airless surface. (NASA photograph.)

Feverishly excited, I sat cross-legged in front of the family television set and watched the fuzzy, indistinct, shapes of Neil Armstrong and Buzz Aldrin moving about amid the grey wash that was the surface of the Moon. The fact that the picture was of poor quality because it had been beamed back to Earth through a quarter of a million miles of space did little to dampen my enthusiasm. I could also make out part of the spidery form of their space vehicle extending from the grey wash into the black stripe that represented the airless sky over the Moon. The sound quality was also poor. The words of those first men on the Moon sounded crackly and wheezy and so were often difficult to decipher, so I listened hard. I was a young boy at the time but my sense of the significance of what I was witnessing was intense. I heard Neil Armstrong's words before he stepped onto the lunar soil. I heard Buzz Aldrin describe the scenery around him as "magnificent desolation". I wished I was there with them to see it.

I was born just after the beginning of what used to be called 'the Space Age'. As far back as I can remember I have been interested in things scientific and technical and have been infected with a particular passion for matters astronomical. I avidly read books about science and astronomy. By the time of that first Moon landing I had acquired an old pair of binoculars and had been bought a very small terrestrial telescope. Whenever I was allowed to go outside after dark I turned these humble instruments towards the Moon and gazed at the dark patches and the craters that they imperfectly revealed. The proper astronomical telescope I yearned for was at that time beyond my means.

Those who were around at the time will remember the feverish excitement and air of expectation that gradually built up through the 1960s as the world's space agencies rapidly made the advances towards that first manned Moon landing. As well as a huge variety of merchandising such as books, booklets, posters, and kits to make plastic models of various rockets, television companies enthusiastically broadcast news items and informative programmes about the 'space race'. Our television screens were also awash with many science-fiction shows – *Doctor Who* and *Space Patrol* (called *Planet Patrol* in the USA) being particular favourites of mine – that featured space travel to other worlds. The fantasy shows reflected the public's yearning for real astronauts to travel through space and walk on real alien worlds. I very much shared that yearning.

The next few years brought further advances and more space missions. The pictures and sound got clearer. The Christmas of 1970 was significant for me in that my parents bought me a 'proper' astronomical telescope. It was a 3-inch (76 mm) Newtonian reflector. Yes, it was still smaller than the size of instrument recommended for useful work but I shall never forget the thrill of turning it to the Moon for the first time and seeing the large iron-grey lunar 'seas' and the rugged mountain ranges and magnificent craters come into sharp focus.

A few years were to pass before I was able to graduate to more powerful telescopes. I was to spend many hours 'learning my craft' at the eyepiece of that first 'proper' one. I didn't know it then but observing the Moon through telescopes was to become an important part of my life. After graduating in astronomy and physics, I was even to spend several years as a Guest Observer of the Royal Greenwich Observatory and so get to use professional telescopes to carry out, amongst other projects, lunar research.

I must have spent several thousands of hours of telescope time observing the Moon. You might have thought that I would be tired of it by now. Absolutely not! I hope to show you why not in the pages of this book. I hope that you, like me, will be thrilled anew every time you view the spectacle of our neighbouring world's "magnificent desolation".

1.1   AN ORBITING ROCK-BALL

Even today there are people (amazingly, even some in our western society) who are unaware of the Moon's true nature and status. I should hope that this does not apply to any of the readers of this book. However please let me, for completeness if for no other reason, state some of the basic facts. The Moon is a solid, rocky, body with an equatorial diameter of 3476 km. It orbits the Earth at a mean distance of 384 000 km. Though often appearing brilliant in the night sky, the Moon does not emit any light of its own generation. It shines mainly because of reflected sunlight, with a very small contribution from fluorescence caused by the re-radiation at visible wavelengths of invisible short-wave solar radiations and absorbed kinetic energy from solar-wind bombardment.

The Moon's diameter is over a quarter of that of the Earth (12 756 km for comparison) and this has led many to consider the Earth and Moon as

Figure 1.1 The positions of the barycentre for bodies of differing masses. The distances of each of the barycentres from the bodies are in the inverse ratios of their masses in each of these cases.

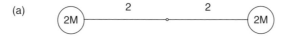

a double-planet system, rather than as a true parent planet (the Earth) and attendant satellite (the Moon). Certainly the statement that 'the Moon orbits the Earth' is an approximate one. In truth both orbit their *barycentre*, or common centre of mass. For two co-orbiting bodies of equal mass the centre of mass of the system lies exactly half way between their centres – see Figure 1.1(a). In the case of one body being more massive than the other, the barycentre is still in mid-space but is shifted towards the more massive body. In fact the ratio of the distances from each body to the barycentre is in the inverse ratio of their masses. This point is illustrated by Figure 1.1(b) and 1.1(c). In the case of the Earth and Moon, the Moon's mass is 7.35×10^{22} kg and that of the Earth is 5.98×10^{24} kg (81 times more massive). This results in the ratio of the distances from the centre of the Moon to the barycentre and from the centre of the Earth to the barycentre being 81:1. Put another way, the barycentre lies 1/82 of the way along a line joining the centre of the Earth to the centre of the Moon; 1/82 of 384 000 km is a little under 4700 km and so the barycentre lies inside the Earth's globe. The Earth may 'wobble' as the Moon orbits but the statement about the Moon orbiting the Earth is approximately true and I, at least, think that this fact qualifies the Moon as the Earth's satellite rather than them both being regarded as a double planet.

1.2 PHASES AND ECLIPSES

Nowadays most people are aware that the Sun acts as the central hub of our Solar System and that the planets orbit at various distances from it.

I have detailed elsewhere the story of how the ancients came to realise this (*Astronomy In Depth*, published by Springer–Verlag in 2003) but suffice it to say here that the researches of Copernicus and Galileo in the sixteenth and early seventeenth centuries were pivotal. Of course, one body was not displaced from its situation of orbiting the Earth, as the ancients had mistakenly believed was the case for all the other bodies of the Solar System: the Moon.

The Moon's *sidereal period*, the time it takes to complete one circuit of the Earth, is 27.3 days. At the beginning of the seventeenth century Johannes Kepler had determined that the orbits of the planets about the Sun were elliptical, rather than being circular in form as had been thought by Copernicus. The Moon's orbit is also elliptical. At the point of closest approach, *perigee*, the Moon's distance is 356 410 km. This increases to 406 679 km at *apogee*.

Figure 1.2 provides the usual elementary explanation of how the Moon's phases are produced over a complete cycle, or *lunation*. What the diagram does not reveal is why it is that the length of the cycle is not 27.3 days, the same as the sidereal period. The reason is that while the Moon is making its circuit of the Earth, the Earth itself is moving along its own orbit around the Sun. Hence the direction of the sunlight changes a little with time, instead of being fixed as implied in the diagram. Consequently, the Moon has to go a little further than one circuit round the Earth to go from one new Moon to the next. So, the length of a lunation, or *synodic period* is 29.5 days.

As well as the phases, *earthshine*, sometimes called 'the old Moon in the New Moon's arms' is another commonly recognised phenomenon. Figure 1.3 shows it well. Most obvious to the naked eye when the Moon is little more than a thin crescent but seen more often with optical aid, this is caused by reflected sunlight from the Earth shining on the Earth-facing part of the Moon experiencing night. Leonardo da Vinci is credited as being first to explain this effect correctly. In part, the earthshine is easiest to see when the Moon's crescent is thin because there is not so much glare from the sunlit portion. Also, when the Moon appears as a crescent from the Earth, the Earth appears gibbous from the surface of the Moon. One could say that the apparent phase of the Earth as seen from the Moon is the opposite of that of the Moon seen from the Earth. So, when the Moon's crescent is thin the amount of reflected light from the Earth shining on the Moon is nearly at its maximum. Apart from the foregoing, the apparent brightness of the earthshine also depends on the amount of cloud cover in the Earth's atmosphere (as seen from the surface of the Moon, the Earth would appear at its most brilliant when largely covered in highly reflective clouds). Finally, the observing conditions local to the observer also have an

Figure 1.2 The phases of the Moon. The upper section of the diagram illustrates the Moon in various positions in its orbit, while the corresponding phases that we see from the surface of the Earth are shown in the lower section.

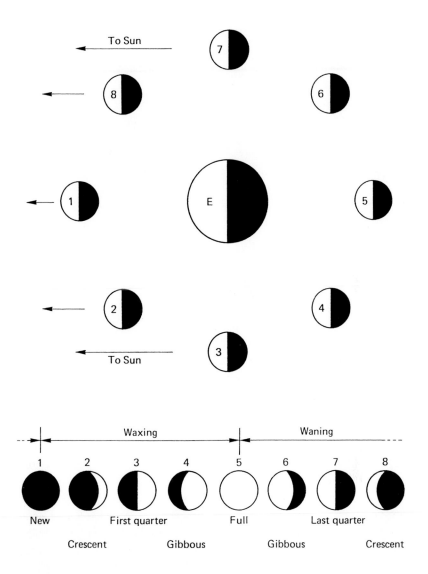

important bearing. Poor transparency and haze both inhibit the visibility of earthshine, just as one would expect.

Another inaccuracy in Figure 1.2 is that it does not represent the true three-dimensional relationship between the Earth, the Moon and the Sun. Realising that the Earth casts a huge cone-shaped shadow into space, one might imagine that every full Moon our satellite ought to pass into this shadow cone (see Figure 1.4). Of course such, *lunar eclipses* do occur but certainly not at the time of every full Moon. Neither do *solar eclipses* occur at every new Moon (Figure 1.5), even though the diagram might suggest that the Moon should pass exactly between the Sun and

Figure 1.3 Earthshine. (a) Photographed by the author with an ordinary camera fitted with a 58 mm f/2 lens on 3M Colourslide 1000 film.

the Earth at these times. What the diagram does not show is that the plane of the Moon's orbit about the Earth is inclined slightly (actually by about 5°) to the plane of the Earth's orbit about the Sun.

A useful concept in astronomy is that of the *celestial sphere*. In this the sky that surrounds the Earth is represented as the inner surface of a

Figure 1.3 (*cont.*)
(b) A close-up,
photographed by Tony
Pacey on 1993 March 26[d]
19[h] 35[m] UT, using his
305 mm f/5.4 Newtonian
reflector. The sunlit
portion of the Moon is
heavily overexposed in
this 12 second exposure
on *Ilford FP4* film.

(b)

sphere, the Earth itself being a tiny dot at the centre of the sphere. All the
stars, celestial bodies and the paths along which any of the celestial
bodies appear to move can be shown as projections onto this imaginary
sphere. Figure 1.6 shows such a celestial sphere on which is projected the
monthly orbit of the Moon. Also shown is the yearly apparent path of the
Sun across the sky, which results from our orbit around the Sun (in effect
the Sun appears to move once around the sky, through the constellations

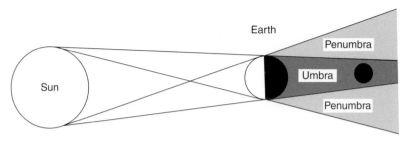

Figure 1.4 Lunar
eclipses. With the Moon
(black disk) in the
position shown, a total
lunar eclipse would be
the result. This diagram
is grossly out of scale for
the sake of clarity.

of the Zodiac, taking one year to complete one circuit). The Sun's annual
path across the sky is known as the *ecliptic*.

The different inclinations of the Moon and Earth's orbital planes are
reflected in the inclinations of the ecliptic and the Moon's path on the
celestial sphere. Note how the Moon's path and the ecliptic cross at two
diametrically opposite points on the celestial sphere. Where the Moon
crosses the ecliptic going from north to south it is said to be at its
descending node. Crossing south to north, it is then at its *ascending node*.

Notice how the only times the Moon and the Sun can appear exactly
together in the sky (put another way, both appearing to be in the same
direction as seen from Earth) are when both are at either the ascending
node, or the descending node, at the same instant. Remembering that the

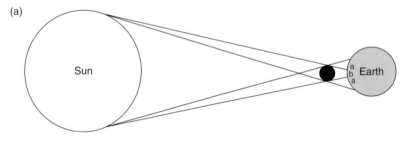

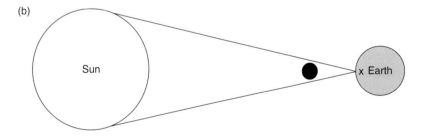

Figure 1.5 Solar eclipses.
(a) An observer stationed
at **b** would see a total
solar eclipse, while
someone in the regions
shown as **a** would see a
partial eclipse. (b) An
observer at position **x**
would see an annular
eclipse. The diagrams are
grossly out of scale for
the sake of clarity.

Figure 1.6 The orbit of
the Moon projected onto
the celestial sphere.

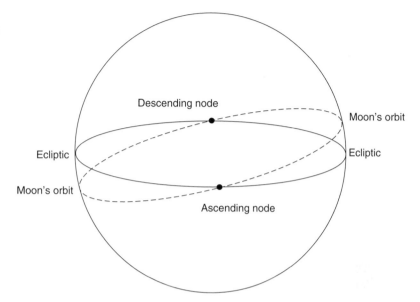

condition for eclipses to occur is that the Earth, Sun and the Moon must simultaneously lie along the same straight line at the time of full Moon (for a lunar eclipse) or new Moon (for a solar eclipse), it is not hard to see why eclipses are relatively rare. For the vast majority of lunations new Moons occur with the Moon appearing just a little north or just a little south of the Sun in the sky. Similarly, the Moon manages to miss the Earth's shadow cone, passing either north or south of it, at the time of most full Moons.

The situation shown in Figure 1.4, very much out of scale for the sake of clarity, is that for a *total lunar eclipse*, where the Earth passes through the full shadow, or *umbra*. First the Moon enters the partial shadow, or *penumbra*. The dimming of the full Moon is only very slight at that time. As the Moon enters the umbra so a 'bite' begins to appear and the direct sunlight is progressively cut off. For a typical total lunar eclipse it will take about an hour for the Earth's shadow to completely sweep across the Moon's surface (see Figure 1.7). Then all the direct sunlight will be cut off. The only light reaching the surface of the Moon then is that refracted and scattered by the Earth's atmosphere. Usually the Moon then looks very strange, bathed as it then is by a copper-coloured glow. For an eclipse of maximum duration, totality lasts about an hour and then the umbral shadow leaves the Moon over the course of another hour or so.

How much dimming there is, and the precise colourations seen, vary from eclipse to eclipse (and can even vary during the course of an eclipse). Also, the size of the Earth's umbral shadow can vary a little from eclipse

Figure 1.7 The lunar eclipse of 1996 April 3^d photographed by Martin Mobberley, using his 360 mm reflector (at the f/5 Newtonian focus) on *Fuji* Reala film. (a) 1/1000 second exposure at 22^h 25^m UT. (b) 1/250 second exposure at 23^h 00^m UT. (c) 3 second exposure at 23^h 20^m UT.

to eclipse, so altering the precise timings and the durations of the eclipses. There is no mystery about these variations. They reflect the state of the Earth's atmosphere at the time of each of the eclipses.

Actually, it might be that for a particular eclipse the Moon is not particularly close to its orbital node and may, as a result, only partially enter the umbral shadow. In that case a *partial lunar eclipse* results. If the Moon misses the umbra altogether, the result is then termed a *penumbral eclipse*, though most casual observers will be hard-pressed to spot the very slight dimming that results. On average, about two lunar eclipses are visible each year from somewhere on the Earth's surface.

The darkness of a lunar eclipse can be rated using the *Danjon scale*. A Danjon 0 eclipse is the darkest. At mid-totality the Moon is almost invisible. A Danjon 1 eclipse is very dark, with a deep-brown or grey umbra, and surface details on the Moon are difficult to make out. A Danjon 2 eclipse is usually deep red, or reddish brown in colour, though near the edge of the umbra the Moon can look bright orange. A Danjon 3 eclipse is brighter still, though the umbra still looks coppery red and its edge is often coloured bright yellow. A Danjon 4 eclipse is the brightest, with the Moon looking bright orange or even yellow at mid-totality.

1.3   SOLAR ECLIPSES

Perhaps I should emphasise that Figure 1.5, which illustrates how solar eclipses are formed, is also grossly out of scale for the sake of clarity. It has always struck me as a remarkable coincidence that the Sun and the Moon both appear to be virtually the same apparent size as viewed from the surface of the Earth. This is approximately ½° – roughly equivalent to a span of a centimetre as seen from a distance of one metre. It just so happens that the ratio of the actual diameter of the Sun to its distance from us is almost equal to that of the diameter of the Moon to its distance from us. As Figure 1.5 illustrates, a *total solar eclipse* can only been seen from a restricted region on the Earth's surface at any given moment. In fact, owing to the Earth's rotation and the relative motions of the Earth and Moon (and their relation to the Sun), this small region sweeps across the globe. A narrow track is generated across the surface of the Earth within which the eclipse can appear as total. All other regions will see, at best, a *partial solar eclipse*.

The maximum duration of totality, as seen from any particular location, is about eight minutes and it varies from eclipse to eclipse. The reason for the variation lies in the fact that the Earth's orbit about the Sun is slightly elliptical, as is the Moon's orbit around the Earth. Totality will last the longest when an eclipse occurs at a time when the Earth is at its greatest distance from the Sun, or *aphelion*, and the Moon is at perigee. In the converse situation, with the Earth at *perihelion* and the Moon at apogee, the Moon's apparent size is actually slightly smaller than that of the Sun. At maximum eclipse the Sun's disk will not be completely hidden by the Moon and a thin ring of sunlight will surround the dark disk of the Moon. This is an annular eclipse and is illustrated in Figure 1.5(b).

A total solar eclipse is a spectacular thing to see. Over the course of about an hour a larger and larger 'bite' is taken out of the Sun as the (invisible against the daytime sky) disk of the Moon passes over it. Then the last sliver of solar photosphere disappears from sight. The sky rapidly darkens and the Sun's pearly *corona* comes into view. Sometimes *solar prominences* can be seen over the edge of the Moon. After just a few minutes the first chink of sunlight peeks once again from behind the Moon and the sky rapidly brightens and the Moon slowly withdraws and the eclipse becomes a cherished memory for those who witnessed it.

As the Moon moves around the Earth, the Earth–Moon system moves around the Sun. Every so often the Earth, Sun and Moon regain very similar positions relative to one another. This happens every 6585 days (a little over 18 years) and this period has been given the special name of the *Saros*. Ancients found the Saros useful in predicting lunar eclipses. A lunar eclipse happening on a particular day will be followed by one 6585 days later. Of course, that is not to say that other lunar eclipses won't happen in-between these times – they will, but each lunar eclipse will be 'paired' with one happening one Saros period later. The Saros is rather less useful in predicting solar eclipses because it is not quite accurate enough.

1.4   GRAVITY AND TIDES

An oft-repeated fable is that Isaac Newton was sitting in his garden one day and chanced to see an apple fall from a tree. Newton's genius was such that he realised that the same force that operated to make the apple fall to the ground was responsible for keeping the Moon in orbit around the Earth. He also reasoned that it was quite likely that the same type of force operates between the planets and the Sun, keeping the Earth and the other planets in their orbits around our parent star. Whether or not it really was the falling apple that gave him his inspiration, Newton explored his ideas mathematically and he published his results in his masterly work, the *Principia*, in 1687.

Newton formulated a 'law' which he thought would be true for anywhere in the observable Universe:

Any two bodies will attract each other with a force which is proportional to the product of the masses and is inversely proportional to the square of the distances separating them.

The law can be expressed in equation form:

$$F \propto Mm/r^2 \tag{1.1}$$

or

$$F = GMm/r^2 \tag{1.2}$$

where F is the mutual attractive force, measured in newtons, M, m are the masses of the attractive masses, measured in kilograms, r is the distance of separation, measured in metres, and G is a constant of proportionality, usually known as the *universal constant of gravitation*.

Historically, getting a precise value for G was not an easy thing to do but by modern times a reliable figure has been arrived at by means of sophisticated laboratory experiments. Its value is $6.67 \times 10^{-11}\,\mathrm{Nm^2\,kg^{-2}}$. Knowing the masses of the Earth and the Moon, one can use the equation to work out the size of the attractive force between them. It amounts to a colossal $2 \times 10^{20}\,\mathrm{N}$. As far as the Earth is concerned, most of this force acts on the solid part of the body, but a fraction of it acts on the Earth's fluid covering and so contributes to the generation of the ocean tides.

The pull of the Moon causes a bulging of the oceans in the direction of the Moon. In effect, the Earth's waters are 'heaped up' because of the

Figure 1.8 The main tide generating process.

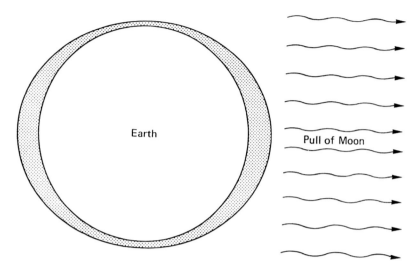

Earth

Pull of Moon

attraction of the Moon. In addition, the Earth is also 'pulled away' from the water on the reverse side, so leaving a bulge of water on the opposite side of the Earth, as shown in Figure 1.8. As the Earth turns on its axis so each position on the Earth experiences two tides per day.

The Sun also contributes its own effect. Though the Sun is very much more massive than the Moon, it is very much further away and so the Sun's tidal force has only about half the magnitude of that of the Moon. Around the times of new Moon and full Moon, the tidal forces act along virtually the same straight line and so at these times the tidal amplitude is greatest, the sea levels rising and falling by the maximum amount. The situation is illustrated in Figure 1.9 and the tides at these times are known as *spring tides*. Near the times of first and last quarter Moon the Sun and Moon's tidal pulls are almost at right angles and so the resultant tides have their minimum amplitudes (see Figure 1.10). These are *neap tides*.

Local topographic features will have their effects on the tides that result at any given location (the situation is often quite complicated in bays and river estuaries, for instance) but the foregoing describes the situation on the global scale.

1.5   MORE ABOUT THE MOTIONS OF THE MOON – LIBRATION

That the Moon always keeps the same face presented to the Earth is obvious even to the casual observer and was well known to the ancients. The explanation for this is both obvious and yet fundamental: the Moon rotates on its axis with the same period that it takes to orbit the Earth. We say that the Moon has a *captured*, or *synchronous* rotation.

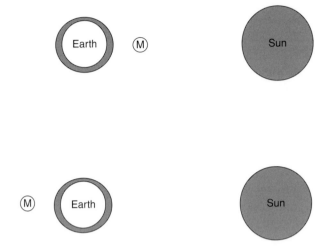

Figure 1.9 Spring tides are formed when the pulls of the Moon and Sun are aligned (even if pulling in opposite directions).

However, the careful observer who is armed with some optical aid will notice that the Moon's topographic features do not quite remain exactly stationary on the visible disk over a lunation. In fact, the Moon appears to slightly nod up and down and rock to and fro over each lunar cycle. Moreover, the nodding and rocking differ slightly from one lunation to the next. This effect is termed *libration*.

If it wasn't for libration we could have mapped only 50 per cent of the Moon's surface before the advent of the space age. We were actually able to map 59 per cent of the Moon, using observations made over a series of years. Three separate effects operate to create libration: *libration in longitude*, *libration in latitude*, and *diurnal libration*.

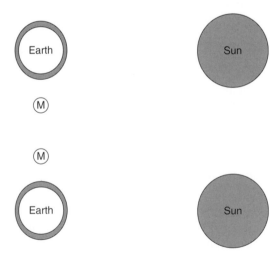

Figure 1.10 Neap tides are formed when the pulls of the Moon and the Sun are at right angles to each other.

Figure 1.11 Libration in longitude. The Moon turns evenly on its axis but the Moon's speed varies around its elliptical orbit. Consequently the two motions are out of step, although the total time taken for one rotation is the same as the time taken for one complete orbit. The result is that the Moon appears (as seen from the Earth) to swivel back and forth in an east–west direction over the course of one lunation.

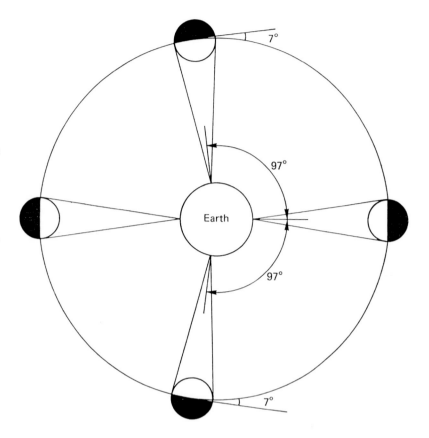

Libration in longitude arises because of the elliptical shape of the Moon's orbit and the fact that its speed changes with its distance from the Earth. When the Moon is close to perigee it moves a little faster than when it is at apogee, the speed changing gradually from one situation to the other. However, the rotation rate of the Moon on its axis remains constant. The result is an apparent 7° east–west rotational oscillation of the Moon's globe during the course of a lunation. This effect is illustrated in Figure 1.11.

The Moon's spin axis is not quite perpendicular to the plane of its orbit. In fact it is canted over at 1½° (by comparison, the inclination of the Earth's rotation axis to the perpendicular to the Earth's own orbital plane is 23½°). Added to this is the already mentioned 5° inclination of the Moon's orbit to the ecliptic (remembering that the ecliptic is, in effect, the projection of the Earth's orbital plane onto the celestial sphere). Taken together, these inclinations mean that we can see, alternately, up to 6½° beyond one pole, then the other (see Figure 1.12). This is libration in latitude.

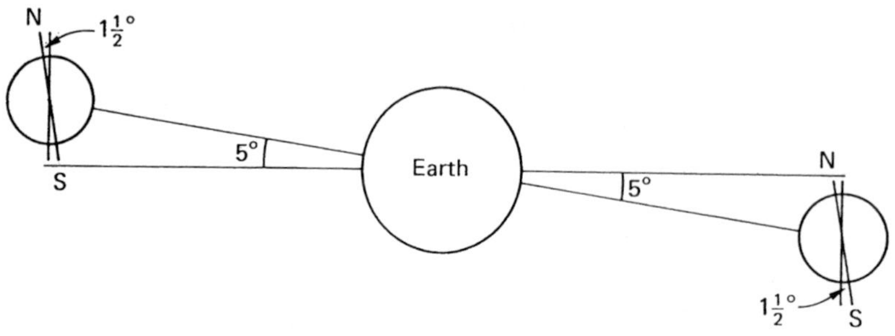

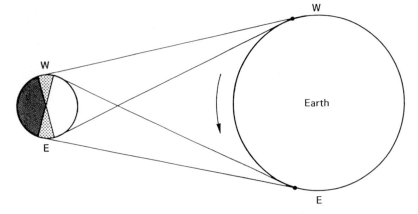

Figure 1.12 Libration in latitude.

Figure 1.13 Diurnal libration.

Figure 1.13 shows how diurnal libration arises. As the Earth rotates, so an observer's viewpoint changes slightly with respect to the Moon. An Earth-based observer watching the Moon rising above the horizon will be able to see a little way further around one limb of the Moon, and then a little further round the other limb when the Moon is setting.

As you might imagine, the way these separate librations combine is complicated, and is made even more so by the fact that the Earth's and the Moon's orbit *precess* (the positions of the nodes shift with time). Consequently, librations differ with each lunation. Figure 1.14 shows well the effect of libration.

1.6   CO-ORDINATES ON THE SURFACE OF THE MOON

Compare a pre-1960s map of the Moon with a modern one and you will notice that east and west are marked on it the opposite way round. On the classical scheme the Lunar 'sea' (dark area) known as the Mare Crisium was situated on the western side. This side of the Moon's face is the east on modern maps. The modern scheme is due to the International Astronomical Union (IAU) and is now the accepted standard.

Figure 1.14 The effects of libration are illustrated well by these photographs taken by Commander Henry Hatfield, using his 12-inch (305 mm) Newtonian reflector: (a) was taken on 1966 May 29^d 21^h 03^m UT; (b) was taken on 1966 November 22^d 18^h 14^m UT. In both (a) and (b) the values of the libration in latitude are close to their most extreme possible, though all three types of libration may be variously prominent at any given time in combination.

Latitudes and longitudes can be assigned to positions on the Moon's globe in the same way that they can on the Earth. Co-ordinates that refer to the surface of the Moon are known as *selenographic*. Of course, libration affects the precise apparent positions of features on the lunar surface but a co-ordinate system has been derived that refers to the mean apparent positions – those that would correspond to zero libration.

The mean centre of the Moon's disk corresponds to a *selenographic latitude* of 0° and a *selenographic longitude* also of 0°. Selenographic latitude is positive going northwards and negative going southwards, being +90° and −90° at the lunar north and south poles, respectively. Selenographic longitude increases eastwards (towards the Mare Crisium) and is 90° at the mean east limb. It further increases (on the part of the Moon turned away from the Earth) to 180° at the mean position antipodal to the Earth and round to 270° at the mean west limb. Now on the Earth-facing side again, the selenographic longitude increases further to 360° (equivalent to 0°) at the mean centre of the disk.

Figure 1.15 shows an outline map, illustrating the modern co-ordinate system. Notice that I have orientated it with south uppermost. This is to make it uniform with the maps and illustrations throughout the book and is because this book is intended to be of use to the practical observer. Most readers of this book will live in the Earth's northern hemisphere and will see the Moon inverted through a normal astronomical telescope (without the use of additional optical elements, such as a star diagonal), that is with this same orientation.

Since the time of publication of the first edition of this book, popular astronomy magazines have increasingly taken to publishing all photographs of the Moon and planets with north uppermost. It is certainly true that observers with modern telescopes (particularly the Schmidt–Cassegrain and Maksutov types now very widely used) often use them with star-diagonals so the old south-up orientation often no longer applies. However in these cases the image is also subjected to a mirror-image reversal (in optics this is properly known as a *lateral perversion*), in which case a mere rotation of the image can never make it match a conventional image or photograph. So, I considered it best in this edition to stick with the south-uppermost orientation that will be commonly encountered in refracting telescopes or reflecting telescopes with an even number of mirrors in the light path.

Just as on the Earth, the lines that pass through both poles and the equator (so forming *great circles* on the surface of the Moon) are known as *meridians*. These are lines of equal longitude. The lines which run parallel to the equator (so forming *small circles* over the surface of the Moon – only the equator is a great circle) are lines of equal latitude.

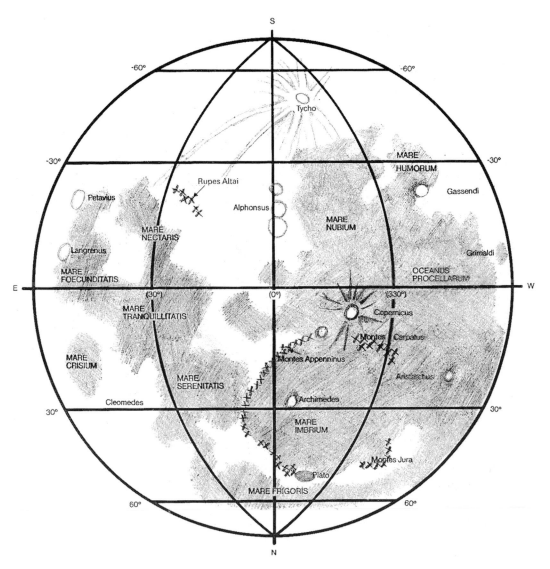

Figure 1.15 Outline map of the Moon, illustrating the modern system of co-ordinates as standardised by the International Astronomical Union.

One can go on to define the co-ordinates of the *terminator*, the boundary between the sunlit and dark portions of the Moon that shifts as the cycle of lunar phases progresses. The *Sun's selenographic colongitude* is the selenographic longitude of the morning terminator on the Moon. Its value is 270° when at new Moon, 0° at first quarter, 90° at full Moon and 180° at last quarter. In ephemerides it is often reckoned with respect to the mean centre of the Moon's disk and so libration can have an effect on the true position of the terminator on the Moon's surface. For instance, comparing a map of the Moon with the ephemeris value of the Sun's selenographic colongitude might suggest that the terminator should run through the middle of a particular feature at a particular time. When you go to the telescope at that time you might find, instead, that libration has carried the feature rather further into the sunlit portion, or alternately has hidden it entirely in the Moon's dark region!

1.7 OCCULTATIONS

As the Moon sweeps around the Earth in its monthly orbit it may pass in front of the planets and stars far beyond. When the Moon hides a celestial body from our sight we say that it *occults* that body. A solar eclipse is an *occultation* of the Sun. Of course occultations of stars are much more frequent than solar eclipses.

Though an occultation is usually quite a simple affair, it is really quite fascinating to watch the edge of the Moon very slowly approach a star until the star suddenly vanishes from sight. Reappearances are also interesting, the once hidden star suddenly snapping into view. Of course, one would normally have to be armed with a prediction that a particular star was going to emerge at that point and time to be able to catch it happening.

The timings of stellar occultations used to be a valuable pursuit because it allowed us to derive knowledge of the Moon's orbit and its surface profile, as well as precise star positions, amongst many other things. In modern times most of these objectives have been better met by other means. However, the long-term nature of occultation-timing data does lend itself to the examination of the dynamical slowing of the Moon in its orbit. This slowing arises because of the Moon's tidal interaction with the Earth.

The binary nature of some stars can be revealed by observing occultations, even if they are too close for resolution by more conventional means. Instead of suddenly snapping out as they pass behind the lunar limb, some stars take a moment to fade. During a casual observation of an occultation, I found one star that had not yet found its way to the catalogues as being a binary. Of course, I reported my find.

CHAPTER 2

The Moon through the looking glass

Who first looked at the Moon through a telescope? The honest answer is that we do not know. We cannot even be sure as to when the telescope was invented, let alone who was first to look at the Moon through one. Until a few years ago most historians had settled upon 1608 as the probable year of invention of the telescope and a Dutch spectacle maker, Hans Lippershey, as its probable inventor. Recently, however, evidence for an earlier invention has come to light. For instance, an Englishman, Leonard Digges, is thought to have produced a form of telescope sometime around 1555.

What we can be certain of is that Galileo heard of the Dutch telescope and, with few clues to help him, he did manage to design and build a small refracting telescope for himself in 1609. Shortly thereafter he built other slightly better and more powerful versions (though still extremely imperfect and lacking in magnification by modern standards) and we know that he used them to observe the celestial bodies, including the Moon. Galileo made sketches of the lunar surface.

An Englishman, Thomas Harriott, had managed to obtain a telescope from Europe and also used it to observe the Moon at about the same time as Galileo. Harriott even produced what was very probably the first complete map of the Moon's Earth-facing side to have been made using optical aid. Despite the imperfections of his telescope, Harriott's map does show features we can recognise today.

You might have expected the coarsest features of the Moon to have been charted before the invention of the telescope. Undoubtedly they were, though the earliest 'map' produced without optical aid that we know of is that by William Gilbert. This was published posthumously in 1651, though it is supposed that he made it in 1600, or at some time close to that date, approximately three years before his death.

Although the very beginnings of lunar study might be shrouded in the mists of time, all that occurred after Galileo's era is quite well documented. The Moon had become a subject for serious scientific study and astronomers set about mapping its surface features. As telescopes improved in their power and quality, so successive observers produced better and better maps.

An essential for any cartographic exercise is the standardisation of nomenclature. Naming systems were devised by Langrenus in 1645 and by Johannes Hevelius in 1647. As an aside, Hevelius's maps were notable because they were the first to take account of, and to represent, the regions of the Moon that were only shown as a result of libration. Despite this advance, Hevelius's system of nomenclature was quickly superseded. Our modern scheme of naming lunar surface features really stems from that devised by Giovanni Riccioli. Riccioli was an Italian Jesuit. A pupil of his, Francesco Grimaldi, had made a telescopic study of the Moon. Riccioli combined Grimaldi's observations into a map, which was published in 1651.

Before taking our story further, it will benefit us to pause to consider the appearance of the Moon through a telescope and to get a brief overview of the modern nomenclature of the main types of surface features revealed by one of these wonderful devices.

2.1   THE MOON IN FOCUS

Even a casual glance made without any form of optical aid reveals that the Moon is not a blank, shining disk. Aside from the phases, the Moon's silvery orb clearly shows patchy dark markings. These give rise to the 'Man in the Moon' (and the variety of animals and maidens which feature in other folklores) effect which is so obvious around the time of the full Moon. Figures 2.1–2.4 show the general appearance of the Moon at successive stages in its lunation, as it is seen through a normal astronomical telescope stationed in the Earth's northern hemisphere – in other words, with south uppermost. Since this book is intended for the amateur telescopist and since it is expected that most of its readers will reside in the northern hemisphere, all the telescopic views of the Moon in this book are orientated with south at least approximately uppermost.

The main bright parts of the Moon are the rocky highlands, known as *terrae*. The large dark areas are known as *maria*, Latin for 'seas'; the singular form is mare. Thanks to Riccioli, we have such charming names as Mare Imbrium (Sea of Showers), Mare Serenitatis (Sea of Serenity), and Mare Tranquillitatis (Sea of Tranquillity) to encounter on the Moon.

In Galileo's time it was widely believed that the patches on the Moon actually were seas. Admittedly, a few scholars considered the darker areas to be the land masses and the rest of the Moon's globe to be ocean-covered. Much later the true, arid, nature of the Moon was

Figure 2.1 The 4-day-old Moon, photographed by Tony Pacey. He used his 10-inch (254 mm) Newtonian reflector at its f/5.5 Newtonian focus to directly image the Moon onto *Ilford* FP4 film, subsequently processed in Aculux developer. The 1/125 second exposure was made on 1991 January 19[d].

recognised and the difference in hue was taken to indicate a difference in chemical composition. In pre-space-age times the dark plains were termed *lunarbase*, while the lighter-hued materials were termed *lunarite*.

As well as the 'seas', we have one 'ocean' (*oceanus*) – Oceanus Procellarum (Ocean of Storms) – and several 'bays' (*sinus* for the singular case), such as Sinus Iridum (Bay of Rainbows). These are the larger dark areas. In addition there are a number of 'marshes' (*paludes*), such as Palus Somnii (Marsh of Sleep) and 'lakes' (*lacus* for the singular case), for

Figure 2.2 The 6-day-old Moon photographed by Tony Pacey. Same arrangement as for Figure 2.1 but he used a 1/60 second exposure on *Ilford* Pan F film, processed in ID11 developer. The photograph was taken on 1992 January 10^d 19^h 00^m UT.

example Lacus Mortis (Lake of Death). These are the smaller mare-type dark plains. They are all easily visible to the user of a pair of binoculars. The lunar equivalent of the Earthly 'cape' is the *promontorium*. An example is the Promontorium Agarum (Cape Agarum) on the south-eastern (IAU co-ordinates) border of the Mare Crisium.

You will find a coarse map of some named lunar features presented in Chapter 7 (p. 154) of this book. In addition, many of the features named in

Figure 2.3 The 11-day-old Moon photographed by Tony Pacey. This time Tony used his 12-inch (305 mm) f/5.4 Newtonian reflector, though with the same technique as he used to obtain the photographs shown in Figures 2.1 and 2.2. The 1/250 second exposure was made on *Ilford* Pan F film on 1992 May 13^d 22^h 14^m UT.

this chapter are discussed in detail in Chapter 8 and images/illustrations of them under differing lighting conditions are included there.

Of course, the view grows more detailed when a proper astronomical telescope is used. Even a small telescope reveals a mass of detail and the sight of the lunar surface in anything larger than a 3- or 4-inch (76 mm or 102 mm) telescope is impressive to say the least. I find that the appearance of the Moon's surface through such a telescope, and using a magnification of the order of ×100, reminds me of plaster of Paris. The waterless 'seas' and other dark plains appear various shades of steely grey and the rougher, crater-strewn, 'highlands' that make up the rest of the surface seem greyish white.

Figure 2.4 The 15-day old Moon photographed by Michael Butcher on 2003 April 16^d 22^h 57^m UT. To take this image he simply held his *Canon* Powershot G2 compact digital camera, set to maximum zoom, up to a 40 mm Plössyl eyepiece which was plugged into his *Meade* ETX Maksutov telescope. The camera was set to automatic exposure and focus and an effective speed of 100 ISO.

When the Moon is close to full, as shown in Figure 2.4, its surface seems dazzlingly bright and covered in bright streaks and spots and blotches. At these times it is difficult to imagine that the Moon is made up of relatively dark rock. In fact the Moon's *albedo* is 0.07, meaning that it reflects, on average, 7 per cent of the light falling on it.

Surface features are difficult to make out near full Moon because the sunlight is pouring onto the lunar surface from almost the same direction as we are looking from. This means we cannot see the shadows, so we see very little in the way of the surface relief as a result.

Away from the times when the Moon is full the effect is far less confusing. Shadowing then makes the lunar surface details stand out. This is especially so close to the terminator, where the sunlight is striking the Moon at a very shallow angle. This is evident even by comparing the wide-angle (and hence low-resolution) views shown in Figures 2.1 to 2.4. Notice how the surface relief along the terminator in Figures 2.1 and 2.2 is virtually invisible in the corresponding positions in Figures 2.3 and 2.4.

Under low-angle lighting even the lunar maria are shown to be less than perfectly smooth. *Dorsum*, networks of ridges crossing the maria,

Figure 2.5 With sunlight illuminating the surface at a low angle even the lunar maria appear far from completely smooth. Patterns of ridges cross the part of the Mare Nubium that is shown in this Catalina Observatory photograph. The instrument used was the observatory's 1.5 m reflector and the photograph was taken on 1966 May 29^d 04^h 41^m UT. (Courtesy Ewen A. Whitaker and the Lunar and Planetary Laboratory, Arizona.)

then become obvious (see Figure 2.5). *Dorsa* are ridges occurring elsewhere than on the lunar maria. They are named after people, for example Dorsa Andrusov and Dorsum Arduino, but the average lunar observer will not have occasion to use these names.

If the lunar 'seas' are the easiest features to see with the minimum of optical aid, then the craters must count as the next-most-dominant surface features on the Moon. These saucer-shaped depressions range in size from the smallest resolvable in telescopes (and smaller, down to just a

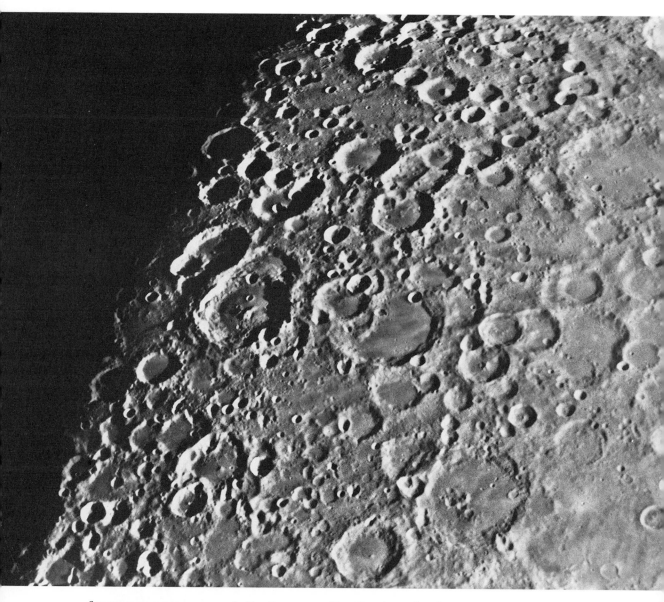

few metres across, as revealed by the manned landings) to a few that are several hundred kilometres in diameter. The smaller craters vastly outnumber the larger ones.

Following the scheme originated by Riccioli, craters are given the names of famous personalities, most usually astronomers. If it strikes you that this is potentially a rather contentious system then you are correct! Over the years many selenographers had taken it upon themselves to modify the nomenclature assigned by the earlier workers, often putting their own names and the names of their friends onto their maps. The result was that a particular crater might have a different name on

Figure 2.6 The crater-saturated southern highlands of the Moon, photographed using the 1.5 m reflector of the Catalina Observatory, Arizona, on 1966 September 5^d 11^h 30^m UT. (Courtesy Ewen A. Whitaker and the Lunar and Planetary Laboratory, Arizona.)

Figure 2.7 The lunar crater Clavius, imaged by the author on 2004 March 01^d 18^h 20^m UT. He used an *S-Big* STV camera in wide-field mode fitted into a ×2 Barlow lens, plugged into the 19½-inch (0.5 m) f/4.8 Newtonian reflector of the Breckland Astronomical Society, stopped to 8 inches (0.2 m) off-axis because the seeing was a rough ANT. IV at the time. The image was subsequently processed using *CCDOPS5* and *Image Editor* software.

different maps. Even more confusing, a particular name might refer to a different crater on different maps! Fortunately, the system has been overhauled by the International Astronomical Union (IAU) in modern times. Under the IAU-standardised scheme, craters are still named after famous personalities (with the proviso that the personality is deceased – the only exception to that being the *Apollo* astronauts) and most of the older assigned names have been retained. The IAU nomenclature is most definitely the one to be adhered to and I would advise caution when using pre-1975 maps.

When seen close to the terminator, craters are largely filled with deep-black shadow and give the impression of being very deep holes. In reality they are rather shallow in comparison to their diameters and can often be quite difficult to identify when they are seen well away from the terminator. Craters saturate the highland areas of the Moon (see Figure 2.6) but there is an obvious paucity of larger craters on the maria. An observer using a typical amateur-sized telescope (around 200 mm aperture) can resolve craters down to about 1–2 km in size and yet many areas of the maria appear craterless. Nonetheless, the photographs sent back by close-range orbiting probes show that even these areas are saturated with small and very small craters. Where there are recognised chains of small craters, these are termed *catena* and are named after the nearest most appropriate named feature. Catena Abulfeda is one example; a 210 km-long chain of small craters near the major crater Abulfeda.

Often the floors of large craters are cluttered with smaller craters and there are many examples of craters breaking into others. In almost all the cases it is the smaller crater which breaks into the larger. Clavius (see Figure 2.7; also Section 8.12), Gassendi (Section 8.22), Posidonius (Section 8.35) and Cavalerius (Section 8.20) are examples of these.

Craters differ in more than their sizes. Some, such as Copernicus, have elaborately terraced walls. Copernicus (Section 8.13) is also an example of one of the many craters to have centrally positioned mountain masses. Other craters, such as Plato (see Figure 2.8; also Section 8.33), have their floors flooded with mare material. Some craters have their walls broken down and are almost totally immersed in mare material. Some craters have bright interiors, such as Tycho (see Figure 2.9; also Section 8.46). This is also one of the best examples of craters which are the source of bright streaks of material, termed *rays*, extending radially from the source crater. Tycho is very easy to see through a pair of binoculars any time close to full Moon, appearing as a bright spot in the Moon's southern highlands. Its rays also seem to extend more than half-way around the Moon's globe. Figure 2.4 shows them particularly well. Other craters have relatively dark interiors and no associated ray systems. All this tells a story and I will have much more to say about crater morphologies and the evolution of the Moon and its various surface details later in this book. For now, we will continue our brief survey of the main types of lunar surface feature and nomenclature.

After the maria and the craters, mountains (generic name *mons*) and mountain ranges and groups of peaks (*montes*) vie for the attention of the telescope-user. They have been named after their Earthly counterparts, so one can find the Apennine Mountains (Montes Apenninus – see Section 8.5) and the Alps (Montes Alpes – visible just below the crater Plato shown in Figure 2.8) on the Moon. The lunar highlands are very rough and hummocky, whereas the maria are much smoother. However, mountain ranges often border a mare. Isolated peaks also exist, sometimes actually on a mare. Examples of this type are Mons Piton and Mons Pico (again close to the crater Plato and shown on Figure 2.8 – see also Section 8.33), situated on the Mare Imbrium. Relatively small blister-like swellings on the lunar surface are termed *domes* but these are not given specific names and are, instead, identified by their proximity to a known major location in the same way as for the crater chains. The easiest domes to locate are those near the crater Hortensius. These are described in Section 8.21.

Figure 2.8 The lava-flooded Plato is the largest crater shown in this view imaged by John Gionis, Mike Butcher and the author on 2005 May 17^d 20^h 45^m UT. They used a *Philips ToUcam Pro* webcam used on the Breckland Astronomical Society's 19½-inch (0.5 m) f/4.8 Newtonian reflector, stopped off-axis to 8 inches (0.2 m). The avi (10 frames per second for 12 seconds) was aligned and stacked in *RegiStax 3* and processed using 'Wavelets' in *RegiStax 3*. The image was further processed in *Image Editor* by the author. Plato is set into the lunar Alps (Montes Alpes). The narrow 'gash' cutting through the Alps to the left of centre is Vallis Alpes. The strip of Mare to the north (below) the Alps is the Mare Frigoris. The expanse of 'lunar sea' occupying the upper part of the image is the northernmost part of the Mare Imbrium.

The closest match to an Earthly cliff on the Moon's surface is an escarpment (a sudden rise in the ground which continues along an approximately linear, or slowly curved path). The generic name for these features are *rupes*, an example being the Altai Scarp (Rupes Altai – see Sections 8.30 and 8.44) on the Moon's south-eastern quadrant.

As well as the craters and the various raised formations, features sunk below the Moon's surface abound. Gorge-like valleys, called *vallis* (in the singular case), such as the huge Rheita Valley (Vallis Rheita – see Section 8.25) are at one extreme of the size range. A smaller version, the Alpine Valley (Vallis Alpes) is shown on Figure 2.8.

Much finer (though often longer) sinuous channels, known as *rilles* (obsolete spelling *rills*; in old books you will also find them often referred to as *clefts*, particularly so the larger examples), also cross the lunar terrain. Several are shown in Figure 2.10. As far as naming them goes, *rima* is used for single examples and *rimae* for networks or groups of rilles. Hence, Rima Hadley (pictured in Chapter 6, Figure 6.4) and Rimae Arzachel. Some examples are detailed in Chapter 8. All the rilles and

most of the lunar escarpments and valleys are named after the closest appropriate major feature. The sole exceptions are: Rupes Altai, Rupes Recta, Vallis Bouvard and Vallis Schröteri.

All the foregoing described features can be seen through small telescopes. Even a humble 3-inch (76 mm) refractor is sufficient to show many rilles, despite their being hard to resolve due to their thinness, when they are seen under low-angle illumination from the Sun (and thus largely filled with black shadow). They were first noted by Christian Huygens with the primitive telescopes of the seventeenth century.

As I indicated earlier, the Moon appears rather monochrome when seen with a small telescope (aside from the prismatic splitting of light through our atmosphere which causes images seen in a telescope often to be spoiled by colour fringing – discussed later in this book). However, if a sufficient aperture is used then some coloured tints can become visible to the observer. Or at least that is the case for many observers. Sensitivity to colours varies enormously from person to person. Some observers fail to see colour in anything they look at through the telescope. For a few lucky individuals the Universe is a very colourful place. Others can see some colours through the telescope eyepiece, perhaps just the strongest hues on Jupiter and the overall colours of Mars and Saturn.

I am fairly fortunate in that I can easily see colours in many objects through a telescope of sufficient size, though I must say that I have

Figure 2.9 Tycho is the prominent crater shown just left of centre in this image by John Gionis, Mike Butcher and the author on 2005 May 17^d 20^h 32^m UT. The other details are the same as for Figure 2.8

Figure 2.10 Systems of rilles situated near the centre of the Earth-facing hemisphere of the Moon. Photograph taken using the 74-inch (1.9 m) reflector at Kottamia, Egypt, on 1965 August 4^{d} 20^{h} 43^{m} UT. (Courtesy Dr T. W. Rackham.)

noticed a significant reduction in my colour-sensitivity as I have got older. I find that I can see subtle coloured tints on the Moon's surface when using a sufficiently low magnification on a reasonably large telescope; for example, ×144 on my 18¼-inch (0.46 m) Newtonian reflector. The overall colour of the rough highlands is still greyish, though perhaps a little 'creamier' in colour than through a smaller telescope, but the large plains of the maria seem tinted with faint blues and greens. In particular, the Mare Tranquillitatis seems especially blue when seen near full Moon. The interiors of some craters, such as Langrenus, appear with a faint brownish or even a golden-yellow tint at these times. Aristarchus appears slightly bluish-white while the raised plateau on which it stands seems particularly brownish to my eyes.

Of course, these colours are very far from accurate. Spectroscopic analysis reveals that the surface of the Moon is really various shades of brown. The human eye has a tendency to normalise the overall colour of

the Moon as white. Hence the different shades of brown manifest as the apparent colours seen. A slightly 'redder' brown produces an apparent yellowish or brownish tint, while a 'cooler' shade of brown seems to the observer to be a greenish or bluish tint.

Figure 2.11 shows a specially prepared photograph on which all the usual grey-scale tones have been obliterated. Instead, the shades of grey represent colour differences. Redder tones show up as lighter, and bluer tones show up as darker. Note the relative blueness of the maria and the relative redness of the interiors of many craters. As far as I can ascertain

Figure 2.11 Colour-difference (610 nm–370 nm) photograph of the Moon created by Ewen A. Whitaker. The normal grey-scale has been eliminated. Lighter regions are redder and darker regions are bluer.

only a minority of people can perceive these subtle tints through even a large telescope. To most users of small telescopes, the Moon is a world of black and white, and steely greys.

2.2   THE PIONEERING SELENOGRAPHERS

As the seventeenth century progressed so refracting telescope object glasses were made which were a little larger than the first, tiny, examples. However, these lenses were single pieces of glass and so suffered badly from chromatic aberration. The remedy for this aberration (and to an extent the other aberrations that arose mainly from the crudeness of the methods of lens manufacture) was to make the lens of larger focal ratio (and hence greater focal length). To reduce the aberrations to a tolerable level, the focal length had to increase out of proportion to the aperture. So, longer and longer refracting telescopes were made. In some cases the focal lengths reached hundreds of feet (several tens of metres). Even then, the sizes of the objective lenses were still less than 9 inches (228 mm)! Despite this handicap, selenography, the charting of the Moon's surface features, steadily improved.

Probably the best map of the Moon made in the seventeenth century was that published in 1680 by Cassini. His 54 cm map (54 cm representing the Moon's diameter), is of remarkable quality considering the cumbersome telescopes he had to work with. Not only is it artistically a fine piece of work but also the positional accuracy of the features it depicts is very good for the time, even if poor by modern standards. It showed unprecedented fine details, such as the minute craters (which we now know as *secondary craters*) around Copernicus. It is also more comprehensive in its depiction of features than earlier works, for instance showing the ray systems that surround many bright craters (de Rheita was, arguably, the first to comprehensively chart the rays in 1645) and something of the variations of hue of the lunar maria.

The later years of the seventeenth century also saw the invention of the common forms of reflecting telescope (the Newtonian, the Cassegrain and the now obsolete Gregorian) which eventually led to more manageable and yet higher-quality instruments, and ever better lunar observations.

In Germany Tobias Mayer produced a small, though accurate, map which was published posthumously in 1775. He was notable in that he was the first to introduce a system of co-ordinates for lunar surface features, having made his measurements with the aid of a primitive eyepiece micrometer.

As far as the 'leading lights' of selenography go, Germans dominated the period from Tobias Mayer's work through to the late nineteenth century. Perhaps the most famous of these was Johann Hieronymous Schröter.

Schröter was a magistrate at Lilienthal (near Bremen, in Germany), where he had enough wealth and leisure time to set up his own observatory. He had various telescopes, including two by William Herschel. His largest (not by Herschel) was a 20-inch (0.51 m) Newtonian reflector of about 8 metres focal length. Completed in 1793, it was the largest telescope in Europe at the time and was surpassed only by William Herschel's 48-inch (1.2 m) of 40 feet (12 m) focal length, though it is thought that the optical quality of the 20-inch was not particularly good.

From 1778 to 1813, Schröter devoted considerable amounts of time and energy to observing the Moon and planets. He set himself the task of making the most detailed map of the Moon to date and he made hundreds of lunar drawings to that end. He used a crude eyepiece micrometer to aid his work, including making measurements of the heights of lunar mountains. He was the first to make a really detailed study of the crack-like rilles. In the end he did not complete his proposed lunar map but instead published the completed sections in a book, *Selenotopographische Fragmente*, in 1791 (a second part was completed and a bound two-volume edition published in 1802). Schröter's work attracted much attention and other selenographers undoubtedly were inspired by the (sometimes controversial) results issuing from Lilienthal.

On the downside, Schröter was not a particularly good draughtsman and he certainly made his fair share of mistakes. In particular he thought he had detected changes on the lunar surface over the years during which he carried out his observations and he was convinced that the Moon possessed a dense atmosphere. Of course, neither are true.

A cruel blow was to befall Schröter when, in April 1813, invading French soldiers looted and then burnt Lilienthal to the ground. His observatory was also looted and then destroyed. At that time Schröter was 67 years old and his health was already in decline. It was too late for him to rebuild his observatory and begin again. Undoubtedly the shock and sorrow he suffered hastened his death. He died three years later.

Wilhelm Lohrmann, of Dresden, also attempted to map the entire face of the Moon in great detail. The first sections of his map were published in 1824 but Lohrmann was eventually defeated by failing eyesight. However, he did manage a general map of the surface of 39 cm diameter.

The quest was taken up by Wilhelm Beer and his collaborator Johann Mädler. Beer had a 3¾-inch (95 mm) refractor at Berlin and, together, they used this telescope to study the Moon in detail for over a decade. They eventually (1837) produced a highly detailed and very accurate map. On it, the whole Moon had a diameter of just over 0.9 m. It remained unsurpassed for decades to follow, a significant achievement given the diminutive size of the telescope they used. Beer and Mädler's map was supplemented with their book *Der Mond*. They portrayed the Moon as utterly dead and changeless, in complete contrast to the picture of it painted by Schröter.

Whereas the Moon of Schröter, with its supposed changes and active weather, tended to excite the interest of others, the Moon portrayed by Beer and Mädler tended to do the opposite. Given, also, the high quality of their map, the general feeling was that 'the last word' had been stated as regards lunar studies. Few others studied the Moon seriously for more than the next quarter-century.

However, one exception was Julius Schmidt. Schmidt had a lifelong interest in the Moon. After posts at various German observatories, he became Director of the Athens Observatory, in Greece, in 1858. He used the 7-inch (178 mm) refracting telescope there to continue his lunar studies. As well as revising the sections of the lunar maps of Lohrmann, and then going on to complete the mapping of the missing sections, Schmidt was eventually to complete one of his own by 1878.

Schmidt's map, 1.9 m to the Moon's diameter (the map was divided into 25 sections) was incredibly detailed as well as being reasonably accurate. It recorded and placed some 32 856 individual features. It took over the torch from Beer and Mädler as the best lunar map. It was to hold this premier position until 1910, when a 1.5 m map of greater positional accuracy was published by Walter Goodacre, the second Director of the Lunar Section of the British Astronomical Association (BAA).

This was not Schmidt's only contribution to selenography. Owing to an erroneous interpretation of his, and other people's, observations, he re-invigorated lunar research. The whole episode concerns a small crater, called Linné, in the Mare Serenitatis. Lohrmann, Beer and Mädler, and Schmidt himself had often recorded Linné as a deep crater. Then, in 1866, Schmidt announced that the crater had disappeared! In its place Schmidt could only find a small light patch. As one might expect, a statement like that was sure to get astronomers turning their telescopes back to the Moon. Many leading astronomers joined in and a vigorous debate ensued. In fact, many astronomers continued to cite Linné as a prime example of an area of the Moon that had changed significantly within the history of Man's observations of it, even to as late as the early twentieth century!

We now know that Linné is really a small crater surrounded by a light area. Under certain angles of illumination it can, indeed, appear in the guise of a deep, apparently larger, crater. It seems certain that Schmidt was mistaken. There never was any change in this lunar feature within the period when astronomers were around to look at it. However, this mistake was just what was needed at the time to counter the view of the

Moon as a dead and uninteresting world that pervaded after Beer and Mädler's epic study of it.

As well as the maps, various other studies of the Moon's topography appeared in the form of books. For instance, there was *The Moon* jointly authored by James Nasmyth (a famous engineer and the inventor of the steam hammer) and James Carpenter. First published in 1874, the authors made serious efforts to understand the origins of the Moon and the evolution of its surface features, though their theories bear little relation to our modern ideas. Much of their researches were based on observations made with Nasmyth's home-made 20-inch (0.51 m) reflector of novel design. Incidentally, the optical arrangement Nasmyth originated is often used in today's largest telescopes and is known by his name. Nasmyth and Carpenter's book also contains beautiful drawings and photographs of sculpted models of regions of the lunar surface (at that time, photography had not technically advanced enough to enable good, detailed, photographs to be taken of the Moon's surface direct through the telescope) along with written descriptions.

Other notable books about the Moon included *The Moon* written by the Englishman Edmund Nevill and published two years after Nasmyth and Carpenter's book of the same name. Actually, Nevill wrote under the name Neison. His book contained a map based on that of Beer and Mädler, along with detailed descriptions of the named features.

If, as a result of the necessary brevity of these historical notes I have given the impression that selenography was only carried out by a few individuals then I must rectify that impression. For instance in England there was the Selenographical Society, formed in the early 1870s specifically for lunar studies. The British Association for the Advancement of Science appointed the Secretary of the Society, W. R. Birt, to head a committee to organise the construction of a new and more detailed map of the Moon. It was intended to be 200 inches (5.08 m) to the diameter of the Moon. Birt was an energetic selenographer and a start was made, though Birt's death and the eventual demise of the Selenographical Society in 1882 meant that the scheme did not bear fruit.

Also, many national and provincial astronomical societies had sections devoted to lunar study. One very active group of the period was the Liverpool Astronomical Society. Its director was T. G. Elger, who became the first director of the Lunar Section of the British Astronomical Association when it formed in 1890. In those early years many people spent a great many hours at the eyepieces of their telescopes studying the Moon. For a detailed account of the early years of Moon-mapping I can do no better than refer you to the book *Mapping and Naming the Moon*, by E. A. Whitaker, published in 1999 by Cambridge University Press.

The last really substantial Moon map to be made using the old-fashioned methods of eye and drawing board to record its finest details was the 300 inch (7.6 m to the Moon's full diameter) colossus of H. P. Wilkins. He published the first version of it in 1946 and made revisions in subsequent years. At the time he was Director of the Lunar Section of the British Astronomical Association. The only version of Wilkins' map I have seen is that reproduced in reduced scale in twenty-five sections in the book *The Moon* by Wilkins and Patrick Moore, published by Faber and Faber in 1955. I was lucky enough to find a copy of this work in a second-hand bookshop some years ago, though it is now very rare. The complexity of the hand-drawn details in the map is mind-boggling. Though it is now recognised that Wilkins' map contains many inaccuracies in its depictions of details (I have stumbled across several, myself, without making any effort to find them), the scale of his achievement still warrants admiration.

Photography, invented in the early nineteenth century, was sufficiently developed to come to the aid of Moon-mappers in the last decade of the nineteenth century and, particularly, those of the twentieth century – but that is a tale for later in this book. Now, after this abbreviated history of the earliest years of lunar study, it is time to consider how you can get the best out of your telescope and enjoy and study the Moon's starkly beautiful vistas.

CHAPTER 3

Telescopes and drawing boards

Why bother to observe the Moon? The answer to that is likely to be different for different people. It is seeing the stunning vistas of an alien world that drives me. What about drawing the lunar surface, though? It is a lot more trouble doing that than passively looking through the telescope eyepiece. Also, the day of the amateur lunar cartographer is now long past. Thanks mainly to Moon-orbiting spacecraft, the various lunar features have now been mapped with much greater precision than can possibly be achieved by an amateur's eye, telescope and pencil.

So, why draw the Moon? The mountain climber's adage "because it is there" might suffice as a reason. The Moon's beautiful orb is every bit as much a part of nature as the mountains and valleys, fauna and flora here on the Earth. Drawing the Moon's surface details is also a powerful way of communing with it. The process of doing so will also create a kinship between you and the selenographers of yesteryear who **had** to draw the Moon because there was no more sensitive way of doing it. You will certainly get to know the parts of the Moon you sketch with great intimacy. You might never get to travel to the Moon but carefully observing it through your telescope and drawing what you see through the eyepiece surely comes as a good second-best. So, if you decide to take up lunar drawing, do so because you enjoy it.

Now we come to the 'nitty gritty' of actually doing it. We must first consider the observer's equipment. At the outset please let me say that if you already own a telescope then use that one for your lunar observing. You will be delighted with the details of our neighbouring world that it can reveal. However, if you are planning to build or buy some new equipment then there is scope (no pun intended!) for making an informed choice. Failing that, there might be ways you can improve your existing equipment to make it more suitable for Moon observing. In either case I hope the following notes might be of help to you.

In many branches of observational astronomy a telescope's light grasp is crucial. In such cases a large aperture is normally an advantage. The Moon is one of the few celestial objects that provides us with plenty of light. It is the various other imaging characteristics of the telescope which are most important for lunar and planetary observation. These can broadly be grouped as resolving power and contrast, though there is a degree of interrelation between them.

The image a telescope makes of a point source (in practice, a star) defines what we call the *point-spread function*, sometimes known as the *instrument profile*, of it. The typical diffraction pattern of a star produced by an unobstructed aperture is represented in Figure 3.1(a). Larger apertures produce smaller diffraction patterns. It is the size of these diffraction patterns that decides whether a telescope has the potential to resolve a close-together pair of stars, or not. This is illustrated in Figure 3.1(b).

So much for stars. Our interest is in resolving details on an extended body: the Moon. The same principle applies. The image the telescope forms of the Moon can be thought of as being composed of a series of overlapping diffraction patterns, each generated by a minute point in the image. A handy way to grasp this is to think of the Moon's image as a mosaic. Obviously the size of the individual tiles determines the fineness of detail that can be represented on the mosaic. If you use a larger telescope the individual diffraction patterns are smaller. This is the same as having the mosaic made up from smaller tiles. The resolving power, R, of an unobstructed optical aperture is given by

$$R = 137/D, \tag{3.1}$$

where R is in arcseconds and D is the diameter of the aperture in millimetres. This formula is derived from Rayleigh's mathematically derived limit and the numerator (137) is true for the mean visual wavelength (540 nm). Many readers will also be familiar with Dawes Limit. The formula takes the same form but the numerator would be 116 in the above equation. In practice, Rayleigh's Limit gives a truer measure of resolution in images of extended bodies. Note this limit is for details with maximum contrast – in other words, blacks and whites. Low-contrast boundaries are less well delineated.

So, we need a larger telescope to see finer details on the Moon. Is that the end of the matter? Actually, no. There is more to it than that. In practical telescopes the point-spread function is influenced by

Figure 3.1 (a) Idealised representation of the diffraction pattern of a star produced by an unobstructed aperture (for example a refracting telescope). (b) In (i) a pair of stars are too close together for a given telescope to resolve them because the diffraction patterns merge. In (ii) the stars are just resolvable and in (iii) they are easily resolvable. The complex extended image that a telescope forms of the Moon can be, albeit simplistically, thought of as being composed of an array of star-like points in order to understand the principle of resolution of it by a given telescope.

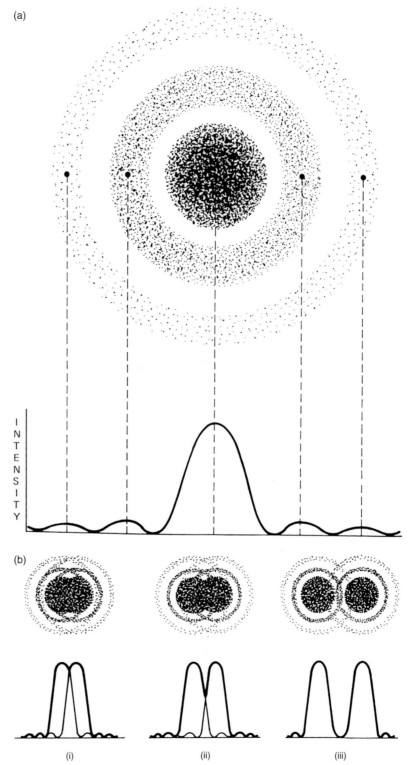

instrumental design and the accuracy of manufacture of the optical surfaces. We also have to contend with the prevailing atmospheric conditions, but more of that last complication later.

On the point about accuracy of manufacture of the optics, all the rays collected by the telescope objective from one point on the object ought to be brought to a coincident point in the final-image plane to an accuracy of within ¼-wavelength of yellow-green light. If the ray fronts deviate by more than this amount, about 135 nm, then the diffraction pattern will be noticeably spoilt. The point-spread function will be changed and both image resolution and contrast will suffer. Even the ¼-wavelength limit is not the ultimate. There would be some improvement in a telescope's performance if it had optics of still greater accuracy. However, **most** of the telescope's performance can be realised at this universally accepted benchmark of quality.

Of the most common telescope designs (and putting aside any subsequent modifications by the telescope-user), the instrument that comes closest to providing the textbook diffraction-pattern structure is the refractor.

Refracting telescopes
The refractors of yesteryear tended to have focal ratios in the range f/12–f/16. The relatively gentle curves on optical surfaces with large focal ratios are easier to manufacture with accuracy and this is one reason why long-focus refractors make good telescopes for Moon and planet observing. Actually, it is quite possible also to make reflectors with long focal lengths, so this is not an exclusive property of refractors. In addition, large focal ratios allow the simpler designs of eyepieces to function better and this is another reason for large focal ratios delivering better images.

Refractors with object-glasses made from two lens elements, typically of common crown and flint glasses, **should** be manufactured to have large focal ratios because of a generic problem with them. While the manufacturer has designed the object-glass to bring the wavelength range to which the eye is most sensitive (circa 540 nm) to a minimum focus, inevitably the correct focal positions for the other wavelengths occupy a range of positions extending further from the lens than the minimum focus. This *secondary spectrum* generally shows itself as a softening of contours in the image, together with a reduction in image contrast, and even visible colour-fringing when the problem is severe.

Of course, what is tolerable depends upon what one is doing. The effect of the secondary spectrum diminishes with the square of the focal ratio for the range of sizes and focal ratios normally encountered. As an aside, the wavefront error (change of focus) for red and violet light at the extreme ends of the spectrum compared to that for yellow-green light can be several times the ¼-wave limit tolerable for the seidal errors (spherical aberration, coma, astigmatism, field curvature and distortion) before becoming objectionable. This is because of the much reduced sensitivity of the eye at wavelengths away from that of yellow-green light.

I have used many telescopes over the years, including a number of refractors. Based upon my experiences with them, I would say that for visual lunar observation 'old fashioned' two-element achromatic object-glasses ought to have a focal ratio of, at the very least, 1.3 times the aperture in inches (0.06 times the aperture in millimetres) in order that the secondary spectrum is not too severe. Even then the image will fall far short of perfection.

As a case in point, I often used the 12.8-inch (325 mm) f/16.4 'Mertz' refractor that was mounted on the 'Thompson' 26-inch (0.66 m) astrographic refractor at Herstmonceux. Seen through the Mertz telescope the Moon's craters were fringed with yellow and the black shadows were filled in with a delicate blue haze! By contrast, the 7-inch (178 mm) f/24 refractor that was mounted on the 36-inch (0.91 m) Cassegrain reflector (the 'Yapp reflector') on the same site gave images of the Moon that were very haze-free and completely free of false colour-fringing.

Right to the end of the 1900s few amateur astronomers owned large refractors. This was because they were very expensive compared to other types of telescope. Almost the only refractors around in the marketplace were those awful camera shop/department store abominations.

That has now changed. Telescope manufacturers have attempted to sate a perceived appetite for refractors amongst the serious amateur astronomical community. There are now a large range of refractors on the market, commonly up to 6 inches (152 mm) in aperture. A few manufacturers even supply larger models. In response to the modern desire for compactness, these instruments are manufactured with much lower focal ratios than the refractors of yesteryear. A focal ratio of f/8 is now quite common, even for a 6-inch refractor. The cheapest of them (2006 prices range from about £230/$350 for a 4-inch rising to about £900/$1300 for a 6-inch) have object glasses of old-fashioned design, that is composed of just two elements, one of common flint glass and the other of common crown glass.

If you buy one of these low-focal-ratio cheaper models please do be aware that the uncorrected secondary spectrum will be visible when you use the telescope to look at anything bright such as a planet or the Moon. You will see lunar craters and other light–dark boundaries fringed with yellow and the image contrast will be reduced by a bluish-purple haze covering everything.

A few companies now market refractors with one component of the two-element objectives made of a special glass (usually known as 'ED glass'). The result is a refractor with a much reduced secondary spectrum compared to one of classical design of the same focal ratio. Again they have focal ratios circa f/8. These are marketed as 'ED apochromatic refractors', and sometimes as 'apochromatic refractors', though that title really belongs to three-element objectives. They should really be called 'semi-apochromatic refractors'.

True apochromatic objectives have the smallest secondary spectrum (about one-ninth the visibility of that of the classical objective of the same focal ratio if made of conventional glass types, and even better if utilising special glass types, for example fluorite).

The two-element 'ED glass' refractors are rather more expensive than those with classically designed objective lenses, typically costing roughly $3000 for a 4-inch rising to $5000 for a 6-inch. True apochromatic refractors are **very** much more expensive, crossing the $10 000 price tag for a 5- or 6-inch refractor. Even then these prices are only for refractors on relatively low-end driven equatorial mountings. Double the price if the telescope is on a top-end mounting with precise tracking and GOTO facility (not necessary for conventional visual observation, though desirable for deep-sky photography).

There are some tricks you can get up to in order to reduce the amount of secondary spectrum you see when looking through the cheapest low-focal-ratio refractors. I describe these in Section 3.6.

Schmidt–Cassegrain and Maksutov–Cassegrain telescopes

In the past couple of decades, the Schmidt–Cassegrain and its close cousin, the Maksutov–Cassegrain telescopes have become extremely popular. Their compactness, and even portability, lend them to the needs of the modern amateur astronomer very well. They are expensive but, for the price, you typically get a computer-controlled instrument of nearly double the aperture of an ED refractor that can automatically set and track on any of thousands of celestial objects.

All very well, but how good are they for observing the Moon? Well, if you need computer control to set your telescope to the Moon then something is very wrong! Of course, the portability and compactness aspects are just as much of an advantage to the Moon observer. The downside is that, aperture-for-aperture, they do not give quite as good lunar and planetary images as some other types of telescope when used for visual observation.

The reasons are two-fold. One is that the steep curve on the primary mirror (typically about f/2 – this is what makes the instruments compact) and the complex curve on the corrector plate (for the Schmidt–Cassegrain, which is the most common type in production; the Maksutov–Cassegrain has a meniscus corrector) are both difficult to manufacture accurately by production-line methods. Inevitably, the optical surfaces will fall just a little short of the ideal accuracy and so the wavefront error might be a little larger than the desired minimum of ¼-wavelength of yellow-green light.

Having said this, the latest models coming off the production lines are **very** much better than the oldest models. This is something to be aware of if you are contemplating buying an older second-hand telescope. In general you can expect the image produced by a Maksutov–Cassegrain telescope to be superior to that from a Schmidt–Cassegrain of the same aperture. However a lot does depend on the manufacturer.

The second reason is common to all telescopes with a central obstruction in the light-path. The central obstruction modifies the diffraction pattern structure. Light is taken from the central disk and given to the rings. This modification of the point-spread function only slightly impairs resolution within an image composed of a pattern of blacks and whites. However, it seriously reduces the visibility and resolution of low-contrast details, especially where those low-contrast markings appear against a bright background. This is usually the case for seeing details on the planets. While it is true that details seen along the Moon's terminator are mostly nearly-blacks and nearly-whites, there are subtle shadings which also form part of the scene.

In addition, the reduction of contrast is worst for the smallest details in the image, making things very hard to discern near the diffraction limit. The central obstructions of Schmidt–Cassegrain telescopes are usually at least one-third of the total diameter, the largest for any type of telescope commonly in amateur hands. Maksutov–Cassegrain telescopes tend to have smaller central obstructions; sometimes less than a quarter of the diameter of the aperture.

The two problems, any slight shortfall in optical accuracy (but you can expect at least fairly good optics in a recently manufactured instrument), and the large central obstruction, each produce additive effects on the point-spread function. The common f/10 Schmidt–Cassegrain telescopes on sale have to have apertures around twice as large as the best refractors in order to show the Moon and planets with equal fidelity under identical excellent conditions; the performance of Maksutov–Cassegrains falls somewhere between them.

However, any lack of image contrast is much less of a problem to the lunar photographer as this can be restored (and even boosted way beyond what is real if so desired) in the subsequent image processing.

The experience of myself and many others is that *Celestron's* line of Schmidt–Cassegrain telescopes, and especially the 9¼-inch (235 mm) model, are particularly good examples of this design of instrument. Damian Peach, possibly the best lunar and planetary imager in the world, uses *Celestron* Schmidt–Cassegrains to produce his finest work.

The 9¼-inch *Celestron* is unusual in that the primary mirror is an f/2.5, rather than the more usual f/2. The secondary amplification is less (×4, rather than the usual ×5) and these factors are quite likely the reason for this telescope's good performance.

My experience of a 9¼-inch *Celestron* Schmidt–Cassegrain telescope (one of the instruments at the Breckland Astronomical Society's observatory) is that it delivers images of high quality, though noticeably down in contrast just as I would have expected from a high-quality set of optics with a large central obstruction. It does, though, need frequent collimation. Even pointing the instrument to different parts of the sky is enough to disturb the collimation. This is because focusing is achieved by small movements of the primary mirror and the mechanism and supports allow varying amounts of drooping of the mirror. The collimation of a Schmidt–Cassegrain is, though, very quick and easy – just a tweak on one of three screws (see Appendix 1 for methods of collimating all the main types of available telescope). The very latest Schmidt–Cassegrains have much improved mirror mountings and so the collimation-drift problem should no longer be an issue if you buy a brand new telescope (but always carefully check the specifications and, particularly, independent reviews); this is something you should bear in mind if you buy a second-hand unit.

Newtonian reflectors

Although they have somewhat fallen out of fashion, the telescope that gives the observer the best value for money is still the Newtonian reflector. Any manufacturer can produce a poor Newtonian telescope but at least it is not very difficult for the manufacturer to produce a good one. The obstruction due to the secondary mirror tends to be about 20–25 per cent of the telescope aperture, higher focal ratios on larger apertures allowing smaller percentage secondary obstructions. Providing the mirrors are of high optical quality the resulting Newtonian telescope will produce better Moon and planet images than will the Schmidt–Cassegrain of the same aperture and maybe roughly equivalent to that from a Maksutov–Cassegrain (at least in the smaller sizes), though still down on what the best refractor can show. However, the Newtonian will be much cheaper aperture-for-aperture than any other type of telescope.

If you decide to buy or build a Newtonian reflector specifically for observing the Moon and the planets, go for a focal length as large as is practical for your situation (size of garden, size of observatory, etc.), while ensuring that it is firmly mounted. A gangling, spindly affair will flutter in the breeze and shudder with every touch of the focuser – not something that is conducive to good observing!

For purely visual work the telescope need not be driven. Of course, a drive is always an advantage as long as it works properly. The requirements for lunar photography are discussed in later chapters in this book.

One endlessly debated question: should the telescope have an open framework tube, or one that is solid? The ultimate in baffling against stray light is not essential for Moon and planet work but warm air from the observer can cause problems if it gets into the telescope's light-path. A solid tube helps prevent this. However, convective tube currents generated by warm optics and fittings can also degrade the image produced by solid-tubed reflectors. In particular, the thermal lag of the primary mirror can give a lot of trouble. Here, an open tube is an advantage. Vents and even electric fans installed near the telescope primary mirror help matters.

My 18¼-inch (0.46 m) reflector used to have an open-framework tube, while my 8½-inch (216 mm) has a solid tube. On some nights the images produced by the 8½-inch are poor and star images betray the obvious presence of a tube current (see Appendix 2). There is a door installed in the tube close to the primary mirror (see Figure 3.2) to gain access to the

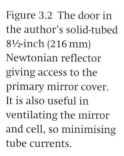

Figure 3.2 The door in the author's solid-tubed 8½-inch (216 mm) Newtonian reflector giving access to the primary mirror cover. It is also useful in ventilating the mirror and cell, so minimising tube currents.

mirror cover. Opening this door significantly improves the images on those nights, since it lets much of the warm air convected from the primary mirror to escape, rather than passing up the entire length of the tube.

To conclude, I think the best design is to have the telescope tube partly closed, especially near the eyepiece. It should, though, be open near the primary mirror (or at least well ventilated) to suppress tube currents. Having the mirrors cool quickly is also desirable to suppress tube currents. A thicker-than-necessary primary mirror and an unventilated cell will work against this. I am currently rebuilding my largest telescope. It will have a solid tube, provided with good ventilation around the mirror cell and across the face of the primary mirror.

Closed optical systems, such as refractors and Schmidt–Cassegrain telescopes, are usually less troubled by tube currents, though both need to cool off for a while if brought out from indoors before observing. However, the sheer mass of the largest examples of Maksutov–Cassegrain telescopes, and Maksutov–Newtonian telescopes (they have thick and heavy corrector plates at one end of an enclosed tube with the heavy primary mirror at the other end, along with the heavy metalwork needed to keep these heavy optics rigidly aligned) makes them prone to thermal problems, though models up to 8 inches (203 mm) in aperture are generally very highly thought of by their users.

3.2   HOW BIG A TELESCOPE DO YOU NEED?

If the optical transmission characteristics of the Earth's atmosphere were perfect and the telescope was in a perfectly temperature-stable environment, then it would be a case of 'the bigger the better' as far as the aperture of the telescope goes. The one caveat is that size must never be at the expense of quality. Of course, the telescope is **not** in a temperature-stable environment and the normal observing conditions are anything but perfect – and that changes matters very considerably.

I still put quality above sheer size when it comes to selecting telescopic equipment. You will have a great deal more success and satisfaction from working with a telescope of good mechanical and optical quality of moderate, or even small, size than you will from a much larger 'light-bucket'. What if one is wealthy enough to have quality and size? Will a bigger telescope always outperform a smaller one of equal quality? Based on nearly forty years of experience of observing with a large variety of telescopes of differing size and design (ranging from a 60 mm refractor to a 0.91 m Cassegrain reflector), my answer to that question is 'No'. In fact, sometimes even the reverse is true. Understanding why this should be so is not very difficult. There are two main reasons.

The first reason is that the column of air through which the telescope is looking is seething with convective pockets, or cells, of air of slightly differing temperatures, and hence differing density and refractive index. The ones that affect the telescopist to the greatest extent are from 10 to 20 cm in diameter and can occur from just in front of the telescope to many kilometres in height. Each of these cells disturbs the passage of light-rays passing though it. The result is that the telescope cannot produce a sharp and steady image at its focus. The blurred and mobile (we say turbulent or bad seeing) image that results is well known to all telescope users.

However, many do not appreciate just how severe the limitation really is. From most backyard sites the seeing rarely allows details to be glimpsed by a human eye looking through the eyepiece that are finer than about 1 arcsecond in extent. For an object situated at the Moon's distance this is a linear dimension of about 1 mile (about 1.6 km).

A good-quality 6-inch telescope will potentially allow you to resolve this level of detail. A bigger instrument will not show you any finer detail on those '1 arcsecond' nights. Of course the image will be brighter when seen through the bigger telescope and the contrast of the image (at least for coarse details) will be greater, the comparison being made at a given, adequate, magnification.

In ordinary conditions, the advantage of large apertures in seeing faint planetary markings (and delicate shadings on the Moon) is not as great as one might expect. While I have experienced rare nights where my 18¼-inch (0.46 m) telescope shows views of Jupiter and Saturn reminiscent of *Voyager* space-probe images, I must say that the normal view is rather fuzzier and lower in contrast. On those normal nights I find that my 8½-inch (216 mm) telescope can show these bodies as well as my 18¼-inch.

Indeed, **sometimes** the bigger telescope will produce significantly poorer images than will the smaller one. If the small-aperture telescope can look through just one convective cell of air at a time, then the image will 'slurp' around its mean position and may distort. However, it will remain quite sharply defined. If the telescope aperture is bigger, then it is looking through a column of air that may include several convective air cells at any one moment. Each cell will produce its own, random, effect and the telescope will combine them. This time, the image will be composed of a number of overlapping components, each one shifting and distorting in a separate way. The end result is a confused and blurred image. It may often be preferable to have the one well-defined, but admittedly gyrating and distorting, image rather than the confused mess.

The second reason why a bigger telescope is not always better was mentioneded in the last section – namely thermal effects. The smaller the mass of the telescope, the quicker it will cool to the ambient temperature and so not be troubled by currents of warmed air convecting from its optics and fittings.

3.3 SO, WHAT TELESCOPE SHOULD I SPEND MY MONEY ON?

All in all, if you can afford it, and if you have the room to house it permanently in some sort of observatory (perhaps a run-off shed), I would say go for a Newtonian reflector of 10-inch–14-inch (254 mm–356 mm) aperture and as large a focal ratio as you can reasonably accommodate. Bear in mind, though, that the number of nights that you will get full performance, even from a 10-inch, will be very few unless your observing site is unusually good. Also remember that thermal problems increase rapidly with increasing telescope size. A particularly good observing site and attention to thermal issues are necessary before you can really benefit from a larger aperture telescope used for Moon and planet observing.

If you can't afford a 10-inch then go for a smaller Newtonian reflector. Remember this type of telescope is the cheapest of any but please do not compromise on quality for the sake of size.

My second choice for an instrument intended for visual observation of the Moon would be a refractor of at least 5 inches (127 mm) aperture. However it would have to be either an achromatic refractor of focal ratio at least f/12 (which would then imply an older second-hand instrument) or a modern (and consequently circa f/8) one with an 'ED apochromatic' objective as the bare-minimum requirement.

Not far behind the refractor in the pecking order on my 'preferred telescope list' would be a Maksutov–Cassegrain telescope provided it had an aperture of at least 6 inches (152 mm) but no greater than 8 inches (203 mm) because of its thermal properties. The Schmidt–Cassegrain telescope would trail in fourth position. I would not be happy with one of those smaller than 8 inches (203 mm) aperture.

Remember this list reflects my recommendation for types of telescope suitable for **visual** observation of the Moon (and, incidentally, the brighter planets). My list would certainly be rearranged were I to consider preferred telescopes for other tasks and other subjects for observation. I would, for instance, be very tempted by a 9¼-inch *Celestron* Schmidt–Cassegrain telescope if I wanted a portable, but still sizeable, instrument for Moon and planet photography, along with deep-sky photography, using a telecompressor with the telescope.

There are other types of telescope that can be used but these are by far the most common currently found in the market place and in amateur observatories. I offer notes on how to adjust various types of telescope to get the best optical performance from them in Appendix 1. Once you have your telescope performing as well as it can, you can evaluate its optical quality under normal observing conditions by means of some simple tests I describe in Appendix 2.

When buying second-hand equipment remember the old adage: **buyer beware**. Unless the price is so low the vendor is virtually making a present of the equipment to you, insist on checking it out under field conditions. If necessary take an experienced friend with you.

You need to be careful even when buying new equipment. For one thing, the quality of a company's advertising does not always reflect the quality of their goods. Make use of the Internet in researching out other people's experiences with the equipment before you buy it. Are there any recurrent problems, or shortfalls in performance? Equipment reviews abound online. Use your favourite search engine, specifying the item of equipment under question and take careful note of what appears.

I have to limit my advice on equipment in this book to just a few aspects of particular relevance to the lunar observer. If you want more let me refer you to my book *Advanced Amateur Astronomy*, the second edition of which was published by Cambridge University Press in 1997. Another book you might like to consult is Martin Mobberley's *Astronomical Equipment for Amateurs*, published by Springer-Verlag in 1999.

3.4 EYEPIECE CHARACTERISTICS

Time and time again someone has enthusiastically shown me the view through their telescope and I have straight away recognised that the instrument is not performing as it should. The most common reason and the easiest to put right is poor collimation (see Appendix 1). The next most common reason is the use of an eyepiece of inappropriate design for the telescope or for the type of observation intended. Surprisingly this inappropriate eyepiece has usually been supplied along with the purchased telescope! So, please forgive the following 'back to basics' notes about eyepieces.

In all eyepieces the quality of the image deteriorates to some extent with increasing distance from the centre of the field of view. Manufacturers mask off the images of unacceptable quality by means of a circular aperture, known as a *field stop*, positioned in the eyepiece's focal plane. As you peer through the *eyelens* of an eyepiece, the angle through which you would have to swivel your eye to look directly from one edge of the field of view to the other is known as the *apparent field* of the eyepiece. The apparent fields of different designs of eyepiece range from about 30° to about 85°.

Another quantity of interest to the observer is the value of the *real field*. This is the actual angular extent of the sky the observer can see when using a particular eyepiece-telescope combination. A rough figure for the size of the Moon is 32 arcminutes (slightly over half a degree). If the real field of the eyepiece-telescope is bigger than this value then all the Moon can be seen at once. If the real field is smaller then the observer will only be able to see part of the Moon in one go.

The value of the real field obtained depends upon the values of the apparent field of the eyepiece (fixed by the manufacturer) and the magnification that it produces with the telescope. The relationship is:

$$real\ field = apparent\ field/magnification. \qquad (3.2)$$

For the equation to work the values of real and apparent field should be expressed in the same units, for instance both in degrees or both in arcminutes, etc.

Eyepieces for amateur telescopes now come in two standard sizes; those which fit drawtubes of diameter 1¼ inch (31.7 mm) and those which fit drawtubes of diameter 2 inch (50.8 mm). Eyepieces with 24.5 mm diameter barrels used to be common, especially in telescopes imported from the Asian continent, and so you might encounter them in the second-hand market.

For an eyepiece of apparent field 57° the field-stop aperture has the same diameter as the focal length of the eyepiece. This means that a wide field of long focal length cannot be accommodated in a '1¼-inch fitting' eyepiece. As one example, an eyepiece of 32 mm focal length and 80° apparent field will have a field-stop aperture of 44.7 mm diameter. So, the manufacturer would build that eyepiece into a '2-inch fitting' body.

Most modern eyepieces have external markings which indicate their type and focal length (for instance MA 20 mm, Or 18 mm, etc.). Those you may well encounter, including some now almost obsolete but which you might come across in the second-hand market, include: Huygenian (H), Huygens–Mittenzwey (HM), Ramsden (R), Achromatic Ramsden (AR or SR), Modified Achromatic (MA), Kellner (K), Orthoscopic (Or), Monocentric and Plössyl (usually named in full). There are also a variety of wide-field eyepieces, which may be useful to you though I would not recommend any of them for the most critical scrutiny of lunar surface details.

One characteristic of eyepieces that varies with design and usually varies with focal length is the *eye-relief*. This is the distance between the eyelens and the *exit pupil*, which is the position you need steer your eye pupil to in order to see the whole of the field of view. A generous eye-relief will be needed if you wear glasses at the telescope, for instance.

As an aside, I should say that you should not need to use glasses at your telescope if you suffer from simple long sight or short sight. The normal action of adjusting the telescope focuser should be enough to compensate. If you suffer from astigmatism then you will either need to wear your glasses or to use a compensating device such as the *Tele Vue Dioptrx*, which is an attachment that fits onto the eyepiece.

All modern eyepieces should have *bloomed* (anti-reflection coated) lenses. Blooming is highly desirable in order to avoid inter-glass reflections causing annoying ghost images and a general reduction of image contrast. The most modern eyepieces have sophisticated multi-layer coatings on most of their component lenses (generally referred to as lens *elements*) and are much superior in that respect to the eyepieces of twenty or more years ago. This is something to be borne in mind if you are buying second-hand equipment.

3.5 SPECIFIC EYEPIECE TYPES AND MAGNIFICATION

In general, if the effective focal ratio of the telescope is f/8 or more, then any of the simpler types will be suitable for lunar observing. Developed from the simple two-element Ramsden eyepiece, the three-element modern Achromatic Ramsden, Modified Achromatic and Kellner eyepieces are commonly available. They generally have apparent fields of about 40–45° and in focal lengths longer than about 12 mm will give good images in f/6 telescopes. A general softening of the image and even colour-fringing will be visible, particularly near the outermost regions of the field of view, when used on telescopes of lower focal ratio. Though the eye-relief of the textbook Kellner eyepiece (recognised by it having an eyelens with a concave surface facing the eye) is uncomfortably small, all modern ones share a similar design to the MA and AR eyepieces and all of these have a moderate eye-relief ranging about half the focal length of the eyepiece.

Like the simple Ramsdens, the two-element classical Huygenian eyepieces are an obsolete type. Again in common with the Ramsdens, they could not deliver good images when used with focal ratios of less than about f/10. However until the end of the 1990s imported Asian telescopes were often supplied with Huygenian eyepieces of modified design. You will recognise one of these by the fact that the eyelens is biconvex, the classical type having a plano-convex eyelens with the flat side facing the eye. These work well with focal ratios down to about f/8. They have apparent fields of view around 30° which are very free of ghost images and scattered light. Their eye-relief is only about a third of their focal length but I find no real difficulty in peering through one of 6 mm focal length. You might come across these eyepieces second hand, most likely with 24.5 mm diameter barrels and so needing an adapter to fit them into your modern eyepiece focuser.

Their main advantage is that they are (or at least should be!) extremely cheap if you can pick them up second hand. Beware, though, any eyepiece marked 'HM'. This denotes the Huygens–Mittenzwey design. It is a three-element eyepiece (doublet eyelens, converging-meniscus-field lens) which delivers a 50° apparent field. It used to be supplied with f/15 refractors and is quite unusable with lower focal ratios.

An eyepiece of yesteryear that used to be a firm favourite with Moon and planet observers was the Monocentric. This cemented triplet gives crisp and false-colour-free images even with f/5 telescopes and is especially notable for its freedom from scattered light in the image, the main disadvantage being a smallish field of view (apparent field about 30°). A Monocentric's eye-relief is usually about three quarters of its focal length.

For many years this design of eyepiece went out of production. Now one optical company, *TMB*, produce a current line of Monocentric eyepieces which they call 'Super Monocentric'. The August 2004 issue of *Sky & Telescope* magazine carries a review of these eyepieces by Gary Seronik. He compared them in use to other high-quality modern eyepieces and concluded that they only just bettered the best of the modern alternatives for image quality at the centre of the field of view. He reports that the image quality remains excellent at the centre of the field of view even when used with an f/4.5 telescope. However the image quality does then noticeably fall off towards the edge of the already narrow (30°) apparent field of view. They cost around $230, which is two or three times the price of a good alternative eyepiece.

If you want modern eyepieces for lunar and planetary observation for about $80 each, then you will be best served by the easily available Plössyl type. They typically have fields of view of around 50–55° and produce high-quality images with any telescope of focal ratio around f/6 or more (in longer focal lengths, say 20 mm, they are fine even with f/4.5 telescopes). They have largely superseded Orthoscopics as the choice four-element eyepiece, even though the finest examples of the Orthoscopics are slightly superior in imaging characteristics to the finest of the Plössyl type, except in their apparent field diameters of 40–45°. The eye-relief of Orthoscopic and Plössyl eyepieces is about the same as that for the Monocentric type. If buying second hand beware the Asian so-called 'Plössyl' eyepieces, which have replaced the Huygenian eyepieces once supplied with their telescopes. These give dismal performance even with f/8 telescopes and the images are noticeably degraded well before the outside of the 40° apparent field of view is reached.

If you can stretch your budget to about $230 per eyepiece then you can have what is probably the best type of eyepiece of all. I refer to the

Radian eyepieces introduced by *Tele Vue* in 1999. As well as excellent image quality they deliver a 60° field of view and 20 mm eye-relief in all focal lengths (the range available goes from 18 mm down to 3 mm).

Apart from the possible exception of the *Tele Vue* Nagler eyepieces (which can also work well with focal ratios down to f/4.5, maybe even f/4), I would not recommend using any of the wide-field eyepieces for lunar and planetary observation, as most lack critical definition even at the centre of the field of view. Many are also troubled by scattered light and ghost images, despite blooming. This is because they contain six, or even more, separate lens elements.

Barlow lenses can be useful. The most common form of this accessory consists of a diverging (concave or plano-concave) doublet lens set into one end of a tube, or series of tubes. This end is plugged into the telescope drawtube while the eyepiece is plugged into the other. The reduction of the convergence of the rays from the telescope objective has the consequence of multiplying the effective focal length of the telescope by a given factor. This is usually ×2 but can be anything the manufacturer desires. All this will be common knowledge to most readers. However, not all may realise that the effective focal ratio (EFR) of the telescope is multiplied by the same factor. Thus an f/5 telescope is converted to an f/10 one, using a ×2 Barlow. As well as the obvious multiplication of the magnification of the telescope by a given eyepiece, using the Barlow lens allows the simpler (and cheaper!) eyepieces to be used if so desired.

One note of caution, though; adding more lens elements into the optical path will increase the amount of light absorbed and scattered. Of course, blooming, and ensuring the lenses are scrupulously clean (not always easy to do) will go a long way to negating this. Also, the Barlow lens must be of high optical quality if it is not to degrade the view produced by the telescope. Regrettably, this is not always the case.

In particular, if your telescope has a focal ratio of less than f/6 I would caution that a common two-element Barlow lens will introduce enough chromatic aberration to produce noticeable colour-fringing when used with powerful (focal length less than a centimetre) eyepieces. Mind you, if you have a cheap low-focal-ratio refracting telescope a cheap short-focus achromatic Barlow lens may actually help it deliver better images, as I describe in the next section.

Some manufacturers supply triplet (apochromatic) Barlow lenses and I would recommend you obtain one of these if you want to use it with a reflecting telescope of low focal ratio. By easing the load on your eyepieces, a first-rate apochromatic Barlow lens may actually upgrade the quality of the images your low-focal-ratio telescope can produce, despite any slight negation due to scattered light. In addition, a few well-chosen eyepieces are sufficient, if used in conjunction with one or more Barlow lenses, to deliver a large range of magnifications with a given telescope.

What about the actual values of magnification? Here I must stress that personal preference must rule the day. Having one eyepiece in which you can view the entire Moon in one go is highly desirable, especially for occasions such as eclipses. The actual field imaged must therefore not be less than about 0°.6. This means a magnification of no more than about ×66 (for an eyepiece with a 40° apparent field) to ×86 (52° apparent field eyepiece).

Having a set of eyepieces that can deliver a series of higher magnifications (and perhaps a Barlow lens to help fill in the steps) is also desirable. If the observing conditions were perfect and the telescope were of good quality then I would normally prefer a magnification roughly equal to the aperture of the telescope measured in millimetres. However, 'fussy' detail, such as the individual peaks in a lunar mountain range, might be better appreciated with higher powers, perhaps up to ×2 per millimetre of aperture. Poorer conditions demand lower powers, of course, as do broad features of low contrast.

Most of my lunar observing is done with powers of ×144 and ×207 with my 18¼-inch (0.46 m) reflector because of the usual, 1 arcsecond, seeing conditions at my observing site. Even then, the 1 arcsecond figure refers to the brief glimpses of fine detail, not the average amount of blurring and distortion which amounts to several times this value. I seldom see any real advantage to using higher powers, though on rare outstanding nights I have had incredible views at ×432.

3.6   MAKING THE BEST OF WHAT YOU HAVE

Let us say that you have a great big 'light-bucket' of a Newtonian reflector with a low focal ratio and mediocre optics. Perhaps it is one of the cheaply produced large Dobsonian-mounted telescopes (though some Dobsonian telescopes have excellent optics). It will probably come supplied with one or more low-cost eyepieces. You find it gives very bright but disappointingly blurred images of the Moon. Can you improve matters without having to buy another telescope? I am glad to say that the answer is 'Yes'.

Let us consider the eyepieces. The set that the light-bucket has, probably Kellners, or Modified Achromats, will undoubtedly work better with a telescope of higher EFR. If your telescope has a focal ratio lower than f/6, you might be advised to obtain a triplet (truly apochromatic) Barlow lens in order to raise its effective focal ratio. A more expensive alternative is, of course, to buy a higher-quality set of eyepieces. Perhaps the foregoing notes might help you make the choice.

Figure 3.3 An off-axis
stop for a reflecting
telescope. Unless the
mirror is thin enough to
distort along its lower
edge (unlikely in
amateur-sized
telescopes) positioning
the aperture to expose
the lower part of the
mirror will normally
provide the best images.
This is because all of the
mirror surface will
normally produce
upwardly moving
convective warm air
currents. Positioning the
aperture over the lower
regions minimises the
amount of convecting air
the light rays have to pass
through.

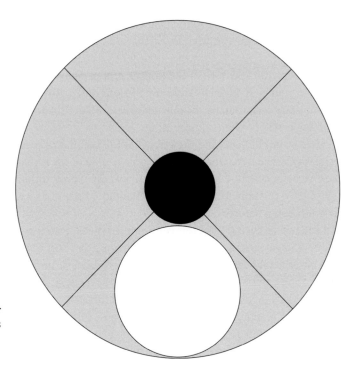

The advantages of using off-axis stops over the telescope aperture is
hotly disputed. My own experience is that stopping my 18¼-inch (0.46 m)
telescope down to 6 inches (152 mm) with a cardboard diaphragm **sometimes** improves the images the telescope delivers. When the seeing is
particularly rough and the images of details tend to be multiple and
confused the stop sometimes 'cleans up' the view, as I described earlier.
I have had the same experiences using aperture stops on other large
telescopes, for instance the 19½-inch (0.50 m) reflector at the Breckland
Astronomical Society observatory.

Certainly if the optics of the telescope are of poor quality, then
stopping the instrument down will usually improve matters, whatever
the seeing. Also, the diffraction pattern produced, using the stop, is more
refractor-like, even if it is broader because the aperture generating it
is smaller. The hole in the diaphragm should be made and positioned
to avoid the secondary mirror and its support vanes (see Figure 3.3). A
16-inch to 20-inch 'light-bucket' may be made to perform like a high-
quality 5-inch to 8-inch refractor in this way. Try this for yourself. All it
takes is a few minutes to mark and cut out a piece of cardboard to suit
your telescope. However, I would caution you only to use the diaphragm
on the occasions you see a definite improvement in the image. Otherwise

the diffraction limit imposed by the reduced aperture might mean that you miss the occasional flashes of fine detail that one usually gets on even poor nights.

The secondary spectrum apparent in the images produced by a cheap achromatic refractor can be reduced by stopping the lens down. However since the refractor is unlikely to have an aperture greater than 6 inches to start with, reducing it still further is undesirable.

There are other ways of alleviating the effects of the secondary spectrum in the cheaper refractors. One way is to try an optical trick that occurred to me years ago, and I have tried with success. Surely other people must have tried the same thing for themselves but I have not seen the method written down before.

My idea (I'll call it that even if I am reinventing the wheel!) is to use a cheap achromatic Barlow lens, best of all one of particularly short focal length, plugged into the telescope before the eyepiece. Its own secondary spectrum will be of the opposite sense to that produced by the telescope's object glass (the Barlow's focus for yellow-green light is furthest from the lens, the other colours reaching focus closer to the lens). Although the Barlow's secondary spectrum is unlikely to be strong enough to completely cancel the refractor's secondary spectrum, it will at least reduce it.

The situation can be further helped by lengthening the barrel of the Barlow lens (by plugging in an extension tube that goes in before the eyepiece). Along with further increasing the magnification of the Barlow lens, this will significantly increase the amount of secondary spectrum generated by it, so cancelling out rather more of the dominant secondary spectrum of the refractor's objective.

However, this cannot be carried too far. For one thing the magnification may become greater than desirable, though one can bring the magnification down by swapping the chosen eyepiece for a longer-focus one. Also the seidal aberrations generated by the Barlow lens (particularly spherical aberration) may then become great enough to spoil the image.

Provided it is also a cheap achromatic lens, rather than a more expensive apochromatic model, a 'dumpy' Barlow will also generate the maximum amount of its own secondary spectrum – and we need as much 'negative' secondary spectrum as possible from the Barlow to counter the 'positive' secondary spectrum of the telescope's cheap achromatic objective.

An alternative way of reducing the effect of the secondary spectrum is to attach a strongly coloured yellow or green filter to the eyepiece before you plug it into the telescope. You will have to put up with a Moon that appears the colour of the filter but at least the image will appear sharp (provided the filter is of good quality) and colour-fringe free.

One step up from that is a filter which is specially designed to reduce the visibility of the secondary spectrum of refractors. Companies that make these include *Baader Planetarium*, *Orion Telescopes and Binoculars*, and *Sirius Optics*. I give contact details for these and other companies in Chapter 7. You might like to refer to an article: 'Curing the refractor blues', in the April 2004 issue of *Sky & Telescope* magazine, where the author compares the effects of a number of different filters on views through his achromatic 6-inch f/8 refractor. The main advantage of these special filters is that they significantly reduce the appearance of the secondary spectrum while only imparting a relatively weak colour-cast to the image of the Moon.

If your telescope is deficient in its mechanical construction then perhaps you can make or purchase some replacement parts to improve matters. If the optics are good you might even think about rebuilding the instrument and just salvaging some of the parts of the original. At the same time you can make any changes that will improve its thermal characteristics. However, that is taking us into matters which are beyond the remit for this book. Let us now, at long last, get down to some actual observing and drawing of the Moon's beautiful vistas ...

3.7 DRAWING THE MOON

Given the telescope, the basic equipment consists of some sort of clip-board with a source of attached illumination. A small piece of hardboard, or a large, stiff dinner mat, a switch, a torch bulb and holder, a small square cardboard or metal box (to mount the bulb in its holder at the head of the board), some wire, a battery (perhaps mounted on the board by means of a Terry-clip) and terminal connections (or soldered joints), or alternative materials, can easily be fashioned into something appropriate. Adding direct shielding from the bulb is also desirable and a small rheostat is also useful for brightness control. However, keep it simple and, above all, keep it lightweight.

Your drawing can be a simple line-diagram. At the other extreme it can be a photographic-quality work of art showing all the half-tones. What is possible depends on your abilities. This will, of course, improve with practice. Even though your work will not be used for pioneering research, astronomy **is** still a science. Therefore your drawing must be accurate. Your fellow astronomers will think little of the most pictu-resque representation that you can produce if it is inaccurate in its proportions and positions. Even the simplest representation that is accu-rate is always vastly preferable.

I ought to emphasise that there are many ways of achieving good representations of the Moon's surface. There is a large array of materials – pencil, pen, ink, charcoal, paint, etc. – to be used on an equally large array of papers, canvases, etc. Each will demand its own techniques. Moreover, each individual will undoubtedly find various ways of working that suit him/her best. Consequently, all I will do here is to offer a few guidelines.

Andrew Johnson is one of a number of amateur astronomers producing superb representations of the lunar surface. He has written an article outlining his own methods, and giving good general advice for the tyro, in the Association of Lunar and Planetary Observers (ALPO) publication *The Strolling Astronomer* (Volume 37, Number 1, May 1993, pages 18–23). Andrew has generously let me reproduce some of the illustrations from that article here.

When you first go to your telescope try not to spend too much time deciding what to draw. At least having an outline plan for the evening will save you a lot of time and effort when you should be actively observing, even though unpredictable weather and observing conditions will demand some flexibility on your part.

Having decided on your chosen target, spend a while scrutinising it with different magnifications before committing anything to paper. Do not attempt to take in a large area in one go. The area you should cover in your drawing should certainly not be greater than about 200 km square on the lunar surface. Better still if it is smaller. When you are about ready to begin drawing, aim for a scale of at least 2 km per millimetre.

Considering just the simplest line drawings, you might generate the sketch entirely from the view through the telescope straight on to a blank sheet of paper. If so, you will achieve the best accuracy by starting with a set of faint pencil guidelines to help you with the proportions. This is illustrated in Figure 3.4(a), which shows the first stage of a drawing of the crater Mairan by Andrew Johnson. Keeping the guidelines very light allows them to be erased easily once the major details are blocked in, as they have been in Figure 3.4(b).

Alternatively, you could base your sketch on a pre-prepared outline of the major features. Your work at the telescope would then consist of filling in the fine details and the shadows. You might make an outline by tracing over a suitable photograph. This technique ought to produce a greater positional accuracy in the drawing, though you must not ignore the effects of libration, especially for features near the lunar limb. The libration value at the time of your drawing is unlikely to be similar to that when the photograph was taken.

The real Moon has shades of grey on it, as well as blacks and whites. There are a number of ways you can represent these on your drawing. One is to add numbers to the areas, representing brightness gradations.

Figure 3.4(c) shows the next stage in Andrew Johnson's drawing of Mairan. On the scale Andrew has used 0 represents black and 10 represents brilliant white. This can be the finished product.

Alternatively, the outline sketch can be carefully traced and, using the numbered original, the shades of grey can be built on in pencil, or what other medium is chosen, on the copy. Most of the really first-rate lunar artists use this approach. Obviously, the finished version of the drawing is made after the observing session. This demands that everything is meticulously noted during the observation. Never rely on your memory of what you think you saw through the eyepiece of the telescope. Get it right at the telescope and you will have no temptation to make subsequent alterations to your drawing.

When you have finished your initial sketch, spend a little time comparing it with the scene through the eyepiece. Any perceived inadequacies of your drawing that you do not feel able to correct can always be noted along with it, e.g. 'small crater should be drawn 20 per cent larger', etc. Do not forget to include all the usual details of date, time, instrumental details, magnifications, seeing conditions.

You might prefer to make your final version at the telescope, complete with all shades of grey and the black shadows filled in. This is only possible if you work with very small, and hence simple, areas of the lunar surface. Close to the terminator the shadows change noticeably in just a few minutes. While it is certainly true that you can lay the outline down and then note the time, subsequently spending time doing the shadings, you will find that there is simply not enough time to do a complicated drawing before the lighting over the scene changes too much. Ideally you ought to complete your sketch within half an hour.

If you do want to do the whole thing, and produce the finished drawing at the telescope, it is highly desirable you take steps to save time and so work with maximum efficiency during the observation period. Having a pre-prepared outline is particularly useful, as is greying the picture area of your paper beforehand. You could use pencil shading for this, or even sprinkling on a little charcoal powder, in either case then smoothing with the finger. At the telescope, a clean rubber can be used for creating the lighter areas and a sharply pointed rubber effectively doubles as a 'white pencil'. Darker shadings can be built up with pencil, and black felt-tip pens, with fine and broad tips as appropriate, can be used to create the black shadows.

If you decide to produce the final version of your drawing after the observing session then, as already stated, the ways you can do it are almost unlimited. However, not all methods will be conducive to subsequent copying. Having back-up copies and copies to send to observing

(a)

(b)

(c)

(d)

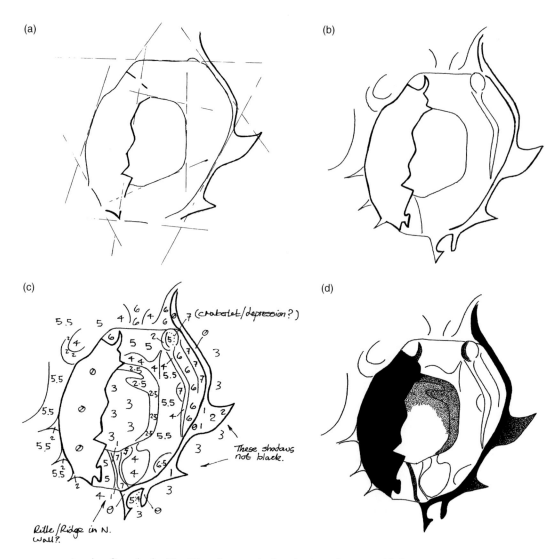

6 (crobertet/depression?)

These shadows
not black.

Rille/Ridge in N.
Wall?

groups, etc., is often desirable. The cheapest, simplest, and most widely available method of copying, these days, is by way of a Xerox or other photocopying machine. These can reproduce any image composed of lines and dots very well. Unfortunately, photocopy machines often do not reproduce half-tone images accurately. Consequently, many lunar artists use stippling as a way of representing half-tone shadings in their finished drawings. In his *The Strolling Astronomer* article, Andrew Johnson suggests using a pen (the choice ranging from the expensive professional drafting pens, to the cheap but less durable felt-tips) with a point diameter of 0.3 mm. With great patience the shadings are built up, dot by dot. Andrew uses a desk-mounted magnifying glass, and provides

Figure 3.4(a)–(e) Stages in the drawing of the lunar crater Mairan by Andrew Johnson. See text for details.

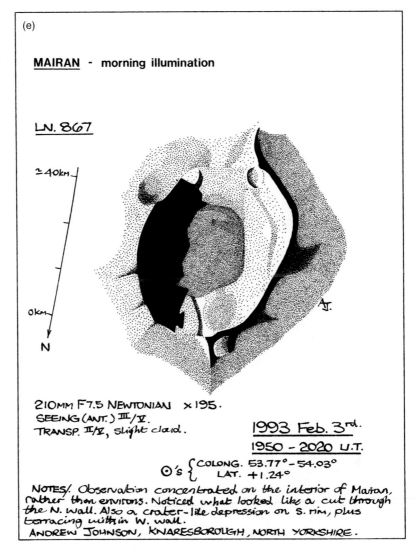

(e)

MAIRAN - morning illumination

LN. 867

≃40km

0km

N

210MM F7.5 NEWTONIAN ×195.
SEEING (ANT.) III/V.
TRANSP. II/V, slight cloud.

1993 Feb. 3rd.
1950 - 2020 U.T.

⊙'s { COLONG. 53.77° - 54.03°
      { LAT. +1.24°

NOTES/. Observation concentrated on the interior of Mairan, rather than environs. Noticed what looked like a cut through the N. wall. Also a crater-like depression on S. rim, plus terracing within W. wall.
ANDREW JOHNSON, KNARESBOROUGH, NORTH YORKSHIRE.

good illumination to ease eye strain. The greater the density of the dots, the darker the shading when the drawing is viewed from the normal distance. Andrew typically takes around two hours to produce the finished version of one of his drawings. Figure 3.4(d) shows the process under way for his Mairan observation and Figure 3.4(e) shows the final result.

I am greatly indebted to Andrew and his co-workers Nigel Longshaw and Roy Bridge for allowing me to feature examples of their fine work in the 'A to Z' section of this book, Chapter 8. You will find there many

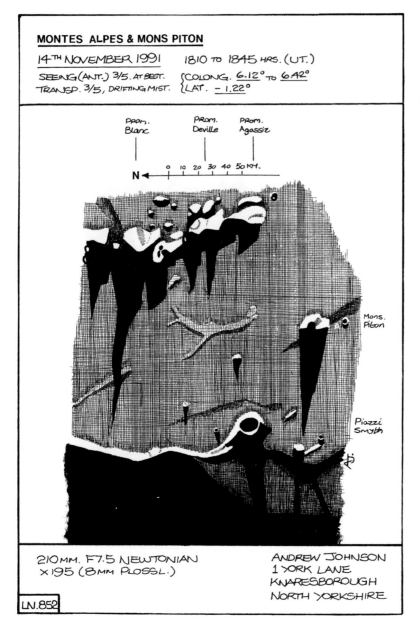

MONTES ALPES & MONS PITON

14TH NOVEMBER 1991   1810 TO 1845 HRS. (UT.)

SEEING (ANT.) 3/5. AT BEST.   ∫COLONG. 6.12° TO 6.42°
TRANSP. 3/5, DRIFTING MIST.   ∫LAT. - 1.22°

PROM. Blanc   PROM. Deville   PROM. Agassiz

0 10 20 30 40 50 KM.
N ◄

Mons. Piton

Piazzi Smyth

210MM. F7.5 NEWTONIAN   ANDREW JOHNSON
×195 (8MM PLOSSL.)   1 YORK LANE
                             KNARESBOROUGH
                             NORTH YORKSHIRE

LN.852

Figure 3.5 One alternative to the stippling technique is cross-hatching, here illustrated in a drawing of Montes Alpes and Mons Piton by Andrew Johnson.

examples of how they represent different types of formations and light-ing aspects.

If the sheer labour of the stippling process puts you off, you might try normal pencil shadings but on stippled paper (obtainable from an art shop), or with normal paper placed over coarse sandpaper. This will create 'pseudo-stippling' which will photocopy much better than plain shading alone. Varying the pressure of the pencil creates larger or smaller dots, producing the required shading variations when viewed from the normal distance.

An alternative to stippling that will still photocopy well is cross-hatching. Figure 3.5 shows a striking example of this method produced by Andrew Johnson. Of course, if appropriate copying methods (laser copier, computer scanner with 'photo-quality' output, etc.) are used, then less-arduous methods can be used to make the final drawing.

One rung up the technological ladder from making drawings of the Moon is to photograph it. That is the subject of the next chapter.

CHAPTER 4

The Moon in camera

The early years of the nineteenth century saw the invention and development of photography. The first processes and photographic materials were clumsy and insensitive but a few determined individuals tried their best to record images of astronomical bodies. J. W. Draper, of New York, is usually credited as the first to achieve significant success in photographing the Moon. In the *Scientific Memoirs* of 1840 he writes:

There is no difficulty in procuring impressions of the Moon by the Daguerreotype. By the aid of a lens and a heliostat, I caused the moonbeams to converge on the plate, the lens being three inches in diameter. In half an hour a very strong impression was obtained. With another arrangement of lenses I obtained a stain nearly an inch in diameter, and of the general figure of the Moon, in which the places of the dark spots might be indistinctly traced.

A decade later J. A. Whipple, also in the USA, succeeded in producing a series of Daguerreotypes of the Moon at various lunar phases. The English amateur Warren de la Rue achieved better results shortly after, as did Lewis Rutherfurd in America. By the close of the nineteenth century the quality of the photographs obtained had improved to the point that the first photographic atlases of the Moon could be compiled.

For example, W. H. Pickering published a complete photographic atlas of the Moon in 1904. The plates were taken at the focus of a specially constructed 12-inch (305 mm) objective. Each region of the Moon was photographed under five different lighting angles, though the scale was not large enough to show the smallest details on the lunar surface. The best lunar photographic atlas of the period was that produced by M. Loewy and P. Puiseux of the Paris Observatory. They employed the 23.6-inch (0.6 m) coudé refractor there between 1896 and 1909 to take the plates for their *Atlas de la Lune*.

As the years rolled on the prospect of space-probes, and even manned missions, to the Moon spurred on further efforts. G. P. Kuiper and his colleagues published the *Photographic Lunar Atlas* in 1960. This was a boxed set of large photographs of the Moon, the best examples of images from Mount Wilson, Lick, Pic du Midi, MacDonald, and Yerkes Observatories.

Under the auspices of the United States Air Force (USAF), Zdeněk Kopal headed the 'Manchester Group' (based at England's Manchester University), carrying out photography from Pic du Midi Observatory from 1959 to the 1970s. Situated high in the French Pyrenees the seeing at the Pic is superb. Initially the chief instrument used was a 23.6-inch refractor, the object glass of which was the very same as that previously installed in the instrument used by Loewy and Puiseux at the Paris Observatory. One member of the Manchester Group, Dr Thomas W. Rackham, had special responsibility for the work with this refractor, though other group members also took part.

Well over sixty thousand photographs were eventually obtained, from which the *Lunar Air Force Charts* were constructed. The photographs were good enough to enable relative heights on the lunar surface to be determined (by means of measurements of the cast shadows) with an accuracy of a few metres in many cases.

The sub-arcsecond seeing typical at the Pic prompted the 'Manchester Group' to procure for the observatory a 43-inch (1.07 m) Cassegrain reflector of design and optical quality especially suited to lunar and planetary imaging. The various members of the Manchester Group, Patrick Sudbury having special responsibility for the use of this instrument, began photographic work with it in 1964. Meanwhile they also collaborated with Professor S. Miyamoto and his colleagues in Japan, (again under the auspices of the USAF and NASA) in further photographic lunar cartography, adding the 74-inch (1.9 m) reflector at Kottamia in Egypt to their arsenal. Dr Rackham very kindly allowed me to reproduce a number of the photographs taken at Kottamia and Pic du Midi in the first edition of this book. Sadly, he has since died but I have kept the same photographs in this edition.

The enormously productive Manchester Group were not the only major players in the lunar mapping game. Gerard P. Kuiper, Ewen A. Whitaker, along with Messrs Strom, Fountain and Larson of the Lunar and Planetary Laboratory, based at the University of Arizona, in America, were also active. They produced their superb *Consolidated Lunar Atlas*, consisting of the best 227 photographs of the Moon taken with the 61-inch (1.54 m) Naval Observatory astrometric reflector and the 61-inch reflector on Catalina Mountain, in Arizona. Most of the work (over 8000 negatives actually exposed) was done with the Catalina telescope between 1965 and 1967. You will find sectional enlargements made from many of the Catalina photographs in this book, thanks to the kindness of Ewen Whitaker and the University of Arizona.

The professional programmes of Moon-mapping were essential to the up-and-coming space missions. Amateur astronomers also tried their hand at photographing the Moon. Many were very successful, even if the typical backyard observing conditions rarely allowed sub-arcsecond resolution to be achieved in their photographs.

During the 1960s one individual even emulated the professionals and produced a photographic atlas that was commercially published. I refer to the very useful *Amateur Astronomer's Photographic Lunar Atlas* by Commander H. R. Hatfield. In it the Moon is divided into 16 sections, there being for each a detailed key map and several photographs taken under different lighting conditions. The photos are mostly at a scale of 25 inches (64 cm) to the Moon's full diameter, with a few supplementary close-ups reproduced at a scale of 36 inches (91 cm) to the Moon's full diameter. Commander Hatfield even built the 12-inch (305 mm) Newtonian reflector and the observatory that housed it himself, as well as much of his photographic equipment. The first edition is now long out of print (it was published by Lutterworth Press in 1968). Springer-Verlag eventually, in 1998, published a new edition of Commander Hatfield's atlas, edited by Jeremy Cook. Now it goes under the title of *The Hatfield Photographic Lunar Atlas*, and is pretty much identical to the first edition, except now having the cardinal points of lunar east and west reversed and the names of some features also updated to the latest IAU (*International Astronomical Union*) edicts.

Jeremy Cook had virtually completed the work on another version of the Hatfield Atlas that was intended for observers using telescopes with star diagonals that give mirror-reversed images when he tragically died. His son Tony put the finishing touches to the work, and *The Hatfield SCT Atlas* was published by Springer-Verlag in 2005.

Georges Viscardy must rank as the premier amateur lunar photographer of the last years of film-based photography because of his very high output of lunar photographs rivalling the professional efforts in their quality and resolution. He used a 20½-inch (0.52 m) Cassegrain reflector set up in the French Alps at a site of superb seeing. His *Atlas–Guide Photographique de la Lune*, published in 1986, contains over 200 plates, with technical details and some descriptions of the surface features.

You will be aware how film-based photography has been in steep decline in these last few years as various forms of electronic imaging have taken over. Also computer processing allied to these methods has allowed the production of lunar images of superb resolution, much superior to the results obtained by film photography.

In the first edition of this book I included a long chapter on how to go about photographing the Moon onto film. If you wish to undertake lunar photography in that medium then I must refer you to that first edition. Do bear in mind, though, that many of the photographic films I referred to are no longer commercially available and all photographic film must surely soon go out of production altogether.

In this edition I only have room enough for the necessary new materials on electronic imaging, though throughout this new edition of I have kept in many of the illustrations that were made using film and provide details in the accompanying captions. I hope that you will feel inclined to try your own hand at lunar photography and I offer the following notes by way of an introduction.

4.1   SOME BASIC PRINCIPLES OF CCD ASTROCAMERAS AND DIGITAL CAMERAS

The imaging devices most commonly used in the arena of amateur astronomy are: the *CCD astrocamera*, the *digital camera* (preferably a *digital single lens reflex – DSLR – camera)* the *video camera*, and the *webcam*. Generally the DSLR is best for wide-angle shots encompassing the whole Moon and the webcam, used with appropriate software, is best for obtaining the highest resolution views of the Moon. In the majority of cases modern imaging devices have at their heart a device called a *CCD*, as the image collector. A cheaper alternative to a CCD is a CMOS (complimentary metal oxide semiconductor) detector. The *Nikon* range of DSLRs use these, as do some other devices such as the cheapest webcams. The *Nikon* cameras are extremely good for their price. In general, though, you get what you pay for and so I recommend choosing devices with CCD detectors wherever possible.

A *CCD*, or *charge-coupled device*, consists of an array of light-collecting units, called *pixels*. Each pixel on the CCD has the same size and shape as its neighbours. That size can range from about 5 μm to about 25 μm (5 micrometres to 25 micrometres) square. Professional astronomers mostly use large CCDs, typically having 2048 × 2048 or more pixels each of around 25 μm square. These are very expensive. Currently amateurs use smaller versions. At the time I am writing these words a typical amateur's CCD astrocamera might have an array of something like 500 × 750 pixels each perhaps 9 μm square. In that case the total imaging area of the CCD would be 4.50 mm × 6.75 mm.

Whatever the size of CCD, the array of pixels are mounted on an 'integrated circuit' or 'silicon chip' type base which has about 20 individual electrical connections to its supporting electronics. The way it works is that photons of light falling on particular pixels liberate electrical charges within each of them. The more light (and so more photons) falling on a given pixel, the more electrical charge is created within it. If an image is focused on the picture-receiving area of the CCD the pixels corresponding to the brightest parts of the image have the greatest amounts of charge liberated in them. The dimmest parts of the image generate the smallest amount of charges in the corresponding pixels.

Charges would continue to build up all the while the light is falling, until each and every pixel is full, or *saturated*. Well before this stage is reached the process, known as *integration*, has to be stopped. Ideally an *integration time* (equivalent to the photographic 'exposure length') is selected so that at the end of it the pixels associated with the dimmest parts of the image have only a small charge while those associated with the brightest parts of the image have lots of charge, though less than the amount necessary for saturation.

When the integration is completed the array of charges are sequentially read off the chip and sent as a representative data stream to a computer, or other electronics, to deal with in order to recreate the image on a monitor/TV screen, or to download it into a computer's memory, or onto a computer disk, to produce a printout, etc.

A CCD is not equally sensitive to all wavelengths of light. The percentage of the number of photons of light falling on the CCD at any given wavelength that is detected by it is known as the *DQE* of the CCD at that wavelength. The term DQE stands for *detector quantum efficiency*. A DQE of 100 per cent is the best that one could possibly have; all the incoming photons then being detected.

The earlier examples of CCDs all tended to have their maximum sensitivity in the near-infrared portion of the spectrum. A typical response might be a DQE of about 40–80 per cent in the 600 nm–950 nm wavelength range falling away steeply at both shorter and longer wavelengths, to become zero at about 400 nm and again at about 1100 nm. This is very different to the spectral response of the eye, the maximum response of which occurs at a wavelength of about 550 nm in the yellow portion of the spectrum and falls to zero at about 380 nm (deep violet) and at about 700 nm (deep red).

Many of the more recent generations of CCDs have coatings which enhance their response to light at the blue end of the spectrum. As an example, the *Philips* FT12 has a response which is closer to that of the human eye. It has a peak sensitivity at 530 nm, falling to half that value at about 400 nm and 700 nm. However, it does so at the expense of some of its sensitivity, having a peak value of DQE of only 30 per cent.

Detector quantum efficiency is a very important factor when imaging deep-sky objects. We Moon imagers are more fortunate because the Moon appears so bright in our skies. Digital cameras have CCDs which are designed to mimic the response of the human eye. They also have a filter matrix layered on top of the CCD so the camera can record images in colour. For these reasons domestic digital cameras have rather lower values of DQE than do astrocameras.

If you are trying to image fine details on the Moon and have a choice between using a digital camera and an astrocamera, then the astro-camera will be better as you can use shorter exposures. This will give you a better chance of securing a reasonably sharp view in average seeing conditions. Your pictures will also be less affected by image drift due to poor tracking or poor polar alignment. By contrast photographic films typically had DQEs of circa 2 per cent and the longer exposures needed made securing fine details much more difficult than is the case with today's digital technology.

So much for the basic principles. There are a number of variations in the design of modern CCD detectors. Look at the literature and you will come across the terms *progressive scan*, *interline transfer* and *frame transfer*. These refer to the way the CCD is structured and, consequently, how the image is read from the chip. Most 'domestic' devices such as video cameras, security video cameras and digital cameras use interline trans-fer chips but there are also a few astrocameras fitted with them, in particular those of the *Starlight Xpress* range.

You will also come across *back-illuminated* and *front-illuminated* CCDs. These terms also relate to the mechanical structure of the CCD. Each type has its theoretical advantages and disadvantages (mainly in sensitivity, free-dom from 'noise', resolution and spectral response). You can do lunar imag-ing with any of these types. In the real world, rather than in the theoretical realm, considerations of cost, pixel size and imaging area are of the greatest importance to you if you are to equip yourself on a limited budget.

One problem afflicting CCDs of all types is something called *dark current*. While an integration is under way thermally liberated charges build up in each of the pixels along with those liberated by the incident light. At room temperature these charges can build up to saturate each pixel fully in just a few seconds. Even before then the charges are reducing the total dynamic range (range of brightness levels) recordable. The effects are negligible for very short integration times, say a fraction of a second. However, integra-tions longer than a few seconds produce images afflicted with a 'noise' of random brightness variations at the pixel level of detail superimposed on the otherwise pure image. This can even build to a snowstorm effect with longer integrations as more and more of the pixels fully saturate.

Practical CCD astrocameras have built-in thermoelectric coolers. An astronomer wishing to record a faint galaxy on a CCD will need to use an integration time of minutes, maybe even stretching into hours. Cooling the CCD is then essential. We photographers of the Moon are fortunate in that it sends us plenty of light. However, if we use optics to provide us with a highly enlarged image, or we are trying to image the Moon during an eclipse, then the integration time needed might stretch to a second or two. In that case cooling, while not essential, is preferable. Some top-range digital cameras are fitted with CCDs capable of reasonable results with exposures as long as several tens of seconds.

4.2   PRACTICAL CCD ASTROCAMERAS AND DIGITAL CAMERAS

Let me say at the outset that if I wanted to perform quality Moon (or planet) imaging on a limited budget then my first choice would **not** be to buy a CCD astrocamera. I would, instead, go for a webcam: more about them in the next chapter. However, if I was interested in imaging comets, nebulae, galaxies and other faint objects then I would purchase a proper, cooled-chip, CCD astrocamera. However, to show that excellent lunar images are obtainable from CCD astrocameras, please see Chapter 2 (Figure 2.7), Chapter 4 (Figures 4.6(a)–(d) and Chapter 8 (Figures 8.13(d), 8.15(b), 8.17(f), 8.33(c), 8.44(e), 8.46(c) and 8.47(a)). Details are given in the accompanying captions.

My attitude is the same regarding digital cameras. If I had one for domestic use, then all well and good. It is not something I would purchase specifically to do my lunar imaging with (with the possible exception of whole Moon portraits – see later). If I had no webcam to use then I would certainly be interested in using whatever I did have, be it digital camera or astrocamera, for imaging the Moon (and planets) as well.

The basic characteristics of a CCD astrocamera for imaging the Moon that we need to concern ourselves with are the size of the imaging area of the CCD and resolution in the image (number of pixels height × number of pixels width comprising the image).

Figure 4.1 shows the camera head of the *Starlight Xpress* SXL8 unit. Actually, I should say that this is one of the company's oldest cameras. *Starlight Xpress Ltd.* have a website at www.starlight-xpress.co.uk on which you can view their latest products. Notice the cooling fins projecting from the back of the camera head. The major part of the mass of the camera head (about 1 kilogram) is associated with the cooling unit. The small grey square that lies within the head is the actual CCD. It is the *Philips* FT12, referred to earlier. Nowadays the ribbon cabling to your computer's parallel port has given way to USB (Universal Serial Bus) connectors and all functions of camera control and image acquisition are controlled from your computer.

The camera head will come with an adapter that screws into the front of it. This is to enable it to be plugged into the telescope drawtube (or

Figure 4.1 The *Starlight Xpress* SXL8 camera head. The CCD can be seen within it (the scale is in cm.).

Barlow lens, or other amplifying system used on the telescope) in the same manner as one would plug in an eyepiece.

One can also use the camera head with a photographic lens if desired. If you have, or wish to buy, second-hand, photographic lenses that formerly were used with single-lens reflex (SLR) film cameras – and there are a lot of them about – then these can be attached to CCD astrocameras with the appropriate adapter ring between the camera body and the lens. Consult the manufacturer/supplier for advice on the specific adapter needed. As well as the obvious thread/bayonet-matching, the adapter has to set the lens at the correct distance from the CCD.

In most cases when using old camera lenses you will also need an infrared-blocking filter to attach to the front of the lens (usually by the conventional 48 mm threaded fitting filter-mount) because the focus for the near-infrared light that the CCD is also sensitive to differs from that of the visual wavelengths the lens was designed for. Photographic films were barely sensitive to deep-red light and were totally insensitive to the infrared.

With the camera plugged into the telescope, the weight of the camera head will necessitate rebalancing the telescope tube, perhaps using additional counterweights and maybe sliding the tube along in its cradles or using strap-on weights, etc. This is not something we have to worry about with the much lighter webcam.

As well as the *Starlight Xpress* astrocameras, there are many others made by other companies, such as the *Santa Barbara Imaging Group* (*S-BIG*). Their web-site address is www.sbig.com. See Chapter 7 for a fuller listing. I recommend searching out advertisements in astronomy magazines current at the time you decide to purchase the camera. Get further information direct from the manufacturers and take the time to make your choice carefully.

When you have made your purchase read the manufacturer's instructions very carefully. The time and effort spent will be more than repaid by how quickly you will be able to achieve acceptable results. There will be variations in operation between one system and the next and so I will confine myself to offering general comments on matters relating to operating the camera with your telescope in order to image the Moon. Firstly, though, we should consider how much of the Moon we can image in one go and just how much detail we can expect to get in our images.

4.3   THE IMAGING AREA OF A CCD CAMERA WHEN USED ON YOUR TELESCOPE, OR WITH AN ATTACHED CAMERA LENS

The image scale (in arcseconds per millimetre) at the focal plane of your telescope is given by:

$$\text{image scale} = 206265/f \qquad (4.1)$$

where f is the *effective focal length* of the telescope. The figure used for the effective focal length must also take into account the effect of the amplifying secondary mirror of a compound telescope and any other additional optics (Barlow lenses, etc.) in the light path before the focal plane. More about that later. The same formula holds true for a camera lens.

When the CCD is placed at the focal plane each pixel will cover a number of arcseconds of the image. Knowing this figure and the size of the CCD's imaging area you can also calculate the actual size of the patch of sky, or part of the Moon, the CCD is going to record in one exposure.

Let us take an example. You have a standard commercial 200 mm f/10 Schmidt–Cassegrain reflector. Its focal length is 2000 mm. You intend focusing the image at the principle focus of the telescope directly onto the CCD chip, with no additional optics being used. Using the foregoing equation you can calculate that the image scale at the focal plane is 103 arcseconds per millimetre. The camera you have is the *S-BIG* model ST-5C. This camera has an array of 240 × 320 pixels, each of 10 μm square (and so the imaging area of the CCD is a tiny 2.4 mm × 3.2 mm).

Hence 100 of the camera's pixels span 1 mm. Remember, this is also 103 arcseconds of image. So each pixel spans 1.03 arcsecond and the total imaging area of the CCD when used with that telescope (and no additional optics) is 247 arcseconds × 330 arcseconds. The Moon on average subtends a diameter of about 1900 arcseconds, which means you will only be able to image a very small section of the Moon in one go with this particular combination of telescope and camera.

The foregoing step-by-step example calculation can be summarised by the following general equations, which you might find useful to have to hand:

$$\textit{Number of arcseconds of image per pixel on CCD} = \frac{206265d}{1000f}, \qquad (4.2)$$

where d = size of one pixel measured in microns (micrometres, μm) and f is the effective focal length of the telescope measured in millimetres.

$$\text{Width of the image (in arcseconds) on the CCD} = \frac{206265\ X \cdot d}{1000f}, \quad (4.3)$$

$$\text{Height of the image (in arcseconds) on the CCD} = \frac{206265\ Y \cdot d}{1000f}, \quad (4.4)$$

where X and Y are the numbers of pixels comprising the width and height of the CCD, respectively. If the CCD pixels are not square (they usually are in modern astrocameras) then the appropriate values of d have to be used to suit, of course.

You can opt to image the maximum amount of the Moon in one go. For instance a telecompressor (usually an optional extra with Schmidt–Cassegrain telescopes but also available for other telescope types) will give you a proportionate increase in the field of view you can image. Do be aware, though, that the standard ×0.33 (delivering an effective focal ratio of f/3.3) telecompressor accessory for *Meade* and *Celestron* Schmidt–Cassegrains will only give a good image across the full extent of a CCD of no more than about 1 cm square.

Another, more expensive, alternative to a telecompressor is to buy an astrocamera with a bigger CCD. I have already mentioned *S-BIG*'s ST-5C camera but the ST-7E and ST-8E cameras are alternatives. They have 510×765 and 1020×1530 arrays of 9 μm pixels, respectively. In these two cases each pixel would span 0.928 arcseconds and the CCD would cover 473×710 arcseconds of the Moon (for the ST-7E) and 947×1420 arcseconds of the Moon (for the ST-8E) when used with the same telescope without a telecompressor. Even 1420 arcseconds in only 23.7 arcminutes, or about four fifths the diameter of the full Moon. However, you could make the whole Moon easily fit on, with sky backround to spare, with the addition of a ×0.63 telecompressor (which could give a good image across the full extent of a CCD of over 2 cm square if needed).

Unfortunately astrocameras with large CCDs cost lots of money. At the time of writing you can purchase the ST-5E for just under $900 but the ST-7E will cost you almost $2700. The ST-8E is priced just a shade under $6000, more than double the cost of an 8-inch Schmidt–Cassegrain telescope.

Of course, a telephoto lens (surplus from a film-type SLR camera, as described earlier) will easily fit the whole Moon onto the CCD by virtue of its shorter focal length.

This is where a DSLR camera really scores. One costing no more than $600 will often have a large CCD, perhaps about 17 mm × 25 mm across covered in about 5 or 6 million pixels. We get a large imaging area along with a very fine resolution. Hence domestic DSLRs are great for imaging the whole Moon in one go (see Figure 2.4, earlier in this book). This is very useful for events such as eclipses, when the one-shot colour imaging is also a great advantage.

A handy rough rule of thumb that will serve to give an idea of the size of the Moon's image on the CCD is:

$$\text{Moon image diameter} = \text{effective focal length}/110. \qquad (4.5)$$

The units have to be the same (both in millimetres, or both in inches, etc.). As an example, a 200 mm telephoto lens will produce an image of the full Moon roughly 1.8 mm across on the CCD. If this lens was attached, via the appropriate adapter, to the *S-BIG* ST-5 camera then the Moon's image would span about 180 pixels (at an image scale of the order of 10 arcseconds per pixel) and would be nicely framed along with some sky background.

Plugging a DSLR camera with a 17 mm × 25 mm CCD into a telescope of 1800 mm focal length will enable the whole Moon to just fit into the imaging area. The Moon's image will then span approximately 2000 pixels at a scale of the order of 1 arcsecond per pixel.

4.4   IMAGE SCALE USING THE SUPPLIED LENSES ON A '35 MM FORMAT' DSLR

If life wasn't complicated enough, the marketplace sports digital single-lens reflex cameras that come with lenses with ascribed focal lengths which, when used with the camera, give fields of view equivalent to the old '35 mm' film cameras.

The imaging areas of the CCDs are not the same as the 24 mm × 36 mm film frames of the old 35 mm film cameras. At the time of writing a typical large CCD in a mid-to-top-range digital camera has an imaging area of 17 mm × 25 mm. So, to get the same field of view the camera lens' true focal length has to be about two thirds of that of the old photographic lens. You should bear this in mind when working out fields of view and image scales.

4.5   PRACTICAL LUNAR PHOTOGRAPHY THROUGH THE TELESCOPE – AT THE PRINCIPAL FOCUS

Some people mean 'principal focus' when they incorrectly say 'prime focus'. I am willing to wager that amateur reflecting telescopes with no secondary mirror and equipment mounted at the true prime focus position are extremely rare. In many cases the principle focus will be the Newtonian focus, though the first focal plane of the refractor, the Cassegrain reflector, or the catadioptric telescope also counts as 'the principal focus'.

The CCD chip's imaging surface is positioned at the principal focus in each case with no additional optics to enlarge the image further. The basic principles already set out in this chapter apply here, equally as well as for conventional photographic lenses. In effect the telescope becomes the telephoto lens.

Image scale, and so the size of the Moon's image on the film, are calculated just as before. Obviously the Moon's image diameter is likely to be larger because your telescope's focal length is probably larger than that of any of your photographic lenses.

The problems encountered with photography at the telescope's principal focus are usually ones of pure mechanics. Just mounting the camera at the correct position can be problematic. Many telescope suppliers provide a wide range of accessories you can purchase, including 'T-ring' adapters that can fit your lensless DSLR camera body into the telescope drawtube. Similarly your purchased astrocamera will normally be supplied with an adapter to enable it to plug into your telescope's drawtube.

However, the CCD's image-plane within the camera will usually be a few centimetres back from the T-ring. You might well find that the focuser will not rack inwards far enough to allow the CCD to reach the focal plane. Making or buying a low-profile focuser might solve the problem. If not, then you might have to take the drastic action of altering the positions of the optics in your telescope's tube. For instance with a Newtonian reflector you might move both the secondary mirror and the focuser a little down towards the primary mirror. Alternatively it might be simpler to move the primary mirror cell a little up the tube towards the secondary mirror.

Sorry, but I must say it: **Do not even think of doing this unless you are sure you are competent to undertake this task, that you can make all of your measurements and alterations accurately, and can safely store the telescope's optics while carrying out the work**. I would hate to think of any enthusiast ruining their expensive telescope because of anything I have written, so please forgive me including this elementary warning.

Remember that if you bring a reflecting telescope's mirrors a little closer together the secondary mirror will then intercept a larger area of the cone of rays delivered from the primary. Make sure that the secondary mirror is large enough to still intercept all the rays. It might work fine in its original position but could easily vignette the field of view (restricting the rays from the primary reaching the outer parts of the field of view) if you move it too close to the primary mirror. It could go as far as to prevent rays from the outermost parts of the mirror reaching even the centre of the field of view, in effect stopping down the telescope's aperture.

Vignetting might also be caused by the drawtube in telescopes of low focal ratio. Vignetting is to be avoided if at all possible but at least it is something that can be corrected in the processing of your image.

Does your telescope need to be driven (at least at the sidereal rate, if not actually at the lunar rate) in order to get sharp photographs? In general, the answer will be "No" for most of your lunar photography at the principal focus. Telescopes larger than about 6 inch (152 mm) in aperture could potentially image details down to about an arcsecond in extent (this is about 1.7 km at the Moon's distance). Another limiting factor comes into play for apertures larger than this: atmospheric turbulence. This might limit the attainable resolution to 1 or 2 arcseconds on many nights, however large an aperture you use.

Consequently, for 'principal-focus' photography using a 75 mm or larger aperture undriven telescope of greater than about 1 metre effective focal length I would impose a blanket ban on exposures of longer than 1/15 second duration. Diurnal motion will smear the Moon's image by about an arcsecond in 1/15 second. Keeping the exposure time less than this is usually possible except when setting your digital camera to a low ISO setting and then using it with an f/10 telescope to photograph a thin crescent Moon. Trying to capture the earthshine or a lunar eclipse almost certainly will require exposures of several seconds, even with a low focal ratio and a fairly high ISO setting. This necessitates driving the telescope during the exposure.

An often overlooked difficulty is that telescope shake is much more likely than diurnal motion to dominate as the cause of blurred pictures for exposure times in excess of 1/60 second. I would say that using a cable release is mandatory for this type of photography. Fumbling to press a button on the camera while it quivers on the telescope is hardly conducive to getting sharp photographs!

There are examples of principal-focus photography in Chapter 1 (Figures 1.3(b) and 1.7(a)–(c)), Chapter 2 (Figures 2.1, 2.2, 2.3, 2.8, and 2.9), Chapter 8 (Figures 8.13 (a), 8.24(b) and 8.46(d)), and this chapter (Figure 4.6(a)–(d)).

4.6 THE POTENTIAL RESOLUTION OF DETAIL IN THE IMAGE

Sometimes we will want to image the whole of the Moon in one go. Even then we desire to see as much detail as possible but we will have to put up with the limits imposed by the image-scale. Perhaps more often we will be photographing the Moon for the very purpose of recording the finest details we can. In that case we will likely need to use a telescope plus some additional optics to enlarge the primary image. More about how to do that later. For now let us consider how much detail we can resolve on the Moon using whatever equipment we have to hand.

There are seven factors that will limit the fineness of detail we can resolve.

1. Deficiencies in the optical system.
2. Deficiencies in the mechanical system.
3. Errors in focusing.
4. Atmospheric turbulence (and atmospheric dispersion may further degrade the image, even if it is not actually a limiting factor).
5. The aperture of the telescope.
6. The size of the pixels on the CCD.
7. The effective focal length of the imaging system.

The first two items on the list might be ameliorated by careful technique, such as stopping down a poor quality set of optics, etc. In mechanical deficiencies I include poor collimation which again is under your control, as is poor polar alignment if your given exposures are long enough for bad tracking to be a factor. Appendices 1, 2 and 3 near the end of this book may be of help with these matters.

Errors of focusing can be reduced by a painstaking approach and a delicate touch on your part, though matters can be helped very considerably by the fitting of a high-quality focuser, and even better a high-quality motorised focuser. Atmospheric turbulence is something you will have to put up with. The best you can do is to ensure things in the environment close to the telescope are as favourable as you can make them. In particular you should do your best to ensure the temperature of your telescope is as close to the ambient temperature as possible at the time you wish to perform your photography. Open up your observatory, or take portable equipment outside well beforehand – an hour or more if at all possible. If you are lucky enough to have any choice, select an observing site of known good seeing. Trying to photograph the Moon through the hot air stream from a bakehouse chimney, for instance, will not prove very successful!

I mentioned atmospheric dispersion along with atmospheric turbulence. It will almost never be a factor that on its own limits the image resolution but may well further degrade an image. One remedy is to purchase an atmospheric dispersion corrector (available from *Adirondack Video Astronomy*, in the USA – www.astrovid.com). Another is to take your photograph through a coloured filter. I recommend a red filter as the effects of atmospheric turbulence appear least destructive in longer wavelengths.

If you really do want a colour photograph of the Moon through your digital camera (surely monochrome is better in most instances?) you could take repeated exposures through coloured filters and combine them (the red, green and blue images will each be slightly separated – that is what causes the image degradation) using software such as *Adobe Photoshop*. However, atmospheric dispersion is usually only troublesome when imaging objects close to the horizon. The effects of image turbulence are also far worse at the same time. Is it really worth bothering to try to photograph the Moon at high resolution when it appears particularly low in the sky?

Next we come to the aperture of the telescope. The reason I have not put this at the head of the list is that one or more of the other factors will likely prove to be the limiting factor. A high-quality telescope of 6 inches (152 mm) aperture, used on the very best nights, should enable you to see (and to potentially resolve in your photographs) detail that is 0.9 arcseconds in extent. On most nights atmospheric turbulence smears the image so that at best you get infrequent brief glimpses of detail that fine. You will be doing well to image sub-arcsecond details however large your telescope may be. As an aside, the video techniques described in the next chapter will give you the best chance of recording the finest detail under any given set of observing conditions. Indeed, they will quite often allow you to break through the 1 arcsecond limit, whereas taking single images will achieve that only rarely.

Next we should consider the interrelation between the size of the CCD's pixels and the effective focal length of the telescope (or camera lens if that is what is used). Formulae for the image scale and the number of arcseconds covered by each pixel of an image focused onto a CCD are given in Section 4.3. Obviously no details finer than the size of one pixel can be represented in an image (think of the tiles that compose a mosaic). In fact, details in the image that happened to fall on the boundaries between two adjacent pixels would actually be recorded on both of them, effectively diluting the detail and spreading it between both pixels.

Hence if you desire to image details down to, for example, one arcsecond in the image then you should amplify the image to such a size that one pixel spans half an arcsecond. Using a lower amplification of the image than that would *undersample* the image, meaning that you would not be realising all of the potential resolution.

This reasoning is encapsulated into what is called the *Nyquist Theorem*. In order not to undersample an image one should choose an image scale such that the angular extent of the finest detail that you desire (or at least can expect to realise in the given conditions and with the equipment you are using) spans **two** pixels. In fact there is some advantage to be gained by enlarging even further. For one thing the diagonal distance across the pixels is 1.4 times greater than the distance measured along one side. Also the best image fidelity is more fully attained when the pixels are no more than one third of the smallest elements resolved in the image.

If you are using a telescope under conditions in which its quality and diffraction limit are fully realised (a rare occurrence when using a telescope bigger than 6 inches aperture) then you might profitably enlarge the image to as much as twice the amount set by the Nyquist limit. This is to wring out the very last bit of the visibility of the finest details, almost all of which would be realised anyway by just attaining the Nyquist limit. Of course the rub is the more you enlarge the image, the more you lengthen the necessary exposure time with the attendant problems that brings.

The image scale at the 2.59 m principal (Newtonian) focus of my 18¼-inch (0.46 m) telescope is 80 arcseconds per millimetre. In perfect conditions the telescope should allow details as fine as 0.3 arcsecond to be seen. Granted, the chances of getting seeing that good are virtually zero on any given night, but that is no reason to forego trying to get the best resolution possible in your lunar photographs. However, a CCD of 5.6 μm pixels (about the smallest available in any common commercial camera) will only provide a resolution, satisfying the Nyquist Theorem, of 0.9 arcsecond on my telescope.

The inescapable conclusion is that to have the best chance of recording lunar images of the highest resolution possible one must enlarge the primary image. In this example, I should have to enlarge the primary image by a factor of at least 3 in order to have any chance at all of resolving details as fine as 0.3 arcsecond in the image. Some methods for enlarging the primary image are discussed in the next section.

4.7 ENLARGING THE TELESCOPE'S PRIMARY IMAGE

There are three main methods of doing this: projection using a Barlow lens (or Powermate), projection using an eyepiece, and afocal (also known as 'infinity to infinity') focusing using the eyepiece and camera lens. The following notes detail each of these methods.

Barlow (or Powermate) projection

Figure 4.2 shows the arrangement, together with the formulae for working out the enlargement factor. The lens must be mounted into a tube with the appropriate bayonet or screw-fitting attachment to your camera. The only real problem with Barlow lenses is that they are usually designed for one specific value of amplification (commonly ×2). They are fine with delivering images a little higher or a little lower than the manufacturer intended. Go too far away from the stated amplification value, though, and the lens will begin to introduce aberrations to the image.

Often the Barlow's lowly amplification may not be enough to match the potential resolution of the telescope to the CCD. More powerful alternatives to Barlow lenses are the range of amplifying lenses marketed as

Figure 4.2 The optical configuration for projecting (and enlarging) the primary image by means of a Barlow lens. Alternative formulae are given for calculating the amplification factor, a. The focal length of the Barlow lens is F. It is entered in the equation as a negative quantity.

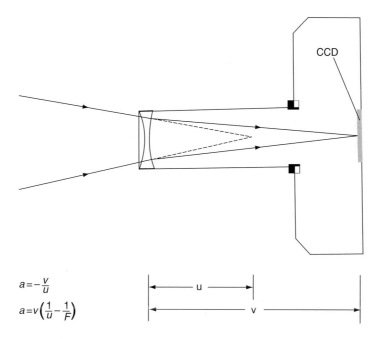

$$a = -\frac{v}{u}$$

$$a = v\left(\frac{1}{u} - \frac{1}{F}\right)$$

Powermate by *Tele Vue*. They come in amplification values of ×2.5, ×4 and ×5 and are the market leaders in terms of quality. Other companies market their own versions, such as *Meade*'s TelexTender series, of ×2, ×3 and ×5.

One should check for vignetting when using a Barlow lens. The minimum diameter, D, the Barlow lens should be in order to fully illuminate the projected image out to a diameter, d, centred on the optical axis is given by:

$$D = \frac{1}{a}\left(\frac{v}{f} + d\right), \tag{4.6}$$

where a is the amplification factor, v is the distance the lens is set inside the telescope's focus, and f is the focal ratio of the telescope without the Barlow in position; f has no units, since it is a ratio, but the units of all the other quantities in the equation must be in the same units for the equation to work. For instance all must be in millimetres, or all must be in inches, etc.

One thing I should point out is that the Powermate or Barlow is designed to produce the magnification it does with an eyepiece plugged into it – and so with the new focal plane situated just a little inside the top of the unit. Attaching your camera to the unit will involve racking the telescope focuser just a little further (maybe by a centimetre or two) inwards in order to bring the focal plane several centimetres out beyond the end of the unit to reach the imaging surface of the CCD. The result is that you will get a little more amplification from the unit used this way than the manufacturer states. For instance a ×5 Powermate used in this way might well actually deliver something close to a ×6 enlargement of the primary image, and maybe even a little more.

With DSLRs and astrocameras with small pixels, a ×5 Powermate will normally provide all the amplification the Moon photographer will need. There are examples of images obtained using Barlow or Powermate projection in Chapter 1 (Figure 1.14(a) and (b), Chapter 2 (Figure 2.7), Chapter 5 (Figures 5.4, and 5.7(a) and (b)), Chapter 6 (Figure 6.4), and Chapter 8 (Figures 8.11(b), 8.13(d), 8.14(e), 8.15(b), 8.17(f), 8.33(c), 8.44(e), 8.46(c), 8.47(a), and 8.49(a)–(e)) and this chapter (Figure 4.5).

Eyepiece projection
As Figure 4.3 shows, the eyepiece is mounted so that the rays emerging from the eyelens converge to form a focused image on the CCD. However, this is not what the eyepiece was designed to do. Sets of parallel rays emerge from the eyelens and enter the observer's eye when the eyepiece is used as the manufacturer intended. Using it as a projection lens involves setting it slightly further out from its normal focused position in the telescope, which is usually not a problem. The problem is an increase in the optical aberrations produced by the eyepiece. Good eyepieces will still produce sharp images within a small area of the centre of the CCD but the outer zones can be very blurred.

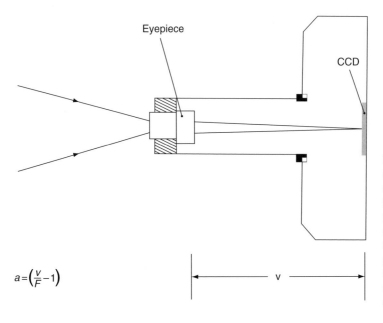

$$a = \left(\frac{v}{F} - 1 \right)$$

Figure 4.3 The optical arrangement for enlarging the primary image by eyepiece projection. The amplification factor, *a*, is found from the equation given. The focal length of the eyepiece is *F*. The equation is approximate as explained in the text.

Figure 4.4 The horrendous blurring of the outer zones of this photograph is a severe disadvantage of the method of eyepiece projection when used to provide low amplification factors to image the Moon. This photograph was taken by the author on 1977 September 2^d 23^h 57^m UT. He used an 18 mm Orthoscopic eyepiece to enlarge the f/5.6 primary image of his 18¼-inch (0.46 m) Newtonian reflector to f/17. The exposure was 1/30 second, made on *Ilford* Pan F black and white film.

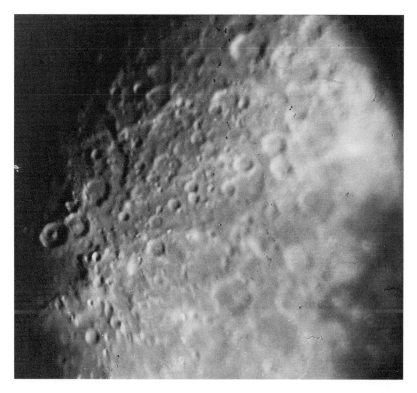

Figure 4.4 shows one of my early experiments with eyepiece projection onto photographic film. Notice the horrendous outfield blurring I got using an 18 mm Orthoscopic eyepiece to enlarge the f/5.6 Newtonian focus of my 18¼-inch (0.46 m) reflector to f/17. This effect is greatly reduced when using the eyepiece to create larger amplification factors. The reasons for this are two-fold. First, the rays emerging from the eye-lens are less convergent (and so the paths of rays through the eyepiece are nearer to what the designer/manufacturer intended). Also, the 'sweet spot' of a sharply focused image on the CCD is increased in size by virtue of the greater magnification. I should also say that outfield blurring manifests when the detector that the image is projected on to is large. In the case shown in Figure 4.4 it was a frame of photographic film of size 24 mm×36 mm. The problem virtually disappears when imaging onto a CCD smaller than a centimetre across.

For lower amplification factors a high-quality Barlow lens or Powermate may serve best. Eyepiece projection would be the best choice for large amplification factors. You will find further examples of images made using eyepiece projection in Chapter 8 (Figures 8.5(c), 8.13(f),

8.14(a), 8.17(a), 8.17(b), 8.24(a), 8.26(a), 8.26(b), 8.30(a), 8.34(a), 8.35(a), and 8.44(a). All details are given in the accompanying captions.

Afocal imaging, using the camera lens and the eyepiece
Though the purist might shudder at the thought of eight or more separate lens elements between the principal focus and the CCD, and have nightmares about multiple reflections and a 'fog' of light swamping the dim image, keeping both the telescope eyepiece and the digital camera's lens in place can actually work really well. With today's multi-coated eyepieces and camera lenses the amounts of light scattered, reflected, and absorbed by them are very much less than was the case for the products of yesteryear.

In this arrangement the eyepiece is focused in the normal way and then the camera, complete with its lens set to 'infinity' focus, is brought up to the eyepiece. The camera is carefully positioned such that it looks squarely, and on-axis, into the eyepiece. In effect, the camera has replaced the observer's eye.

Given that there might be some uncertainty in the precise 'infinity' focus of your eye caused by the natural accommodation of focus and/or any long- or short-sightedness, one can expect better and more consistent results by directly viewing the digital camera's display screen and focusing manually. It is obviously then an advantage if your camera has a display with a 'enlarge part of the picture for fine focusing' facility.

The amplification factor, *a*, is given simply by the ratio of the focal lengths of the camera lens, F_c, and eyepiece, F_e:

$$a = F_c/F_e. \tag{4.7}$$

Remember the focal length of the camera lens has to be its true focal length in this equation, not its 'DSLR 35 mm film format equivalent focal length'. One advantage of this method is that both eyepiece and camera lens are being used well within their intended parameters and so can deliver images of the highest quality.

While it is always best to attach the camera firmly to the telescope, another advantage of the afocal method is that tolerable results, sometimes very good results, can be obtained by hand-holding the camera to the eyepiece. My first photographs of the Moon were taken this way through my 3-inch reflector. That was back in 1972. In the years since I often used this method when I wanted to take quick photographs without going to the trouble of setting up the equipment to attach the camera to the telescope. You will find examples of pictures taken using afocal imaging in the following chapters of this book: Chapter 2 (Figure 2.4), Chapter 5 (Figures 5.1 and 5.2), Chapter 8 (Figures 8.20(a) and (b), also 8.34(b)). The details in each case are given in the accompanying captions.

In order to get consistently sharp photographs, hand-holding the camera demands that the exposures should certainly be shorter than 1/30 second, better 1/60 second or less. Using this technique it is possible to photograph terminator details in the first or the last quarter Moon with an exposure of 1/60 second with a digital camera set to ISO 800 provided the effective focal ratio is no higher than about f/20. With the Moon at a fuller phase you will be able to either increase the effective focal ratio a bit further, or set the camera to a lower ISO, or give a shorter exposure. The Moon's brilliance increases very rapidly as its illuminated phase expands. The full Moon is fully twelve times as bright as the Moon at first or last quarter.

If you feel that attaching your camera to your telescope and arranging to project the image onto the CCD is just too much like hard work, then I urge you to try some hand-held 'snapshots' using the afocal technique. You may be surprised at just how good your lunar photos will turn out. Go on – have a go! You might even be spurred on to more serious efforts afterwards.

4.8   IMAGE PROCESSING

One great advantage of today's electronic imaging technology is that the results can be instantly viewed. In the century and three quarters of what one might call 'chemical photography' the results were far from instant. Even though I had my own amateur darkroom and used to process my own films and print them, I never felt energetic enough to do the processing the same night as I took the exposures. For people who had to send their films away for processing the delay was even longer and there was the additional disadvantage of not being able to control the process yourself.

Getting instant feedback from your digital camera's screen, or the monitor of the computer controlling your astrocamera, enables you to make as-you-go adjustments, such as to the length of exposure. You can then try again and again, making changes until you get a reasonable result.

Invariably, though, even this final image will not satisfy you. Indeed, you might think it rather disappointing in its quality. Instead of a sharp and contrasty picture showing incredibly fine detail, you see a rather blurry and washed-out image. This might even be the case with the results from your digital camera despite it having some clever onboard electronics to enhance the image its CCD naturally obtains. This will almost certainly be the case for what you see delivered to your computer by your astrocamera.

This brings us on to an even greater advantage of electronic imaging. It either already exists in, or is naturally converted to, a digital format in order that one may perform image processing on it. With some software – and it doesn't have to be complicated or expensive – you can turn your fuzzy and foggy-grey image of the Moon into something very impressive, showing fine details that you thought were never even present in the first version.

Way back in 1996 I bought my 133 MHz PC with, the addition of a *Hauppage* video capture card. The software suite that came with this card, *Ulead*'s *MediaStudio VE 2.5* included *Image Editor*. This is a delightfully simple and easy to use, yet very effective, image-processing program. Modern versions are still available, packaged along with other software – see Chapter 7 for *Ulead*'s web-site address. I think of *Image Editor* as a simplified version of the other, often expensive, image processing programs you might have heard about (such as *Adobe Photoshop*, *Paintshop Pro*, *AIP4WIN* and *MaximDL* – see Chapter 7 for web-site addresses). The point is that this simple little program has almost all the functions I have ever desired to use in finishing off the processing of Moon and planet, and even deep-sky, images. To show you it in action on lunar images, please see Figure 4.5(a)–(e), where I am making improvements to an already good image (initially processed in *Starlight Xpress*'s proprietary software). Of course, you can do this and much, much, more if you do use *Adobe Photoshop* or one of the other highly sophisticated programs – but be prepared for a steep learning curve. Also the various versions of these and other commercial image-processing software can be expensive, often costing hundreds of dollars.

Of course, the first piece of software in the chain is that required to operate the camera (in the case of an astrocamera), download the camera's output to your computer and (especially in the case of astrocameras) to do some initial processing and to convert the image file to one of the formats that will allow it to be outputted to a printer, or other device, and maybe to some more software for further work on the image.

As an example please take a look at Figure 4.6 (a)–(d). 4.6(a) shows the state of the image as it was captured by an *S-Big* STV astrocamera. Well, it almost does. In truth I cheated a bit. Actually it was rather dark on my computer monitor so I artificially brightened it solely for the purpose of reproducing it clearly in this book. The next stages in the computer processing were done on the original image, not this brightened version. The software suite is called *CCDOPPS5* and this was supplied with the camera.

Figure 4.6 (b) shows the result of me *re-sampling* the image, ×2. That sets it into a new grid of four times as many pixels as the original (the pixels have half the diameter so there are four times as many of them covering the area as before). Doing this allows me to do things like applying algorithms to sharpen the image without the 'blocky' nature of the pixels becoming apparent. This does, though, also make the size of the file four times as big. I also adjusted the brightness range of the image (leaving the black level untouched – I did not want to extinguish any faint details that might be lurking in the image) to the point that the brightest parts of the image did not fully saturate as seen on my computer's

Figure 4.5 (a)–(e) Further manipulation on the author's computer of an image initially processed in *Starlight Xpress* software.
(a) 'Original' image.
(b) 'Tone Adjustment' histogram for the 'original' image (having first 'zoomed-in' on a selected area).

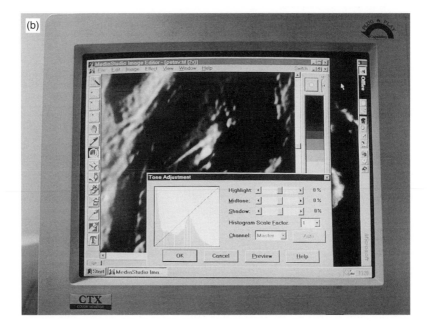

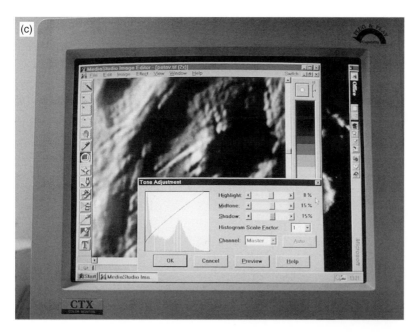

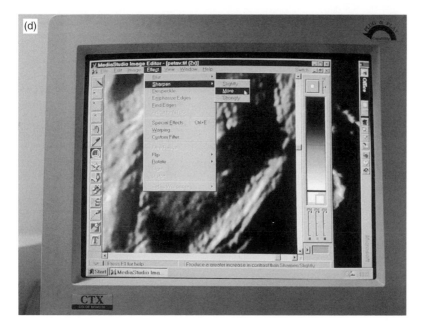

Figure 4.5 (*cont.*)
(c) Effect of altering the parameters in 'Tone Adjustment' (by dragging the slide-bar controls in the window) in order to improve the visibility of the shadow details within the crater. Note the change in the parameters and the new histogram.
(d) Calling up the appropriate window in order to sharpen the image.

Figure 4.5 (*cont.*)
(e) The final result.

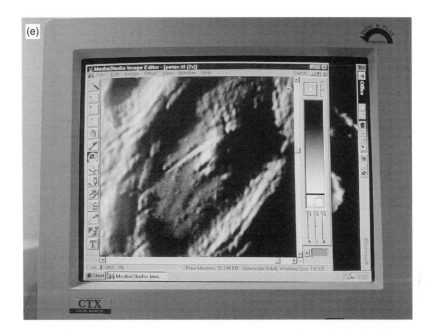

monitor (I suspect, though, they might appear saturated as printed in this book due to the limited dynamic range of the printing process).

Figure 4.6(c) shows the image, still in *CCDOPPS5*, after I have applied sharpening. The options available were for the type of sharpening ('planetary' or 'nebulae') and the level of sharpening ('soft', 'medium', or 'hard'). I chose 'planetary' and 'hard'. So far the image file has been in the camera's proprietary file format. My next job was to save it as a TIF file (by clicking on 'File', then choosing 'Save' from the menu that appears, then clicking on the file-type widow and selecting 'TIF' from the list of available file types). I usually like to save the file to a floppy disk (or other storage device) at this stage.

Figure 4.6(d) shows the result of me importing the file to *Image Editor* and using 'Tone Adjustment' to alter the brightness levels, my aim being to show better some of the detail in the dark terminator region while preserving as much of the contrast in the rest of the image as possible. Finally I have converted the image to a JPG file format (high-quality setting) just to show you that a JPG file of only 90 Kb can still produce a good-looking image. This compressed file format is useful if you are intending to e-mail images to your friends or colleagues.

A word of warning about JPG files – **never** convert to a JPG file until you have finished all the processing you intend to do on the image. If you do you will find that the amount of processing you can do will be very limited.

Figure 4.6 (a)–(d) Region of Copernicus imaged by the author on 2004 March 4^d 18^h 44^m UT, using an *S-BIG* STV camera on the 0.5 m Newtonian reflector of the Breckland Astronomical Society. (a) The original image captured by *CCDOPS5* software. (b) The image resampled × 2, and the brightness range decreased while leaving the black level undisturbed.

That brings me to the mention of a possible difficulty with image processing images from digital cameras at the cheaper end of the market; JPG files are all that you get from them. Admittedly these JPG files are usually multi-megabyte in size and that does help a lot. You can get away with doing some image processing on large JPG files.

As time goes on the provision to save images in alternative file formats is becoming available in cameras of lower cost, so maybe there is no problem with the camera you already have. If you are choosing a digital camera you intend to use for astronomy, I advise stretching the budget to a good DSLR. Before your purchase make sure that you will be

Figure 4.6 (*cont.*)
(c) The image after applying a sharpening filter.
(d) The image after converting to a TIF file format, importing into *Image Editor* software and making alterations in 'Tone Adjustment' in order to improve the visibility of details in the terminator region.

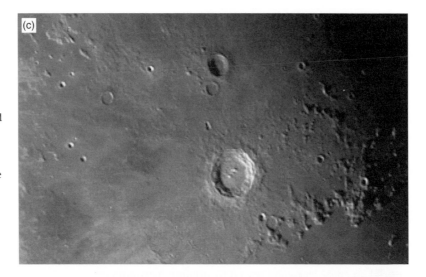

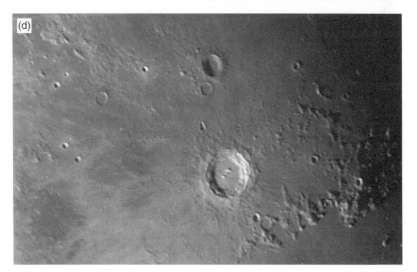

able to export your images to whatever processing software you intend to use. Cameras that save images in RAW format can be opened, worked on, and converted to other file types in *Adobe Photoshop CS2*, for instance.

Astronomers using CCDs frequently take *dark frames*. These are to cope with irregularities in the response of pixels in the CCD and particularly with the effects of 'noise' build up due to thermal effects. The procedure is to make an integration with no light being allowed to reach the CCD. The length of the integration is the same as that for the actual exposure of the subject. I have never needed to bother with dark frames in any of my Moon or planet photography but I always do when deep-sky imaging. You might use a dark frame only if using a poor quality CCD camera or for some reason giving an unusually long exposure. If your camera has provision for taking dark frames – an astrocamera certainly will – simply follow the instructions.

Astronomers also frequently take *flat fields*. These are exposures made of the twilight sky background or of an evenly illuminated screen or wall. The purpose is to effectively map out the variation in sensitivity of the whole optical-CCD system across the field of view. Again this is something I have never needed to do in Moon and planet imaging. Vignetting is corrected by applying a flat field to the image in processing. As ever, follow the instructions.

I list some software packages that might be of use to you in Chapter 7. There are also some books that may be of help to you listed in the same chapter. I would give special mention of *Photoshop Astronomy*, by R. Scott Ireland and published by Willman-Bell in 2005. This is an invaluable tutorial on *Adobe Photoshop* geared to amateur astronomers.

Perhaps here I should also mention *Images Plus*, a program especially designed for DSLR users who wish to undertake astronomical imaging. Other devices can also be used with this software. You will find a review of version 2.75 of this product in the July 2006 issue of *Sky & Telescope* magazine. As the reviewer points out, the program is very complex in its operation and you will need to prepare yourself for a steep learning curve. You certainly do not need anything this complicated if you only intend to do lunar imaging with your camera.

Indeed, if you are using a CCD astrocamera you might well be satisfied with the processing tools that are included in the manufacturer's supplied software package and might not want to get involved with other software suites. If you are a beginner at image processing then I strongly urge you to keep things as simple as possible. Avoid getting bogged down with the intricacies of complicated software until you are comfortable with the basics. You surely want to have fun from the start with your lunar images, not frustration!

CHAPTER 5

Stacking up the Moon

If you want to obtain images showing the very finest lunar detail possible with your equipment in any given observing conditions then you have to put in just a little extra effort and go for a more sophisticated approach than taking single frames. Most of this chapter is given over to describing how you might do that. Let me first, though, show you how to get impressive results by adopting a very simple approach with a piece of equipment you might already have and have never thought of putting to astronomical use . . .

5.1   THE MOON AND YOUR DOMESTIC VIDEO CAMERA

Take a look at Figure 5.1. It shows the great crater Plato and part of the Montes Alpes. I took this photograph using my own telescope and a mid-priced domestic video 'palmcorder' of early 1990s vintage that recorded onto its own miniature videotape. I re-recorded the footage onto VHS tape in my 1980s-vintage domestic VCR, played it back on my 1980s-vintage television and paused the tape at a moment of good seeing. I then photographed the TV screen using my old film camera. Using modern digital devices at each of these stages would undoubtedly produce superior results.

The seeing was very ordinary that night and yet I estimate the resolution of the image to be approaching the 1 arcsecond level. Figure 5.2 shows another part of the Moon, videoed in the same session showing similar resolution. I have plenty more images from this and other video sessions which are of similar quality. The telescope I used was my 18¼-inch (0.46 m) reflector but the seeing was the limiting factor in the resolution of the images, not the size of the telescope. You could have done just as well as I have with a telescope of less than half the aperture of mine. In better seeing conditions you could easily improve on these images. In the following notes I explain how.

The video camera

If you already own a video camera, then you might like to put that one into service with your telescope. I am sure that you will be delighted with the images you will get from it. However, if you are wishing to buy one and the main reason for doing so is to use it with your telescope then I recommend considering a few factors before making your choice.

One important consideration is the camera's size and weight. Though it is sometimes possible successfully to arrange the camera on a separate tripod and have it peering into your telescope, you will normally have to have the camera riding on the telescope. Way back in 1985 I achieved some success in hand-holding a borrowed (and rather heavy) camera to the eyepiece but the results were decidedly hit and miss.

You will have to construct something to do the job of firmly attaching the camera to the telescope. The heavier the camera is, the firmer – and this usually means heavier – this contraption will have to be. Will your telescope remain shake-free with all this weight hanging close to the eyepiece? Do not forget that you will also need to add counter-weighting

Figure 5.1 Video image of the Moon taken by the author, using his 18¼-inch (0.46 m) reflecting telescope on the date and time shown. The large crater at the lower right is Plato and the mountain range extending from the upper left to Plato is part of the Montes Alpes. Further details in text.

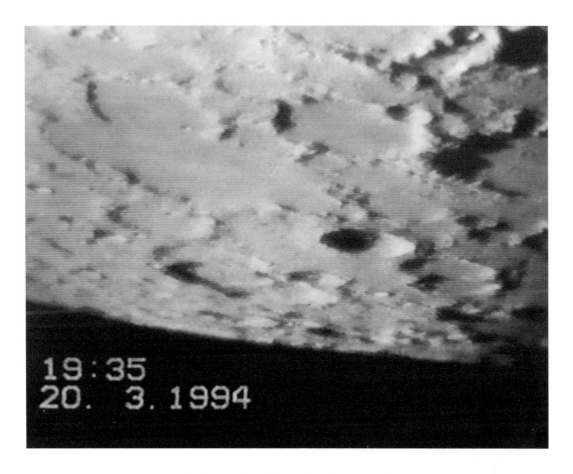

19:35
20. 3. 1994

Figure 5.2 The north polar region of the Moon. A video image taken by the author at the date and time shown. Further details in text.

to re-balance the tube and/or the mounting. Consequently, unless you own a telescope which is constructed like a battleship, I recommend selecting a video camera which is as lightweight as possible.

The next thing to worry about is the resolution of the camera. Obviously it should be as great as possible. A low/medium-quality camera will have a '⅓-inch CCD image sensor', more expensive models having a '½-inch' version. You will normally find the figures for picture resolution, and size of CCD in the technical specifications section of the instruction manual. I recommend that you insist the storekeeper opens the box so that you can examine the manual prior to purchase. The stated picture resolution is often that in the horizontal direction. The vertical resolution is usually slightly better. A ⅓-inch CCD camera ought to have a horizontal resolution of more than 200 lines across the full width of the TV frame.

On the low light-level exposure setting my camera records in illumination levels as low as 1 lux. Using it with my 0.46 m telescope, I was able to successfully record the dust shells around the nucleus of comet Hale–Bopp in 1997. You should not have any trouble in recording the Moon's vistas through the camera, even when using a telescope of quite small light-grasp. On the contrary, you might even have to take precautions against too much light if you use a low magnification on a large-aperture telescope.

There is one major pitfall I must warn against: do not purchase a video camera which has no manual override for focusing. Using a video camera on 'automatic focus mode' to try and image the Moon through the telescope is doomed to failure. Do so and you will find the camera's electronics will not like the image your telescope delivers one little bit. The camera focusing mechanism will restlessly zoom and jitter about the mean focus setting. You must be able to focus the camera manually.

My old camera's ⅓-inch CCD sensor has a horizontal resolution of 'more than 230 lines', according to the specifications given in the back of the instruction manual. The camera has the facility for on-screen recording of the date and time, if desired, as well as a selection of 'shutter speeds' which allow one to optimise the image quality for different lighting conditions. The camera lens can 'zoom' between effective focal lengths of 5 mm (the widest-angled view in normal use) to 40 mm for close-ups of things in the distance. We need the lens set to 'maximum zoom' ('×8', 40 mm effective focal length in the case of my camera) when using it with the telescope.

Mounting and shooting

I certainly do not recommend attempting to remove the lens of your video camera. So, you are forced to leave the telescope eyepiece in place and use the 'afocal' focusing method, as described in Section 4.7.

The camera needs to be mounted so that it looks squarely into the telescope eyepiece. Figure 5.3 shows how I attached my video camera to my 0.46 m telescope. There are as many solutions to the problem of mounting the camera on the telescope, and ways of achieving the necessary rebalancing of the telescope, as there are different telescopes. It is purely a matter of mechanics, which I must leave to you.

At least Figure 5.3 might provide some inspiration. Note how I have provided some adjustment for 'squaring on' the camera to the eyepiece – necessary to get the best-quality images. I have also made some provision for racking the camera back and forth, which is a great convenience when using different eyepieces and for initially setting everything up – particularly changing eyepieces and focusing, when some clearance will be needed in front of the camera lens.

The Moon is in the sky and we have just set the camera into its mounting, fitted to our telescope, and have also attached the counterweighting. The telescope is balanced and ready to turn towards the Moon. We have

Figure 5.3 The arrangement the author uses to attach his video camera to his telescope. Counterweights (not shown) attach to the bottom end of the telescope tube to restore the balance.

already made sure that the camera focus is set to 'manual' and that the focus ring is moved all the way to 'infinity' (long distance) focus position, also that the 'zoom' setting is at maximum (more about this shortly) and that the camera is otherwise ready for operation.

With the camera set back as far as possible from the telescope, plug a low-power eyepiece into the telescope drawtube. Adjust the rackmount so that the eyepiece is close to its normal position for infinity focusing. Perhaps a predetermined felt-pen mark on the drawtube would be a useful guide, since you will not be able to get your eye to the eyepiece with the camera in position. If the eyepiece is fairly close to its correct infinity focus position, then the camera can be safely brought up so that its lens is a centimetre or two from the eyepiece but do leave room enough for any necessary fine adjustment.

Now it is time to point the telescope towards the Moon. At the stage where the moonlight is dropping into the telescope drawtube you will probably be able to see it emerging from the eyepiece and illuminating the camera lens. Powering-up the camera and looking through its view-finder, you ought to be able to easily find the Moon's image and set the telescope on the part of it that you want to record. At this point you can make any fine adjustment necessary to the telescope focuser but please do be very careful that you do not drive the eyepiece into collision with the camera lens. If desired, you can now bring the camera a little closer to the eyepiece, though a gap of a centimetre will make no difference; the camera's front lens is big enough to capture all the emergent rays without it having to be in dangerously close contact with the eyelens of the eyepiece.

Having brought the image to as fine a focus as possible, you should now be seeing an impressive view of the Moon's mountains and craters. Try the various camera exposure settings. The shortest on my camera is the so-called 'sports' setting. This produces the best result with my telescope. The image is sharpest at this setting because the effects of turbulence and any tremors of the telescope are virtually 'frozen' on individual frames.

Also, I find that my 0.46 m telescope gathers too much moonlight for the camera to cope with on the other settings. On my first attempt, I found that a black-curtain effect (caused by severe overload) descended over the image when the Moon entered the field on all but the 'sports' exposure setting. Even if your telescope is rather smaller than mine, you will probably find that the fastest exposure setting will give the best recorded image.

Of course, the phase of the Moon, the magnification of the image, the transparency of the air, etc., will all determine the correct exposure. At least you can see what is happening through the viewfinder while you select the different exposure settings on the camera. You might be satisfied with the view you already have through the camera and can set it recording. Alternatively you might wish to change the eyepiece to give a higher magnification. More on this later, though here is the place to explain why the camera ought to be left on a setting close to full zoom.

The reason is that on lower settings (at which the camera lens has a shorter effective focal length) not all of the rays emerging from the eyepiece can find their way unobstructed through the camera lens assembly. If you experiment with the camera on the telescope you will find that on a low-zoom setting the image appears as a small island in a sea of surrounding blackness. Press the zoom button and you will find that this island expands in size. Nearly full zoom will be needed before the image fills the field.

If you can successfully mount your video camera onto your telescope, and get it balanced properly, you will find this technique very forgiving as regards any other requirements of your telescope. An equatorial mount is a great convenience, while a drive is very much a luxury. You could even get superb results from an undriven altazimuthly mounted telescope. There are other advantages in using a domestic video camera: the images are in full colour and one can record a commentary at the same time as recording the view through the telescope.

Field of view and image scale
The effective focal length and effective focal ratios are calculated in the same way as for a conventional camera in the afocal configuration (see

Section 4.7). By way of an example, the photographs shown in Figures 5.1 and 5.2 were taken with my 0.46 m reflector, which has a focal length of 2.59 m. My camera, with its lens adjusted to 'full zoom' (effective focal length 40 mm) was set looking into an 18 mm focal length Orthoscopic eyepiece. The amplification factor was 40/18, or 2.22. Thus the effective focal length of the combination was 5.76 m.

Knowing the effective focal length, the image scale on the CCD can be calculated. In this case it was 206265/5760, or 35.8 arcseconds per millimetre. However, I do not know the precise size of the CCD so this figure is only useful in making a rough prediction of the area imaged. The '⅓-inch' size of the CCD is only a rough guide. It is a category, rather than a precise figure, and refers to the approximate (and usually exaggerated!) length of the diagonal between corners of the CCD's imaging area. If your video camera has a ⅓-inch CCD you can expect it to have an imaging area of something like 4.0 mm × 5.3 mm in the old standard aspect ratio. The area covered in each of the photographs presented in Figures 5.1 and 5.2 is approximately 140 arcseconds × 190 arcseconds. These reproductions are virtually the full frames photographed from my TV screen.

Maximising the resolution

You might not know the size of the pixels in your camera but you can still ensure that you do not under-magnify the image and so limit the potential resolution. The manufacturer's specifications sheet will provide you with a figure for the resolution in terms of the size of the full frame. My camera has a resolution in the horizontal direction of 'more than 230 lines'. With the arrangement of my telescope, camera and eyepiece already described the width of the full frame comes out as 190 arcseconds. Therefore the potential resolution of the image is 190/230, or 0.8 arcseconds.

On most nights I find that I can only fleetingly glimpse arcsecond-level details, so this choice of amplification is about right. Increasing the magnification would serve only to reduce the field of view and produce an apparently blurred image. However, on nights of perfect seeing my telescope ought to resolve down to about 0.3 arcsecond. I could match this by exchanging my 18 mm focal-length eyepiece for one of 6 mm focal length. The size of the field of view would then be 48 arcseconds × 63 arcseconds.

Other video devices

Closect circuit television (CCTV) security cameras, specialised CCD video cameras designed for the amateur astronomical market – such as those produced by *Adirondack Video Astronomy* (a company in the USA – the web-site address is given in Chapter 7), and almost any device that can record video – including some domestic digital cameras – can be used with your telescope.

Unfortunately I have not enough room here even to begin to go into their use in adequate detail. You can always search the Internet to see what other astronomers are doing and there is a book which may help you: *Video Astronomy, Revised Edition*, written by Steve Massey, Thomas A. Dobbins and Eric J. Douglas and published by *Sky Publishing Corporation* in 2004.

Playback and hard-copy
Videoing the Moon is an easy way of getting high-quality images of it. Your friends will be wowed by the impressive views of the Moon you will be able to show them on your television screen. With more than two-dozen images recorded every second, you can use your DVD player's frame-by-frame advance function to search out the few best images in each session. These will occur at a rate of one per hundred to one per several hundred of the recorded frames, so a degree of patience is needed. At least you can concentrate on the sections of the recording in which you see you get flashes of sharp imaging and individually search just those. Using the slow-motion playback on your DVD player will also be a help.

You can even photograph the TV screen and so obtain hard-copy of those 'paused' best images. Set your camera to its lowest ISO setting and exposure of at least ⅛ second, better ¼ second if your television has a cathode ray tube (a short exposure will only give you a partial picture, or banding across the picture, because of the way the picture is created on a CRT). Setting the camera on a tripod and darkening the room (to avoid reflections on the screen spoiling the image) will give you the best chance of obtaining a good reproduction of the image on your television.

Take note that we have collected images from our telescope, shown them on a television, and even obtained hard-copy photographs without the otherwise ubiquitous computer even getting a look-in. Of course, we can involve a computer if we want to ...

Linking your video or DVD to your computer
There is a way of sending the output of your DVD recorder or video camera to your computer. Your computer must be fitted with a piece of circuitry (in computer jargon a *card*) known as a *frame-grabber*. If you are up to the job you could install this hardware (and then install the operating software) yourself. If not, then you will have to have a computer specialist do the job for you.

In any case you will have to consult your computer specialist dealer in order to find out what products are available at the time you go shopping, as well as their specifications. Tell the dealer what it is you want to do with the frame-grabber and its software. Do tell the dealer that you do not need a package that will allow you to make your own special-effects-laden science-fiction movies. You most certainly do, though, need it to store selected full quality uncompressed images from your video camera. One important feature your frame-grabber package must have is the ability to save images in file formats other than JPG. A JPG file may be acceptable if all you want to do is send it to your printer. It most certainly will not do if you intend to perform image processing on your saved images; TIF or bitmap (BMP) files will prove satisfactory for image-processing requirements.

5.2 THE BENEFITS OF STACKING SELECTED IMAGES

Before the mid 1980s the most successful way of recording fine details on the Moon and planets available to an amateur astronomer was to study critically the view through the eyepiece and make a drawing. The finest details on that drawing were gleaned from the occasional few flashes of fine detail that occurred in a session of something like ten to twenty minutes long for drawing the planets, or maybe half an hour for sketching a small part of the Moon. It was difficult for the users of photographic film to get such detailed views because atmospheric turbulence usually blurred the image during the exposure. This was made worse by the relatively insensitive films needing exposures of a large fraction of a second, to maybe even a few seconds, when the image was dimmed by being enlarged enough to have any chance of resolving fine details.

The sensitive CCD and onboard electronics in modern video cameras are able satisfactorily to record images of the Moon (even those dimmed by enlargement) at a rate of 25 or 30 frames a second. As an aside, the American video and television systems work at 30 frames per second. European devices work at a frame rate of 25 per second. The individual exposures can be rather less than $\frac{1}{25}$ or $\frac{1}{30}$ second each.

This is why one in a hundred to one in a few hundred of the recorded video frames will show details much finer than the others recorded in the same session. These frames are the equivalent of the 'few flashes of fine detail' that the visual observers of old had to do their best to utilise when making their drawings.

I have already described how these best frames can be searched out in any recording and hard-copy made of them. I have also described how individual frames can be 'grabbed' by the computer for the purpose of being worked up using image-processing techniques. However, there is a problem. If you try processing a single grabbed frame you will find that the already slightly grainy-looking image will not stand much enhancement before it looks ghastly; like a picture first printed on coarse sandpaper and then sprinkled with salt and pepper. This happens because the individual images are afflicted with electronic *noise*. Each pixel in the CCD should ideally have an amount of charge liberated in it that is directly proportional to the amount of light falling on it during the exposure. In fact most of the pixels are afflicted with additional, seemingly random, amounts of additional charge.

This noise has a number of root causes. One major one is that the CCD is not cooled, as is mandatory for a proper CCD astrocamera. Another is that the process of reading the image off the CCD and into the cameras electronics also generates noise. Some pixels have no additional charge but most do and a few may have enough to fill them almost to saturation. The effect is only mildly visible in the frames as they come from the camera but most enhancement techniques applied in image processing will cause this noise to spring to the fore.

Fortunately there is a solution to this: combine a number of frames together so that the random effects of the noise in each individual frame is diluted. In the normal way you watch a video playback, persistence of vision does this combining for you. You are presented with either 25 or 30 frames per second but the effects of noise are averaged over several of these in any one moment. That is why the quality of the picture seems better – rather smoother – in normal playback than it does when you study individual frames.

So, wouldn't it be great if we could select out the best of the images we get from our telescope in any one session and combine just those to form a new image, which will then have much of its random noise averaged out? There would be a further advantage. Even the best individual frames will be affected by some atmosphere-generated blurring and distortions which vary across the image. By combining frames we could average out these afflictions, reinforcing the 'true' fine details and positions and diluting the 'untrue' ones.

The composite image would have the effects of atmospheric turbulence greatly minimised as well as having a much better *signal to noise ratio*. We could then be more aggressive with our enhancement tools and so really drag out all the fine detail that it is physically possible for our image to contain.

Well, yes it most certainly would be great. In fact we really can do that. This is what all the best contemporary Moon and planet imagers actually do!

5.3   MANUALLY STACKING INDIVIDUAL FRAMES

How many images of a given lunar scene **can** you combine? The answer to that depends upon the file size of each image and the available memory space on your computer. However, with today's computers with multi-gigabyte hard drives memory space ought not to be a problem when combining only a few dozen images, even if they are the multi-megabyte images produced by digital cameras.

How many images **should** you combine? The answer to that depends on the quality of each image and the amount of electronic noise it contains. Statistically, the noise content is proportional to the reciprocal of the square root of the number of separate images combined (noise $\propto 1/\sqrt{N}$). An image blended, or *stacked*, from four others should have only half the noise content compared to any one of the four original images. A final image assembled from sixteen separate images ought to have only a quarter of the noise content of any one of the originals.

There is another factor that may limit the number of images you can combine: the time elapsed between the first and the last of the component images must not be too long. Otherwise the lighting of the lunar scene will have changed sufficiently to alter the shadows perceptibly. At the terminator this time should not be more than a few minutes, though this time can be increased for areas away from the terminator.

How do you go about combining the separate images? You need an image software package that will allow you to accurately stack the images one on top of the other and produce a true average of the combined result. There are automated packages that will let you stack up to several thousand frames if you want to. More about those later.

In this section I will concentrate on manually combining a relatively small number of frames. This will prove quite adequate for combining multiple images from digital cameras and astrocameras, as these are relatively noise-free. The same software can even be used for combining manually selected images from video cameras and webcams. However the best results from these devices, particularly so the 'noisy' webcams, will be obtained from combing large numbers of images and this is where the software packages detailed later come into their own.

Major image-processing software packages ought to have image-stacking facilities. One such is the highly popular *Adobe Photoshop*. Whatever the package you use, you will have to resort to the supplied instructions to carry out any of its myriad of processing operations. At least *Adobe Photoshop* has a helpful book, *Photoshop Astronomy*, written by R. Scott Ireland and published by Willman-Bell in 2005, whose target audience is amateur astronomers.

For instance in *Adobe Photoshop* you would begin the stacking process by pasting one image roughly on top of the next until your stack is complete. The bottom image is known in *Photoshop* jargon as the 'background'. The images on top of the 'background' are described in *Photoshop* as 'layers'. You can select one layer at a time, leaving the others temporarily 'hidden' (invisible).

Next painstakingly align each layer in turn with the background (the layers you are not working on all being hidden), working one at a time through the stack. One way of doing this accurately is to select the 'difference' blending mode and then move that layer's image about until it exactly cancels out the background layer to leave a dark screen.

Non-overlapping details stand out bright against the darkness. You can expect a few bright bits on the general field of darkness even on a properly aligned frame. This is because of the image distortions due to atmospheric turbulence, which are effectively frozen in time on each individual frame. Once you are happy with the alignment switch the blending mode back to 'Normal' and the result you now have is the image in the first layer in tight alignment with the background image. Repeat the whole process until you have the background image and all of the images in each of the layers fully aligned.

An alternative method of alignment is to create your own, initially empty, new layer on top of the topmost one containing the image. Next select the 'background' and your new layer. Then 'paint' some markers on your new layer in positions that align with points on the background image that you decide to use as reference. Then you deselect the background and select, one at a time, each in turn of the layers above the background as well as your newly created layer. Making sure that you do not disturb the position of your new layer adjust just the layer containing the image until your chosen points in it coincide with the markers. Work your way through the stack until you are satisfied you have got all the images precisely aligned. Make use of the 'zoom' tool as an aid to precision.

If all the layers are selected (unhidden) in turn at this stage we would still not see a properly averaged image. We would only see the image in the top layer. We have one more job to do. This is to set the opacity of each layer to such a value that it makes an equal contribution to the final image. I find this operation in *Photoshop* rather counter-intuitive. I would have thought that if I had, say, 10 images to combine I would set each at 10 per cent opacity and the result would be a true equal blend. This is not the case in *Photoshop*. Perhaps a better way of thinking is to imagine each of the images in a stack of clear acetate sheets and how it is that you have to look through a particular sheet to see the sheets below. The top sheet would have to have the weakest image, lower sheets progressively stronger images, and the bottom sheet the strongest if each is to make an equal contribution to the final effect.

In the *Photoshop* world what has to be done is to set the lowest, 'background', layer to 100 per cent. The next layer is set to 50 per cent. The next is set to 33 per cent and the next to 25 per cent, and so on. The rule is to divide 100 by the number of the layer (starting with the 'background' as 1). The tenth layer's opacity should be set to 10 per cent.

Of course, you can always deviate from these values if you do want to change the relative contributions of the layers and so favour one or more frames that you judge to be better than the others. However do be aware that you might then not reduce the noise component to its lowest possible value.

As an aside, using the opacity control gives you a third alternative way of getting the images in the layers all accurately aligned. The procedure would be to set the 'background' at 100 per cent. Next select each layer in turn (with all the others deselected), setting that layer's opacity temporarily to 50 per cent. Then move the layer about until you get the images aligned. As, before, using the 'zoom' tool will be a great help in achieving an accurate alignment. Work image by image through the stack until each one is aligned to the background.

Whatever alignment method you used, afterwards go through the stack setting the opacities so that all images give equal contribution as I described a couple of paragraphs ago. Finally save the composite image, perhaps in TIF or BMP format. Then you can either stay in *Photoshop* or export the file to another program, in order to perform your favourite image-enhancement routines (such as those described in Section 4.8 of the last chapter).

5.4 THE WEBCAM REVOLUTION

It would have been practically impossible for an amateur astronomer using film photography to obtain the image shown in Figure 5.4. This is the sort of image that, in the old days, a professional astronomer might have obtained during a session with the 43-inch (1.07 m) reflector at Pic du Midi. In fact it was taken by Damian Peach with an 11-inch *Celestron* Schmidt–Cassegrain telescope from his UK back garden! Yes, Damian is legendary for the effort he puts in to get his great images but even the rest of us ordinary mortals can follow his lead. You might care to take a look at Damian's website: *Damian Peach's Views of the Solar System*. The section devoted to high resolution lunar images can be found at www.damianpeach.com/lunar.htm.

I regard Damian's work as the gold standard but it is also true that many people are not very far behind him in the quality of the results they get. Other lunar images taken by various people using the same basic techniques are displayed in Figure 5.7(a) and (b), and in Chapter 2 (Figures 2.8–2.10), Chapter 6 (Figure 6.4), and Chapter 8 (Figures 8.11(b), 8.13(a), 8.14(e), 8.49(a)–(e)). Of these, Figures 8.11(b) and 8.14(e) are two more examples of Damian's superb work. I find Figure 8.14(e) the most remarkable of any because to get this incredible image of Cleomedes he used his portable 9.25-inch (235 mm) *Celestron* Schmidt–Cassegrain telescope!

You too can take images that are superior to those of even the best of the film-based photographers of yesteryear. Moreover, you do not need

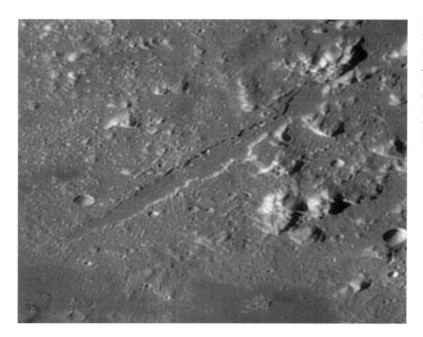

Figure 5.4 Image of the Vallis Alpes taken by Damian Peach, using an *ATiK* 1HS camera and his 11-inch (280 mm) *Celestron* Schmidt–Cassegrain telescope on 2004 March 1[d].

the most expensive imaging devices to do it. A webcam costing less than a couple of hundred dollars and a computer of post-1998 vintage, perhaps most conveniently a modern laptop, fitted with a USB port will be enough for you to potentially get nearly diffraction-limited images from your telescope.

Several recent advances have come together to make this possible. One is the technical development of 'video conferencing devices', or *webcams* as we now call them. In particular the webcam's sensitive detector (a CCD in the best ones) and fast USB download (maybe helped by a little onboard compression) allow us to capture images with exposures of one tenth of a second or less with good resolution in quite low-light conditions.

Another factor is the ever-expanding speed and memory capacities of today's personal computers and portable laptops. This enables us to take large numbers of individual frames in any one session and store them all in the computer.

This brings us onto the final reason for 'the webcam revolution': software packages that automatically, or semi-automatically, can sort the very best images from as many as a few thousand saved ones and then align and stack them. Add a bit of final image processing and voilà: a picture of a planet or small part of the Moon that would have astounded even the best amateur or professional astronomers of not that many years ago.

5.5 YOUR WEBCAM AND COMPUTER

At the time of writing there is one webcam in the 'under $200 price bracket' currently widely favoured by the amateur astronomical community for imaging the planets: the *Philips* ToUcam Pro II PCVC 840 K. You might even be able to obtain this webcam for under $100 in the time it remains available after this book is published. The camera comes complete with a CD-ROM containing the driver software you will need to install on your computer. This is called *VLounge*. However, many people find this piece of software a bit unstable on their computers and they resort to using *QCFocus* (the version current at the time of writing is *Version 2.0*); *QCFocus* can be downloaded free from www.astrosurf.com/astropc. The screens, menus and layouts of *VLounge* and *QCFocus* are well-nigh identical, which is a help if you start with *VLounge* and decide to resort to the other if you find its operation troublesome on your computer.

In normal 'conferencing' use with its lens in place the webcam is capable of producing images in a rather dimly lit room (with ambient lighting of brightness less than 1 lux). This is possible because it has as its detector a *Sony* ICX098BQ, a fairly sensitive, good quality, and relatively low-noise CCD. This makes it also good for imaging the Moon and planets.

This chip has an array of 480×640 pixels, each 5.6 μm square, with an overlaid *Bayer Matrix*. This is a filter grid that sits over the pixels and allows the device to record images in full colour. In any square grouping of four pixels, two of them are filtered green and one each is filtered blue and red. This is where the colour information comes from.

It might strike you that the resolution of the colour component of the image (known as *RGB bands* from the filtered colours) must be much poorer than the resolution that could be synthesised from all of the pixels – and you are right! The human eye and brain are critical of the spatial resolution in a monochrome image but not so critical of the RGB resolution. By means of some artful processing what the computer does is to construct a full resolution 'grey-scale image' using all the pixels and overlay it with a lower-resolution coloured version of image based on the information from the R, G, and B channels.

Thanks to its successful use as a device for imaging the planets in full colour, the *Philips* ToUcam Pro II PCVC 840 K is also the webcam amateur astronomers most frequently plug into their telescopes for imaging the Moon at high resolution. You will find examples of images taken using this webcam in this book: in Chapter 2 (Figures 2.8–2.10), Chapter 6 (Figure 6.4), Chapter 8 (Figures 8.13(a) and 8.49(c)) and in this chapter (Figure 5.7(a) and (b)).

Of course, developments in the computer world are rapid. At the time of writing a new version of this webcam, the *Philips* ToUcam SPC 900NC, is beginning to make an appearance in the market place. As far as I can determine, though, it is pretty well identical to the 'Pro' version except in that it has the possibility for a faster download (90 frames per second as opposed to 60 frames per second – which we do not really need) and is USB2 compatible. I have yet to find out the detailed specifications for this new camera but I wonder if it can deliver its faster download speed without compression? If so, then this camera will turn out to be a step up as compression degrades image quality. By the time you are reading this the information should be readily available.

Most of the best planetary imagers in the world have used the ToUcam webcam and then progressed to other small-chip (typically 480×640 pixels) cameras capable of fast download. In particular they have moved away from 'one-shot' colour imaging devices to cameras with CCDs that don't have an overlaid Bayer filter matrix. To get colour images they take separate video sequences each exposed through coloured filters and one exposure without any coloured filter (the *luminance*, or *L-band* image) and combine them in the computer after the event. More light gets to the CCD without the Bayer filter and consequently the L-band images have a much better signal-to-noise ratio.

You might eventually consider doing the same. If you do not require full colour imaging of the lunar scene, the monochrome high-resolution L-band images are just as much 'one-shot' with one of these cameras as are the full colour images captured by the ToUcam webcam. For lunar observing the benefits these cameras offer are greater sensitivity and an improved signal-to-noise ratio. In addition, they allow a much faster download speed without any of the image-degrading compression that afflicts normal webcams used faster than ten frames per second. They tend to be more expensive than the humble webcam, typically somewhere in the region of $1000, though prices vary greatly from supplier to supplier.

I don't have the space in this book to be able to go into these various alternative cameras but suffice it to say that Figure 5.4 and a couple of images in Chapter 8 in this book (Figures 8.49(b) and (e)) were taken with the *ATiK* 1HS camera and another four images in Chapter 8 (Figures 8.11(b), 8.14(e), also 8.49(a) and (d)) were all taken using the *Lumenera* Lu075M camera.

If you are new to imaging then I would advice you first to serve an apprenticeship with a ToUcam webcam (or equivalent, see further on in this section) before considering moving on to more elaborate and expensive cameras. Get everything right with your telescope and your ToUcam and your images could be of world-class quality.

Figure 5.5 Michael Butcher demonstrates his *Philips* ToUcam webcam with its lens removed and a 1¼-inch (31.7 mm) telescope adapter screwed in its place.

For use with our telescopes the webcam's lens must be removed. A removable lens is an important requirement in any webcam we intend using for high-resolution astrophotography. Please do bear in mind that if you need to disassemble the casing in order to remove the lens you straight away void the manufacturer's warranty. This is not a problem with many webcams, including ToUcams, as the lens simply unscrews from the front.

Next on our shopping list is an adapter that will screw into the hole left by the now-removed lens and which will be a proper slide-fit into the standard 1¼-inch (31.7 mm) telescope drawtube (see Figures 5.5 and 5.6). Many astronomy suppliers keep webcam adapters in stock, particularly so for the most popular models. One important point: **Never leave the webcam chip exposed to dust and moisture**. Keep the webcam in a sealed clean box, or keep a cap over the hole, or keep the webcam plugged into a Powermate or Barlow lens or other sealed device. If dust does get on the CCD, then use only an aerosol dry-spray cleaner (such as that used by photographers) to remove it.

Figure 5.6 Michael Butcher's equipment ready for a night's lunar imaging. The webcam with its adapter (as shown in Figure 5.5) is plugged into a ×5 Powermate and this is in turn plugged into his 10-inch (254 mm) *Skywatcher* Newtonian reflector.

One more item might be regarded as an optional extra, though I regard it as almost essential: an infrared-blocking filter. Again this is something your astronomy retailer should have in stock. The wide end of the adapter usually has an internal thread to take this filter (or any other screw-in filter). The webcam's lens had an infrared-blocking filter built in. Without it colours are a little washed-out and the colour balance is slightly disturbed. This might be a problem in planet imaging but will probably not bother you for lunar work.

However, there is another problem. If your optical system includes lenses then the focus for the near-infrared wavelengths will not be coincident with that for the other visual wavelengths to which the webcam is sensitive. To get the sharpest pictures you should really have an infrared-blocking filter screwed in. The filter your supplier will provide will likely also block near-ultraviolet wavelengths in order to get a good colour balance in normal use.

There is an all-in-one package, *Celestron*'s NextImage: Solar System Imager, which contains an identical device to the ToUcam webcam (even built around the same CCD), supplied with a telescope adapter and a CD-ROM of all the operating software and *RegiStax*, which is the most popular of the image-stacking and processing programs. *RegiStax* is available to download free of charge from the Internet but the other items would otherwise have to be purchased separately.

Celestron introduced this package in 2004 at a price of $150 but many retailers now sell it as cheaply as $100, making it a very attractive product. The infrared-blocking filter is, though, one of the optional accessories, and usually retails at about $55. Other useful optional accessories on offer with this package include a 'reducing lens', costing about $25, which will double the field of view your telescope can fit onto the tiny (2.7 mm × 3.6 mm) imaging area of the CCD. This is useful for the occasions when you desire to image a larger area of the Moon's surface in one go.

Meade also offer a planetary imaging camera along the same lines, though their detector is a cheaper CMOS chip. These are less sensitive and more noisy than CCDs. I reckon the camera package from *Celestron* is your best bet if you are new to webcam-style imaging. Do be aware, though, that this is a fast-moving field and there may well have been developments in the time between me writing these words and you reading them. Please do at least some research before you go shopping for your own camera system.

As far as the computer requirements go for the ToUcam webcam or NextImage cameras, you will need a personal computer later than 1998 vintage (at least 333 MHz, Pentium II standard), fitted with a USB1.1 socket or higher, running Windows ME or later software. The computer should have at the very least 128 Mb of RAM (random access memory) and multi-gigabytes of free hard-drive space. Also please do not skimp on the screen resolution. Go for at least 1024 × 768 pixels in order to achieve accuracy when focusing.

If you go for one of the 'higher level' cameras, such as the *Lumenera* mentioned earlier, then you will need a computer of significantly higher specifications. To get the full benefit from these cameras you will need a system with something like a 2.6 GHz processor and a 7200 rpm. hard drive (though some additional software packages, such as *StreamPix* will allow you to effectively synthesise these requirements on a slower machine).

Most amateur Moon and planet webcam users prefer to use laptop computers. Although these devices are much more expensive than desktop PCs with the same specifications, and have a rather shorter average working life, their portability is a great advantage. Most people prefer to be close by the telescope in order to make fine adjustments. However if the telescope is on a fully driven mounting equipped with remote slow-motions and an electric focuser then this is not strictly necessary.

The cable supplied with the webcam is usually only about 1.5 m long., which is another reason for wanting your computer close to the tele-scope. Alternatively you will have to use USB extension/repeater cables but in practice this might lead to a loss of stability in the link-up. Check with your computer dealer to see what will work well with your machine and webcam. (hubs, powered cables, etc.). Let me also recommend that if you buy a laptop, you spend a bit more and get one with the highest-quality display possible.

5.6   THE WEBCAM'S FIRST NIGHT ON YOUR TELESCOPE

Having purchased your camera and installed its controlling software (supplied with it), please take a little time to experiment with it before your first intended night of Moon imaging. A little familiarity will make things go much better on that first night. Find the control box that switches the camera from 'auto' to 'manual' and play with all the controls. If you have purchased a conventional webcam, keep its lens in place and use it in a dimly lit room. If you have purchased an astronomical version of the unit, such as a *Celestron* NextImage and don't have a conventional webcam lens with it, then plug it into a telescope, which is set to look at some distant scenery (you will not be able to rack the focuser far enough out in order to focus on a nearby object). You will probably also have to fashion a cardboard diaphragm to reduce the light entering the telescope.

A recorded video sequence is known as an AVI. The part of the software suite supplied by *Philips* that does this is called *VRecord*. Practise using your camera to obtain AVIs. Once you are reasonably familiar with the camera and how to control it, then it is time to think about using it to obtain your first Moon images.

First set up your telescope with its sidereal drive switched on and an eyepiece plugged into it. Make sure that you have your PC/laptop safely connected to the mains, or that your laptop's batteries are fully charged. Also check that your webcam is plugged into the USB port **before** powering up the computer.

Select 'Video Properties' and set the webcam's resolution to maximum (in most cameras this will be 640×480 pixels). In the *Philips* software this is in a box called 'Output Size', to be found in a menu that goes under the heading of 'Stream Format'. Calling up the various camera-control and image windows, make sure that the camera control is set to 'Manual' and adjust the gain setting to somewhere between half and full. Make sure the 'Audio' box is unticked in the 'Audio Control' menu which you will also find under 'Properties'. Leaving it on may cause problems later – don't worry about this, just make sure 'Audio' is turned off.

In the 'Image Controls' menu, set the 'Brightness', 'Gamma' (this is contrast) and 'Saturation' (this is how strong the colours are) all to their mid-settings. Next select the 'Frame Rate'. In the *Philips* program you have a choice of 5, 10, 15, 20 and a few higher speed settings. Choose 10 frames per second. Set the 'Exposure' setting to $\frac{1}{25}$ second (however, with the camera on 'manual' and a chosen frame rate of 10 per second the duration of each exposure is actually $\frac{1}{10}$ second; don't worry about this – just accept it as one of life's little oddities!).

If necessary click off the various control settings menus until you can see the camera display screen. Next aim the telescope at the centre of the bright portion of the Moon and exchange the eyepiece for the webcam. Do not worry about enlarging the image for this first attempt. You will see the camera display screen awash with white light. If the screen is virtually brilliant white, call up the control window and adjust the 'Gain' until the screen looks light grey. Leave the 'Brightness' control alone at this stage – control the screen brightness by means of the 'Gain' setting. It is always an advantage to turn down the gain because this reduces the electronic noise the camera generates.

Now have a go at trying to focus. At some point you should see the lunar vista come into focus on the camera display screen. You might need to adjust the 'Gain' setting once more, and then further adjust the focus. Keep going until you are satisfied. Then you might like to call up the 'Image Controls' menu again and adjust the 'Gamma' setting, and maybe just tweak the 'Brightness' setting, until you are pleased with the image. A final check of the focus and all should be ready.

Finally call up the 'Capture' menu and set the duration and frame rate of the AVI, (choose 10 frames per second and 30 seconds for your first go) and a filename ('Moon#1' – not very original but it will do!). When you press 'Return' off it will go – image files will be created from the camera's output and will be rapidly filling up your computer's hard drive.

5.7 STACKING THE IMAGES USING *REGISTAX*

After your session at the telescope you will have one or more AVI files of images to process. You will normally want to do this in comfort indoors – and this brings me to an important warning: **NEVER power up computer equipment, or any mains-driven equipment, that has been recently brought from a cold environment into a warmer one**. Condensation could breach the electrical insulation on any mains-powered equipment causing danger to you. Also delicate electronics (even if low voltage) and disk drives can be wrecked, I repeat **wrecked**, if operated when damp. An observing session in sub-zero temperatures certainly demands that you leave all your equipment until the next day before powering anything up. This might not be such a bad idea, even for occasions when the temperature difference between inside and outside is not so extreme.

Now we come to the stacking, alignment and processing software. The most popular package is the wonderful *RegiStax*, by Cor Berrevoets. This is free to download from: http://aberrator.astronomy.net/registax/. This product can stack and align up to 5000 frames (or up to a 2 Gigabyte limit), which is plenty. It also includes some powerful processing software to use on the stacked image. *RegiStax* is also bundled along with *Celestron*'s NextImage camera, a very attractively priced all-in-one package at the time of writing.

Other programs, some of which can also control the webcam for making and saving the AVIs, include: *IRIS*, *K3CCDTools*, *AVIedit*, *AstroVideo*, *AstroStack*, and *Astro-Snap*. You will find the full set of website addresses for them in Chapter 7. They all come with descriptions, Help files, and/or tutorials and I recommend you study these well. However, do prepare yourself for a steep learning curve. In particular, you will find that there is a number of different choices to make at each stage of the process. There are different ways of doing things and a wide array of settings you can alter to your preferred values.

Perhaps it would be some help if I present here a simplified procedure for getting some good results from your very first attempts at using *RegiStax*. Lack of space prevents me covering anything more about this program, let alone details about all the other software packages. Stick to using this procedure for the first few nights of your imaging and glory in the images you will get. Then, if you so desire, you can begin to experiment with the software package in tandem with rereading the software author's advice (who, after all, is the expert!) in order to improve your results.

I should also warn you that each successive version of *RegiStax* differs somewhat from the last. You may very well find the layout and operations a little different to how I describe them in the following notes, when you come to download the latest version to your own computer. Please allow for that possibility and adopt a flexible attitude of mind. Hopefully the following notes will still be of use in pointing the way to quickly and easily getting good results on your first night of trying.

You have *RegiStax* downloaded onto your computer and you start it running. Then:

1. Click 'Select Input' and load the AVI file. As you work through the program please leave all the various control buttons and boxes at their default settings for the first few times you use it (except where I indicate otherwise in the following steps). The default settings can potentially give you great results but you can so easily make it impossible to get good results if you meddle while still inexperienced. Get some consistently good results **before** you start experimenting!
2. Make sure that the 'Colour' box is ticked (near the upper-left on your screen). Set the 'Processing Area' to 512 pixels square (this is usually the default setting – only consider making this smaller if your machine has limited memory capacity) and select the 256 pixels 'Alignment Box'

button. Check that the 'Quality' box is set to 80% (the normal default setting), you will find this towards the upper-right of your screen.

3. Click 'Show Frame List'. This creates a window in which you will see listed all the individual frames that make up the AVI. These are all numbered. Selecting any one of these makes that image appear in an inspection window. To go through inspecting the images you have a choice of clicking on the listing, frame by frame, or clicking on one frame and using the up and down arrowed buttons on your keyboard, or you can use the slider bar on the side of this window to move rapidly through the sequence of images.

4. We need to create a 'reference frame' to be used as the anchor for aligning the others to in the AVI. There are a number of ways of going about this. You will find the following the most straightforward, even if not the quickest or best, method until you become familiar with this program:

Begin to work manually through the frames until you find a good (sharpest and least-distorted image) frame. Make a note of its number. Keep going for as long as you feel you want to, looking for a better frame. If you find one note its number. When you have had enough of that activity (after two or three minutes, perhaps, depending on your boredom threshold!) scroll back and select whichever frame you decide was the best of those you looked through.

At this point you drag the 'Alignment Box' (remember stage 2) over a particular strong and well-defined feature in the frame, preferably somewhere near the middle of the frame. A large crater would be ideal. Right click your mouse and what comes up next is the 'Align and Stacking' menu.

5. As before, ignore all of the setting controls and boxes on this screen. Simply look for the 'Align and Stack' button. Found it? Good – just click it and sit back.

The program is now busily aligning all the images in your AVI. When it has done this it will commence going through the lot again, discarding those it decides are too poor in quality. Then it will stack the saved best images. All this can take a while, depending upon the speed of your computer. A few minutes is typical but stacking a few thousand frames on a slow machine can take hours!

At the end of this you will be presented with an image that looks quite bright and is noise and speckle free. It will probably look slightly soft of focus – see Figure 5.7(a). At this stage it is a good idea to save the image as a BMP file. We have still not quite finished, though. We can next make use of a very powerful image processing routine while still in *RegiStax* . . .

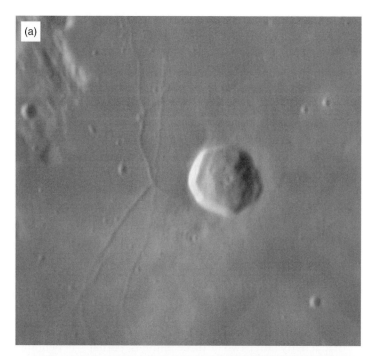

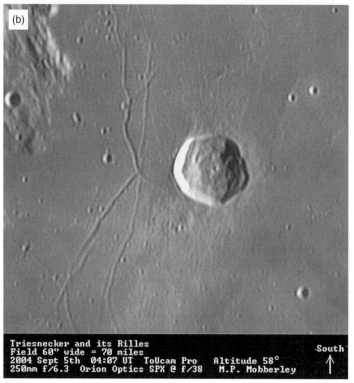

Figure 5.7 (a) A raw stack of the best 150 frames out of an AVI of 2000 frames taken by Martin Mobberley.
(b) The stack now processed in *RegiStax*, using 'Wavelets'. To finish the job the image has been converted to greyscale and the accompanying caption created and stitched on.

Triesnecker and its Rilles
Field 60" wide = 70 miles
2004 Sept 5th 04:07 UT ToUcam Pro Altitude 58°
250mm f/6.3 Orion Optics SPX @ f/38 M.P. Mobberley
South

5.8   PROCESSING THE STACKED IMAGE IN *REGISTAX*

If you click on the 'Wavelets' tab you will open the wavelets processing page, with your stacked image still in the picture window. Wavelets allows certain ranges of spatial frequencies, in other words details on different scales, to be enhanced. As before, I simply do not have the room to explain everything, so please use the following advice as merely a starting point that will quickly and easily get you some initial results. Later resort to Cor Berrevoets' supplied 'About and Help' materials and your own experimentation to take things further.

Find the tick-boxes grouped under 'Options' and make sure the box labelled 'Autoprocessing' is ticked. Look for the two tick-boxes grouped under the heading 'Wavelet Scheme' and make sure the box marked 'Linear' is the one that is ticked. Finally find the two boxes marked 'Wavelet Filter' and make sure the one marked 'Default' is ticked.

Experiment with the six slider controls just one at a time. Starting with the top one, slowly drag it to the right and notice the effect on the image. You will also notice the number on the slider's indicator increasing from its initial '1' setting. Then take the slider back to the '1' setting and next try the same thing with the next slider down. Try each slider in turn. The top one enhances the finest details (the details with the highest spatial frequency), including the unwanted electronic noise. Each of the sliders in turn enhances larger and larger scale features (lower and lower spatial frequencies).

Your aim is to balance the application of the slider controls in order to achieve a combination of settings that best exploit the spatial frequency information in the image. Those actual best settings will very much depend on what is in your image. However a good start is to begin with all the sliders set to '1'. Then move the third slider until the image looks as good as it can without looking artificial due to pushing things too far. Next start moving the fourth slider but only keep going as long as you see any improvement in the image. Then do the same for the second slider – you probably will not want to move this one very far, if at all. Then turn your attention to the fifth slider. You might then try the first slider but I think that the noise that springs to prominence will persuade you to leave this one set to '1'. Then do the same for the sixth slider. You might like to save the image at this point as a BMP file. Then you can further tweak the sliders to see if you can improve the appearance of the image further. For instance, you might try backing off the setting on the third slider and see if that allows you to put a little more on the second and/or fourth ones, etc.

The point of saving the files at the end of the stacking phase and after you have got a preliminary result from Wavelets processing is that if your further work goes wrong then you do not have to go back to the beginning and repeat everything all over again.

Finally you will have the image that best pleases you. Click the 'Final' tab and the last *RegiStax* page appears. The controls on it are very obvious. These allow you to tweak the hue, saturation, and brightness of the image, as well as rotate it and crop it if desired.

Then save this image. Choose to save it as a JPG only if you are sure that you will not want to do anything more to it. I recommend a BMP format if you wish to export it to another program. In fact, I think it is well worth you doing so, as you may well be able to improve the image sharpness a little, especially if you have not quite managed to get your wavelets sliders set to their optimum values for your image. Also you can adjust the tonal values with rather more flexibility in many other software suites ('Curves' is good for this in *Adobe Photoshop*, for instance).

Figure 5.7(b) shows the same image as in Figure 5.7(a) but after Wavelets processing has been applied, and the image has been finished off by being converted to greyscale and a caption has been created and stitched on to the bottom of it.

5.9   STRIVING FOR THE BEST RESULTS

Is your telescope accurately collimated? If in doubt please refer to Appendix 1, near the back of this book, for details on how to check this and the procedure for making any necessary corrections. Poor collimation is a real killer of a telescope's optical performance. The world's best lunar and planetary imagers are all obsessive about precise collimation – and with good reason!

Is your telescope sited in the best place? Looking over rooftops, long lengths of concrete and patio slabs, warm buildings, etc., will all cause bad seeing. I realise that you may have little choice in the matter but do whatever you can to give yourself the best chance of obtaining quality images. Taking a warm telescope out into the cold night air and trying to image straight away is usually pointless. Do whatever you can to minimise the temperature difference between the telescope and the night air. Take the portable telescope outside, or open up the observatory as long as possible before you begin.

If your telescope has a cooling fan (very useful on larger reflecting telescopes) then switch it on as early as possible before you begin. If you have a choice of what equipment to use then a small high-quality telescope may well be better than a large mediocre-quality one. In particular the thermal characteristics of the smaller telescope may well allow you to obtain images superior to those you get from a larger one on many, maybe even most, nights.

When taking the AVIs make sure that you and any heat-generating equipment, such as your laptop, are well away from the front end of the telescope. If at all possible position yourself and your equipment downwind of the front of your telescope.

Take the time to focus the image as precisely as possible. The normal state of the Earth's atmosphere will make this a trying exercise. The image will shift in and out of focus as well as gyrating and distorting all on its own, even without you touching the focus adjuster. It can be quite difficult judging where the true focus point really is. A good quality electric focuser is a real boon. If you have to touch the telescope to focus it you will likely throw the image into jitters, making the task of achieving a good focus even more difficult.

To allow the possibility of resolving the finest possible details in the image, the principle focus image needs to be enlarged. I discuss ways of doing this in the previous chapter (Section 4.7). The formula for predicting the finest details physically resolvable by a given aperture of telescope – the diffraction limit of the telescope – is given in Chapter 3 (Section 3.1).

For a CCD having 5.6 μm pixels, as is the case for most webcams, including the ToUcam Pro, the Nyquist sampling limit and diffraction limit are both satisfied when the effective focal length of the telescope is 17 times its aperture. In other words, the Nyquist theorem predicts that the '5.6 μm-pixeled' CCD is just capable of recording the finest detail it is theoretically possible for the telescope to show when it is used at an effective focal ratio of f/17.

While most of the potential detail will indeed be recorded at this focal ratio, there are some gains to be had by enlarging the image further. Indeed, I would generally advise any webcam apprentice to use an effective focal ratio of around f/20–f/25. The premier Moon and planet imagers tend to use effective focal ratios in the range f/30–f/50 under the very best observing conditions. However, it is also true that they have graduated beyond the ToUcam to cameras like the *ATiK* and the *Lumenera* which are more sensitive.

Of course the ambient seeing conditions and the size of telescope have a big bearing on how much image enlargement is profitable. The bigger the telescope, the less often will it be able to resolve anywhere near its theoretical limit. Once you have mastered the basics, experiment to see how far you can push the enlargement factor with your equipment in given seeing conditions before that enlargement becomes unprofitable.

Experiment with the length of AVIs. Do more frames get you a better final result? Generally the optimal number of frames increases with image enlargement. This is because the image dims with enlargement. Find out what works best for you.

Finally, get to know the camera-control software and the stacking and processing software and experiment with a view to improving the results you get. I must again emphasise that the brief notes I have given in the last two sections are only intended as instructions for the easiest way to get some results on your first few nights of using *RegiStax* while it is still unfamiliar to you. Make use of any Help files, About notes, and tutorials the software authors provide. Also find out what other people are doing.

There are a couple of books: *The Lunar and Planetary Webcam User's Guide*, by Martin Mobberley (Springer-Verlag, 2006) and *Introduction to Webcam Astrophotography*, by Robert Reeves (Willmann-Bell, 2006) which have my highest recommendation. There are also a couple of *Sky & Telescope* articles which may be of help: *Processing Webcam Images with RegiStax*, written by Cor Berrevoets (the author of the software, here describing his version 2.0) in the April 2004 issue; and *Planetary Processing with RegiStax 3*, a brief review of version 3.0 of this software by Sean Walker, in the December 2005 issue.

CHAPTER 6

The physical Moon

While it is true that the Moon's stunning vistas can provide many hours of entertainment of the 'sight-seeing' kind, I would argue that observing the Moon is ultimately a sterile and pointless exercise unless one is attempting to understand and know it better. If you accept that premise then it follows that having some knowledge and understanding of the Moon, including knowing what mysteries still remain to be solved, will expand, and give some meaning and purpose to, your observations of it.

In that spirit I offer the following highly abridged account of the space-borne missions to the Moon together with some of our modern ideas about the physical nature and evolution of the Moon that arose because of them.

6.1   THE FIRST LUNAR SCOUTS

In 1903 Orville and Wilbur Wright made their first powered flights at Kitty Hawk. Astonishingly, it was only 66 years later that Neil Armstrong and Edwin 'Buzz' Aldrin stepped from their space-going vehicle onto the Moon's alien surface. The pace of progress at that time was breath-taking. Indeed, it was only in 1957, a mere dozen years before that first manned Moon-landing, that the Earth's first man-made satellite – *Sputnik 1* – was launched into orbit, marking the true beginning of the 'Space Age'. The many elements of progress – such as in launch-vehicle design, probes, satellites, telecommunications, and much, much, more – all form part of a complex story. Here, though, I can mention only the main highlights.

The first Moon mission successes came with three Russian probes in 1959. *Luna 1* (at the time the *Luna* probes were called *Lunik*) was the first to achieve a flyby, passing less than 5000 km from the Moon and revealing that it has no significant global magnetic field. *Luna 2* made further measurements as it headed towards the Moon, eventually impacting with the lunar surface in the Mare Imbrium. It was thus the first man-made object to make physical contact with the Moon. *Luna 3* was much more ambitious. As well as making a full range of scientific measurements, its trajectory carried it about 4600 km beyond the Moon so that it could look back and photograph the Moon's rear side. The images it transmitted to us might have been of poor quality judged by modern standards but the Earth-averted hemisphere had never before been seen. We learned much from those first blurry photographs.

The next few years brought forth a mixture of successes and failures. The continuing *Luna* series of probes, and a *Zond* probe (*Zond 3* in 1965, another mission to photograph the Moon's averted face while it was on its way to Mars – two objectives for the price of one!) were joined by the American *Ranger* series of probes. *Ranger 7* was the first to photograph the Moon at very close quarters. As it hurtled to destruction in the Mare Nubium in July 1964 it took and transmitted back to Earth over four thousand photographs. The area around the crash site was afterwards re-named Mare Cognitum, the 'Known Sea'.

Altogether, nine probes bit the lunar dust and seven others either missed the Moon or were not intended to hit the surface before *Luna 9* became the first soft-lander in February 1966. It touched down in the Oceanus Procellarum, near the great crater Grimaldi.

Luna 10 became the first lunar-orbiting satellite in April 1966 and in the next ten years 38 further lunar satellites and soft-landers were sent to the Moon by the Russian and American space agencies (including the manned missions, described in the next section). Among them the American *Orbiter* series of lunar satellites, in the late 1960s, were particularly valuable in mapping much of the Moon to a finer resolution than was ever possible from the Earth (see Figures 6.1, 6.5 and 6.6; also see Figures 8.7(c), 8.13(g), 8.17(e), 8.22(h), 8.33(e), 8.37(e), and 8.46(d) in Chapter 8). Of course, the mapping also included areas that were either poorly seen, or totally hidden on the lunar farside (see Figures 6.1 and 6.6).

The American *Surveyor* craft also produced particularly valuable results, as they were in effect soft-landing laboratories, sending back photographs from their landing sites, as well as testing the mechanical properties and chemical composition of the lunar soil. Meanwhile the Russians continued with their *Luna* probes. *Luna 16*, launched in 1970, was the first robot vehicle to return a lunar soil sample to the Earth. *Luna 17* (better known as *Lunokhod 1*) did even better, in that it was the first robot rover vehicle to explore the surface of another world. It spent over 10 months exploring the Moon's surface, covering about 10.5 km of the Mare Imbrium in that time. *Luna 20* was another mission to recover lunar soil, this time from the Apollonius highlands, in 1972.

Figure 6.1 The Moon viewed from an angle impossible from the Earth. This *Orbiter IV* view of the Mare Orientale, clearly shows the multi-ring structure of this vast impact basin. The inner, basalt-lava-flooded, section has a diameter of about 320 km while the outermost ring spans about 930 km. The south pole of the Moon appears at the top of this photograph and part of the nearside feature of the Oceanus Procellarum appears to the lower left. The small patch of dark mare material between the Mare Orientale and the Oceanus Procellarum (but closest to the Oceanus Procellarum) is the basalt-flooded crater Grimaldi (see Section 8.20 in Chapter 8). (Courtesy NASA and Ewen A. Whitaker.)

The next year saw another Russian roving vehicle, *Lunokhod 2* (*Luna 21*), put down in the Mare Serenitatis – but of course the main glory in the years 1969–72 belongs to the Americans. For it was in those years that science fantasy became science fact – and men walked upon the surface of another world.

6.2   MEN ON THE MOON

While the unmanned probes were doing their work, preparations were underway to send men into space. Major Yuri Gagarin was the first man sent above the Earth's atmosphere in the Russian *Vostok* capsule on 12 April 1961. Gherman Titov was next but the Americans were not far behind. Colonel John Glenn was launched into space on 20 February 1962 in *Friendship 7*, one of the *Mercury* series of manned capsules. The Americans eventually overtook the Russians in what unofficially became known as 'The Space Race' – and what a thrilling race it was!.

The American *Gemini* missions gave way to the *Apollo* programme. The hardware to enable men to get to the Moon was designed, built and tested in stages. The Christmas of 1968 was memorable because the astronauts Frank Borman, James Lovell and William Anders became the first people to travel beyond the Earth's realm as their *Apollo 8* spacecraft went into orbit around the Moon. Only two more *Apollo* missions were required to further refine and rehearse everything. *Apollo 11* would be the one to achieve the great goal.

On 16 July 1969 the mighty *Saturn 5* three-stage rocket launched from *Cape Kennedy* (previously known as *Cape Canaveral*, the name was changed back again after the *Apollo* missions) with the astronauts Neil Armstrong, Edwin 'Buzz' Aldrin and Michael Collins riding in the nose-cone capsule of *Apollo 11*. Stages one and two were released to fall away from the rocket when their fuel loads were expended and their work was done. The third stage finally put the astronauts and payload into orbit. The payload consisted of the Service Module, at the top of which was the nose-cone capsule (Command Module), and the Lunar Excursion Module, or LEM, which was stored inside the upper part of stage three.

Three hours after launch the Command–Service Module was separated and turned and the nose-cone attached to the LEM. The LEM was then pulled out of stage three of the rocket. The Service Module engine was then fired and stage three of the *Saturn 5* rocket was left behind as the Command–Service Module and LEM together moved out of Earth's orbit and headed towards the Moon.

On 19 July the spacecraft was driven into lunar orbit. Armstrong and Aldrin in the spidery-looking LEM (which had been named 'Eagle') separated from the Command Module, which was to stay in lunar orbit with

Collins keeping a lonely, if busy, vigil. Armstrong and Aldrin prepared for landing. On Sunday 20 July the astronauts fired the LEM engine, slowing it and allowing it to drop out of Lunar orbit and descend to its chosen landing site on the Mare Tranquillitatis. Television viewers watched the drama unfolding. The world held its breath until Armstrong's words: 'Houston – Tranquillity Base here – the Eagle has landed' were received with relief and jubilation.

The astronauts peered through the spacecraft windows at the unearthly scenery. They could see the long shadow of the LEM cast onto the Moon's dusty surface. Seven hours later Armstrong opened the hatch and carefully climbed down the ladder. With the words 'That's one small step for a man – one giant leap for mankind' Neil Armstrong became the first human to set foot on another world.

As an aside, I have given here the official wording of Neil Armstrong's famous declaration. However, I have listened to the recording a number of times and I always hear '. . .step for man', missing the 'a', but I suppose I must be wrong.

A little later Aldrin also stepped out onto the moonscape. For about 2½ hours they busied themselves setting up a television camera, planting the American flag, taking photographs, collecting rock samples, and setting up experiments on the Moon's surface. Here on Earth millions of people watched the astronauts go about their business and listened to the conversations between themselves and Mission Control at Houston. The astronauts moved with a bouncing, almost slow-motion, gait in the low surface gravity (on the surface of the Moon objects have only one sixth of their earthly weight). All too soon it was time to climb back into the LEM.

After a night's sleep the astronauts blasted off, leaving the lower part of the LEM still sitting on the surface of the Moon, and they successfully re-joined the Command Module. Then the journey home. On 24 July the *Apollo 11* capsule (just the nose-cone section) ploughed into the Earth's atmosphere. Hung under three large parachutes for the last part of its descent, the capsule splashed down in the Pacific Ocean. The first great adventure was over. Men had been sent to land on the Moon and returned safely to Mother Earth.

Subsequent *Apollo* missions followed the same basic mode of transporting men to the Moon (*Saturn 5* launch vehicle plus payload, including LEM) but with longer and longer periods spent on the Moon's surface and progressively greater quantities of Moon rock and scientific data collected. Instrument packages were left to monitor conditions (temperatures, seismic activity, solar wind data, etc.) and radio the information back to Earth long after the astronauts had left.

The *Apollo 12* astronauts Charles Conrad and Alan Bean landed their LEM, 'Intrepid', in the Moon's Oceanus Procellarum on 19 November 1969, while Richard Gordon orbited the Moon in the Command Module. They had landed only a couple of hundred metres from the *Surveyor 3* probe which had been landed there 2½ years before. As well as carrying out their other activities, they photographed the vehicle (see Figure 6.2) and added a few parts of it to the booty they brought back to the Earth.

Apollo 13, launched on 11 April 1970, nearly resulted in tragedy as a mid-flight explosion onboard the Service Module put an end to the mission. The only option for the astronauts James Lovell, John Swigert and Fred Haise was to carry on to orbit the Moon and hope that the Service Module motor could be fired to get them back to the Earth. They made it back despite the crippled state of their spaceship – a sound reminder of the dangers of the enterprise. There had been previous fatalities in both the American and Russian space programmes.

Apollo 14 put the programme back on track. The LEM carried Alan Shepard and Edgar Mitchell down to the Fra Mauro region of the lunar surface – the first manned landing in rougher terrain – on 31 January

Figure 6.2 Clearly *Surveyor 3* bounced before it came to rest on the lunar surface, as revealed by this photograph of one of its feet taken by an *Apollo 12* astronaut. (Courtesy NASA.)

Figure 6.3 *Apollo 15* astronaut James Irwin pictured by the Lunar Roving Vehicle, with Mount Hadley providing a spectacular backdrop. (Courtesy NASA.)

1971. They hauled a small hand-cart about in two traverses of the lunar surface while Stuart Roosa orbited in the Command Module.

The *Apollo 15* LEM was landed at the foothills of the Lunar Apennine mountain range (Montes Apenninus) on 30 July 1971. David Scott and James Irwin drove over the lunar surface in their 'rover' vehicle (see Figure 6.3), covering 27 km in three separate traverses, including driving almost to the edge of the winding Hadley Rille (see Figure 6.4). Alfred Worden orbited in the Command Module.

Apollo 16 touched down in the Descartes region of the southern highlands of the Moon on 21 April 1972, with astronauts John Young and Charles Duke in the LEM. Thomas Mattingly orbited in the Command Module.

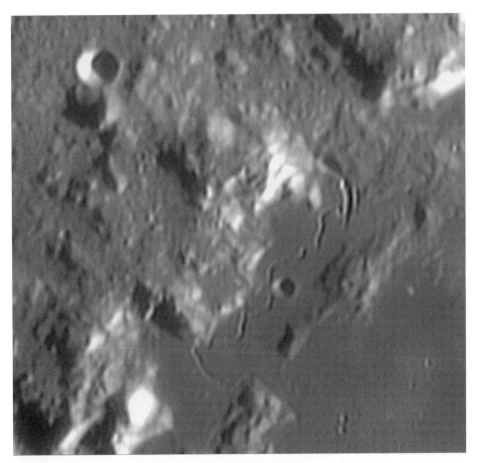

Figure 6.4 Hadley Rille (Rima Hadley) – a webcam image by Martin Mobberley taken on 2004 September 5^{d} 03^{h} 57^{m} UT with a ToUcam Pro camera, 250 mm f/6.3 Orion Optics SPX @ f/38.

They also had a rover vehicle to aid in their exploration, experimentation, and sample-collecting activities in the stunning rock and boulder strewn environment. They also used equipment to make the first astronomical observations from the Moon. Of particular importance were ultraviolet photographs of the Earth's atmosphere, interplanetary gas and stars.

Apollo 17 was the grand finale as the LEM, with Eugene Cernan and geologist Harrison Schmidt aboard, touched down in the Taurus–Littrow region on 11 December 1972. They went even further in their rover vehicle (see Frontispiece). Each mission bettered the last with the astronauts spending longer periods outside their LEM, doing more photography and making increasingly sophisticated geological examinations and scientific experiments, as well as setting up more intricate remote telemetry packages. As a result of the space-probes, and particularly the *Apollo* programme, our knowledge of the Moon grew many-fold. When Cernan and Schmidt rejoined Ronald Evans in the Command Module and they headed back to Earth they brought to a close a period unprecedented in human history.

6.3   THE POST-*APOLLO* MOON

Originally, the *Apollo* missions were to go beyond number 17 but the public grew bored, and some vociferously objected to the money spent on the project. Vote-conscious politicians cut back NASA's budget. The programme came to a premature end. Three successful Russian probes, and one failure, went to the Moon post-*Apollo*, the last of these in 1976, as well as one Japanese orbiter in 1990. To date there have been no further manned missions.

We had to wait two decades for the next American probe – *Clementine* – which was injected into a polar lunar orbit in February 1994. By circling from pole to pole as the Moon turned under it, the probe was able to photograph the entire Moon in twelve different wavebands spanning ultraviolet through to infrared. Some of the images it obtained were of particularly high resolution (see Figure 8.41(e) in Chapter 8), the best showing details as fine as 100 m across.

Clementine also carried a laser-ranging system which built up a laser echo map of the Moon with a horizontal resolution of the order of 200 km, though with a vertical sensitivity of about 40 m. The advantage of the laser echo technique was that the picture of the Moon it built up was three-dimensional, containing as it does height information. For the first time scientists realised the true extent of a depression in the Moon's south polar region, known as the South Pole–Aitken Basin. It turns out to be about 12 km deep and 2500 km across, and holds the record for being the largest impact basin known in the Solar System. Other ancient basins were also studied.

The multi-waveband images provided valuable information about surface composition. *Clementine* was a multi-purpose probe, having military as well as scientific goals. For two months it did its work orbiting the Moon. It was then to be dispatched to a rendezvous with the asteroid Geographos but a technical hitch caused it to be sent spinning off into space, instead. As in the words of the ballad, *Clementine* was 'lost and gone forever'. Allied to the previous space-mission results, *Clementine* provided another leap in our knowledge of the Moon.

In January 1998, the *Lunar Prospector* probe arrived at the Moon. Other probes had passed by the Moon on their way to other targets but I will pass over these, even though some telemetry and lunar photography was undertaken. Like *Clementine*, *Lunar Prospector* was sent into polar orbit, following up on the investigations undertaken by the earlier probe. Mapping, studies of surface composition, including the search for ice deposits at the lunar poles, magnetometry, radioactive-particle counts and gravitational data all were carried out.

The biggest news headline from the *Lunar Prospector* results was the apparent discovery of ice in the Moon's polar regions. NASA's public relations team announced they had found about 6 billion tons (later downgraded to 300 million tons) of the stuff! The news was breaking at the time I was putting the finishing touches to the first edition of this book and in it, amid speculations in the Press about astronauts setting up bases at the Moon's poles to utilise all the ice laying about on the surface, I sounded a note of caution. For one thing I doubted that there could be any solid deposits of ice laying around even in the parts of the lunar surface permanently shielded from sunlight. Also it occurred to me that the probe had detected the signature of hydrogen released from chemical combination and the source of that only **might** be water ice. I thought at best what ice might be present was probably all due to accumulated materials from impacting comets and would be mixed up in the upper regolith.

It turns out I was right to be wary. NASA scientists decided to sacrifice the probe a little prematurely in order to give it a chance of providing conclusive evidence for surface ice deposits. On 31 July 1999 *Lunar Prospector* was crashed into the shadowy interior flank of an anonymous 51 km diameter crater at the Moon's south pole. A number of telescopes, including the *Hubble Space Telescope*, looked at the exact time and location of the crash with spectrographs set to detect any of the hydroxyl species or water molecules that should have been thrown up if the probe encountered any ice. Nothing was detected.

While it is true that a crash at one particular location does not rule out surface ice elsewhere, more negative evidence arrived in 2003. Bruce Campbell and his colleagues of the Smithsonian Institute used the Arecibo Radio Telescope in Puerto Rico to fire radar signals at several specific wavelengths at the Moon's polar regions. The waves penetrated the lunar regolith to a depth of several metres, as well as seeing into the depths of the craters. The reflectance data showed that there could be no substantial concentrations of ice. By contrast the same observational methodology had revealed thick ice sheets at the poles of the planet Mercury. Lunar ice is not ruled out but it must be well mixed into the regolith and rather spread out. Since the Moon's chemistry is starkly anhydrous (more on that in the next Section), I remain convinced that whatever water got to the Moon and remains frozen in at the poles was delivered by comets.

At the time of writing another probe, *SMART-1*, is orbiting the Moon. Its main purpose is to test a new ion-drive propulsion system. Launched in September 2003, its 233 looping orbits of ever increasing radius of the Earth (getting some gravity-assist from the Moon as it passed our satellite with each Earth orbit), eventually resulted in it passing into lunar orbit in November 2004. It began imaging the Moon's surface at medium resolution soon after and it is equipped with an infrared spectrometer and an X-ray telescope in order to carry out mapping of the chemical composition of the Moon's surface. It also has particle and field detectors. [Note added during production: the *SMART-1* probe was deliberately crashed into the Moon on 3 September 2006.]

6.4   NOT GREEN CHEESE BUT . . .

The Russian *Luna 16*, *Luna 20*, and *Luna 24* craft returned a total of about 0.3 kg of Moon rock. We already have about 4½ kg of identified lunar material on the Earth, in the form of meteorites blasted from the Moon's surface by impacting meteors, but it was the six *Apollo* manned landings that provided us with the greatest quantity and variety of samples; some 381.7 kg in all.

The chemical composition of the Moon's surface covering is very different to that of the Earth. It is made up of a variety of igneous rock types. In the main these are complex silicates. Unlike earthly rocks, there is a complete absence of water in the chemical make-up of any of the samples so far examined. As far as we can ascertain there is no water natural to the Moon itself. Apart from anything that might have been deposited onto its surface by external means, the Moon is a completely arid world. Nor did we find any signs of life, past or present. In fact, the Moon has only trace amounts of the carbon compounds that are needed as building blocks for the genesis of life as we know it. Most of these compounds originated from outside the Moon, delivered onto it by the solar wind plus meteorite and comet impacts.

However, there are some similarities between the lunar rocks and the exposed **mantle** materials we find on Earth (these rarely get to the surface but are occasionally found amongst volcanic ejecta). Even more significant is that the ratios of the proportions of the three common isotopes of oxygen we find in the Moon-rock samples match closely with those we find in terrestrial rocks.

The highland rocks and mare rocks are themselves very different from each other. Considering first the lighter-coloured highland material, there are at least three major categories of distinct subtypes of the silicate-type rocks. *Ferroan anorthosites* are particularly rich in calcium and aluminium and largely composed of a mineral known as *plagioclase feldspar*. The so-called magnesium-rich rocks are also plagioclase-based but also have mixed in with them various magnesium-rich minerals such as *olivine* and *pyroxene*. The minerals *norite*, *dunite* and *troctolite*, are found in highland samples but these are really subtypes of the magnesium-rich rock types, each having various proportions of pyroxene, plagioclase and olivine.

The other major class of rock found in the lunar highlands is the so-called *KREEP*. This is an acronym for potassium (chemical symbol K), rare-earth elements (REE) and phosphorus (chemical symbol P). Rocks with enhanced concentrations of these components are known as KREEP rocks – an example being *KREEP norite*.

The lunar maria are composed of iron- and titanium-rich volcanic rock types known as *basalts*. They also have a diversity of compositions within the main type – pyroxenes, plagioclase, *ilmenite* (iron–titanium oxide), and olivine, plus many others. The lunar basalts have one important physical characteristic: when they are heated to melting point their viscosity becomes very low, much less than is the case for earthly volcanic lavas. To give you some idea, the lunar lavas might have been as runny as 20/50 W engine oil. This fact has an important bearing on how the Moon got to look as it does to us today.

6.5   GENESIS OF THE MOON

Four main theories of the Moon's origins were developed by theorists over the years. The first of these is that the Moon was once part of the Earth but broke away from it, leaving a hollow which became the Pacific Basin. We now realise this idea is dynamically untenable. Also, the Moon's chemical composition is so unlike the Earth's that any simple separation would be out of the question. This idea is now only of historical interest.

The second theory is that the Moon and Earth were formed at about the same time, 4600 million years ago, but the two bodies were formed in completely different parts of the Solar System (which was also being 'born' at the same time). At some later date, the theory goes, the paths of the Moon and the Earth crossed and the Moon became gravitationally captured by the Earth. Tidal interaction then caused the Moon to settle into a stable orbit around the Earth. This idea is not totally out of the question but the dynamical difficulties are very great and most theorists of today have little confidence it. Further evidence against this idea comes from the oxygen isotope ratios we measure in the lunar rock samples. These vary with location in the Solar System but, as already noted, the ratio is the same as that which we find for terrestrial rocks.

The mathematical difficulties associated with this second theory disappear if we assume that the Moon and Earth formed in the same region of space and at about the same time – the third theory. The isotope abundance ratios would then also be explained. However, even allowing for the chemical differentiation that would take place inside a condensing cloud of protoplanetary matter – heavier elements sinking to the core, leaving the lighter elements at the top – it is very difficult to explain the sharp difference in the chemistries of the two worlds.

The fourth, and currently the most popular, theory we have today is that at some time very early in the history of the Solar System, perhaps before the Earth had developed a solid crust, our planet had a glancing collision with another planet or protoplanetary body (perhaps it was even as large as the planet Mars is today). This would have 'chipped off' a sizeable chunk of material from both bodies. While much of the resulting shoal of material and the remains of the second planet/proto-planet would be lost, some of this debris would settle into orbit around the Earth and eventually form a new body – our Moon. If the Moon was mostly formed from the material that originated in the impacting proto-planet and from the **mantle** of the Earth then this might explain the compositional differences between the materials of the Earth's and the Moon's surfaces.

6.6   THE MOON'S STRUCTURE

The seismic detectors left on the Moon as part of the *Apollo* missions gave us data for years after the last men walked on the Moon. Just as seismologists have built up a picture of the Earth's structure by studying earthquakes, so planetary scientists have been able to build up a picture of the Moon's structure from the results of the very gentle *moonquakes* that trouble the Moon's globe. About 3000 of these were detected each year the seismic detectors were in operation. Lest you should get the wrong idea, though, I should add that the total seismic energy output of the Moon is less than one ten-billionth the energy the Earth expends in earthquakes in the same period – feeble by any standards. Moonquakes seem to originate about 600–800 km below the lunar surface.

Also, a few spacecraft were deliberately smashed into the Moon's surface to generate much more violent tremors and some moonquakes have been triggered naturally because of meteorite impacts to provide further information about the Moon's structure. In addition, the *Apollo* astronauts also conducted seismic sounding experiments. Meanwhile the various heat-flow experiments left by the *Apollo* astronauts indicate a heat-energy flux of about one third that of the Earth. Models suggest that most of this heat energy is accounted for by radioactive decay deep in the Moon and so the Moon must have lost most of the heat generated by its formation. These results have helped in the refining of the theoretical models.

The crust of the Moon extends to about 60 km depth and exists as three distinct layers. The top layer is known as the *upper regolith* and is made of fragmented and impact-welded rocks of the type geologists call *breccias*. Ranging from 1 to 20 metres deep on average, the upper regolith is the result of aeons of bombardment by meteorites large, small and minute (micrometeorites), together with the crumbling stresses caused by the diurnal heating and cooling. It could only form as it has on a world which for a long time has been devoid of a protective atmosphere. The soil is also churned and intermixed with materials from underlying layers due to the same forces. This process has the delightful name of *gardening*.

Extending down to a depth of about 20 km is the *lower regolith*, composed mainly of basaltic rocks. The bottom layer of the crust is chiefly made up of the rock type known to geologists as *anorthositic gabbro*.

Below the crust is the *mantle*, rich in certain minerals such as olivine and pyroxene. The mantle becomes less rigid with depth. Gravity data, especially that obtained from the *Lunar Prospector* space-probe, indicates the presence of a small iron-rich *core*. It might be about 600 km across if composed mostly of pure iron, increasing to about 900 km if it is composed mostly of iron sulphide. It is likely to be molten, perhaps at a temperature of about 1700 °C if it is pure iron and ranging down to about 1000 °C, or even a bit less, if mostly iron sulphide (nickel is another likely ingredient). The core could not be more massive than that because the density of the Moon averages a mere 3.34 times that of water, too low for there to be a larger core. Gravimetric data (obtained by precisely monitoring the paths of orbiting space-probes) and even studies of lunar libration confirm the small size of the core.

Overall, the Moon's globe is very iron-depleted compared to the Earth. Given that the Moon has no significant global magnetic field (though there are concentrations of weak magnetism 'frozen' into the surface rocks), it used to be thought that the core has now solidified. The newest data, though, do suggest it is at least partially molten.

The presence of an iron-rich core is significant in that the most popular theory of the creation of the Moon (see the previous section) should produce a completely iron-free Moon if the impact had occurred after the two colliding worlds had become chemically differentiated. Assuming the theory is correct, perhaps either or both were still so 'newborn' as to be largely undifferentiated when the collision occurred?

I mentioned that the lunar crust is about 60 km thick. Actually, this is only an average figure. It is much thinner on the Earth-facing hemisphere, being only 20 km or so in places. However, on the reverse side the crust is over 100 km thick.

The *Luna 3* probe of 1959 had revealed there to be an almost complete absence of maria on the Moon's reverse side, yet about one half of the

Earth-facing hemisphere is mare-covered. This was a real surprise to scientists at the time. The reverse side of the Moon is covered with the same sort of rough, cratered terrain we see in the highland areas of the near side. We now understand the reasons for the asymmetry. The explanation is bound up with the evolution of the Moon after its formation, so let us briefly review our modern ideas on this subject.

6.7 THE EVOLUTION OF THE MOON – A BRIEF OVERVIEW

When the Moon was still a molten body, about 4600 million years ago, its own gravity operated on the components making it up and caused the heaviest to sink towards the core – the chemical differentiation I referred to earlier. This separation continued until the lightest elements floated to the top. These lightest materials formed the basis of the lunar highlands. At the time when the Solar System was still young, space was cluttered with debris left over from the formation of the various planets and moons.

During these early times massive lumps of material were smashing into the Moon and the other planets. Great basins and smaller craters were created by these huge, explosive, impacts on the now solidified lunar surface. The gravitational fields of the Moon and the planets acted as 'celestial vacuum cleaners', gradually disposing of the Solar System leftovers. All the biggest pieces were used up first, only the progressively smaller pieces of debris being left as time went on.

The ferocity of this lunar Blitzkrieg abated until it was all but over by about 3800 million years ago. The Moon's surface was then heavily scarred and saturated with craters of all sizes, though with the greatest numbers of them being the small ones.

I ought to mention that for many years a great controversy raged among astronomers concerning the origin of the Moon's craters. Setting aside those with really wacky ideas, of which there were many, astronomers dichotomised into two camps. Many maintained that the craters were formed by endogenic (internal) processes. Endogenicists' theories ranged from violent volcanism, through to more quiescent mud-bubble scenarios. The evidence they drew on to support their views included a perceived non-random distribution of craters on the Moon (north–south-going crater chains), and the fact that smaller craters almost always break into larger ones – and virtually never the other way round.

In fact, the primary craters of the Moon **are** pretty well randomly arranged. Much of the perceived north–south alignment is due to the direction of the incident sunlight, which throws features along the lunar meridians into prominence. This is particularly so when the terminator

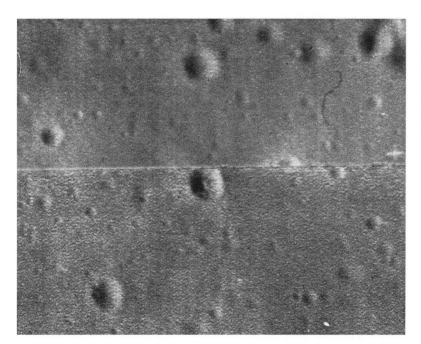

Figure 6.5 A man-made lunar crater! The 13½ m diameter crater at the centre of this *Orbiter IV* photograph was created by the impact of the space-probe *Ranger 8*. It even has a central peak! (Courtesy NASA and Ewen A. Whitaker.)

looms close. In my view, the extreme rarity of larger craters breaking into smaller ones is a little more convincing but even this factor is explainable under the *exogenic theory*, otherwise known as the *impact theory* – the idea that the craters were created by meteorites impacting with the lunar surface. I cannot resist showing you the crater pictured in the centre of Figure 6.5. There can be no controversy about the origin of this one – it was created by the impact of the spacecraft *Ranger 8*!

At an early stage after the formation of the Moon, tidal drag between the Earth and Moon locked it into a synchronous orbit with the Earth. Hence it always keeps the same face to the Earth. Moreover, their mutual gravity produced some asymmetry in the internal structure of the Moon, for instance it bulges towards the Earth a little as well as having the thinnest part of its crust on the Earth-facing hemisphere.

The creations of the biggest basins were real Moon-rocking events, leaving a heavily fractured crust, some of the fissures extending down to the still-molten mantle. Low-viscosity lavas flooded out of the fissures to fill the basins and so form the maria.

Where the crust was thickest the fissures could not reach through to the mantle and the basaltic lavas could not escape to flood the surface. That is why the maria are predominantly on the Earth-facing hemisphere. The crust was too thick to allow the process to happen on the reverse side.

As the interior of the Moon cooled, so the lava-flooding activity dwindled and eventually stopped about 3200 million years ago. Some small-scale volcanism probably continued for a little longer and the Moon certainly continued to receive impacts thereafter but all the really major activity was over and the Moon then became a much more sedate place.

6.8   LUNAR CHRONOLOGY

The 4.6 billion year history of the Moon has been divided into a number of periods, or eras, marked by specific events. These have been dated by means of the laboratory testing of soil and rock samples brought to Earth. Various techniques have been used to date other lunar features/events with these as primary benchmarks.

The first of these events was the formation of the Nectaris Basin – which later lava-flooded to form the Mare Nectaris (the subject of Section 8.30 in Chapter 8). This occurred some 3.92 billion years ago, according to modern determinations. The first lunar era is thus the *Pre-Nectarian Period* (4.6 to 3.92 billion years ago). The massive amount of bombardment the Moon suffered at this stage has obliterated much of the earliest-formed surface features, though an undefined number of Pre-Nectarian structures have survived.

The second benchmark event is the formation of the Imbrium Basin some 3.85 billion years ago. This basin was later lava-flooded to form the Mare Imbrium – the subject of Section 8.24 in Chapter 8.

The period between the formation of the Nectaris and Imbrium basins (3.92 to 3.85 billion years ago) is known as the *Nectarian Period*. Here the determinations of the ages of lunar formations become much more clear-cut. The Moon was still suffering a very heavy bombardment but this had reduced enough so that most of the basins, craters, and other formations created at this time were not completely obliterated by subsequent impacts. About a dozen of the basins we recognise today were created by gigantic impacts in the Nectarian Period, along with thousands of craters. The lunar soil was heavily churned and mixed by the pounding it received during this time.

The 'carpet-bombing' continued to abate during the Nectarian period. After the Imbrium impactor had done its work, only the projectile that created the Orientale Basin (see Figures 6.1 and 6.6) a few hundred million years later remained as the last really massive piece of debris to hit the Moon. Smaller fragments, though, continued to rain down.

The next accurately dated event was the formation of the crater Eratosthenes, some 3.2 billion years ago. This crater is described, and pictured, in Section 8.5 of Chapter 8. The period between the formation of the Imbrium Basin and Eratosthenes defines the *Imbrium Period*.

Figure 6.6 The north-eastern sector of the Mare Orientale. This is an enlargement of the view shown in Figure 6.1. The outermost rings are just visible on the limb of the Moon as seen from Earth (as a series of mountain ranges – the basin being virtually edge-on to us), when the libration and lighting angles are most favourable. In fact Julius H. G. Franz recognised the presence of a farside lunar sea in 1901. In his book of 1906 he named it Mare Orientale, meaning the 'Eastern Sea' because it lay on the side of the Moon we then called the east. Of course, by the IAU convention this is now the lunar west!

The impactor that created the basin, the last really big chunk of debris to hit the Moon, was probably a piece of rocky debris several tens of metres across. The concentric rings are, in effect, 'frozen' shock waves in the lunar crust. Radial features are also apparent, especially in the outer parts of the structure, including secondary impact scars. (Courtesy NASA and Ewen A. Whitaker.)

Materials ejected from the enormous explosion site of the Imbrium Basin are scattered over a substantial portion of the Moon's globe and the shock waves that permeated the Moon caused much restructuring of the lunar topography. I discuss some of the physical evidence that remains of this Moon-shaking event in Chapter 8. It is during the Imbrium Period that most of the basaltic lava-flooding of the basins occurred.

The formation of the crater Copernicus (see Chapter 8, Section 8.13), 0.81 billion years ago, provides the last of the key chronological markers. The *Eratosthenian Period* (3.20 to 0.81 billion years ago) betwixt the formations of Eratosthenes and Copernicus saw the last vestiges of the episodic lava-flooding of the maria and the continuing diminution of the meteoritic bombardment of the Moon.

This brings us to the *Copernican Period*, which spans 0.81 billion years ago to the present day. Very few of the large lunar craters are younger than Copernicus and only the most minor volcanic happenings have disturbed the Moon's quietude in the last billion years.

6.9 FILLING IN THE DETAILS

Planetologists are fairly confident about the main details of the story of the Moon's nature and evolution, such as I have given them here. However, even if there are no major errors, there is plenty of uncertainty and even ignorance about many of the fine details.

For instance, the way space-probes behaved in lunar orbit led to the early discovery that there are distinct concentrations of dense material situated some way beneath the lunar surface. These are known as *mascons*, a contraction of 'mass-concentrations'. The first results suggested that these all coincided with the lunar maria. It was assumed that the lunar maria were made of denser-than-average materials, which they are (being mantle material brought to the surface), and this explained the anomalies. In recent years the picture has grown more complicated. In particular, *Lunar Prospector* found several new ones, including four on the lunar farside, but only some of these coincide with lunar maria. The latest thinking is that the mascons result from dense plumes of convected material from deep in the Moon, rising into its upper mantle. It is only relatively recently that geologists have appreciated just how important mantle plumes are in explaining earthly tectonic and volcanic structures and activity. That is a salutary reminder that we still have much to learn and some of today's accepted 'truths' might well be replaced with new ones in the future!

The entire spectrum of lunar features and all the myriad pieces of evidence – those obtained from remotely viewing the Moon as well as actual samples of lunar material – have had to be woven into a coherent scenario. The overall scheme might appear simple but the details are rather complex. Given that the main purpose of this book is to be an observer's guide, I have incorporated many of our modern ideas into the accounts of selected lunar features I provide in Chapter 8.

What is the nature of the brilliant ray-systems that some craters possess and why do not all craters have ray-systems? Why do some craters have much brighter interiors than others? What causes the wrinkle ridges on the lunar maria? What are the nature of the lunar rilles and how did they come about? You will find answers to these questions, and more, in Chapter 8.

What about more up-to-date information, such as the knowledge we have learned from *Clementine* and *Lunar Prospector*? The next chapter, which in part serves as resource guide, will help you locate a selection of materials available at the time of writing (spring 2006), as well as incorporating a key map which you will find useful in finding your way to the selected lunar features examined in Chapter 8.

Lunarware

This chapter is mostly a resource guide of books, maps, atlases and websites that I hope will help you to take things further. I finish the chapter with a key map for locating the lunar surface features/areas explored in the next chapter.

7.1   OUT-OF-PRINT BOOKS

I would like to give special mention of a couple of major works about lunar science which are no longer in print. If you want a large (over 700 page) single-volume guide to the science of the Moon as we knew it post-*Apollo* then you can do no better than to locate a copy of the *Lunar Sourcebook – a User's Guide to the Moon*. It is edited by G. Heiken, D. Vaniman and B. French and was published by Cambridge University Press in 1991. It is chock-full of data, information and explanations about the physics, chemistry and geology of the Moon and how that information was obtained. It includes a list of hundreds of references to scientific papers and the details and contact addresses of many sources of lunar databases, imagery and archives. It is a superb springboard to further studies as well as being a mine of information itself.

Another excellent book about the exploration of the Moon and our knowledge of it post-*Apollo* is *The Moon – Our Sister Planet* by Peter Cadogan, which was published by Cambridge University Press in 1981. Of course we have learned much since the times these two books were published but the major advance in our understanding came about as a result of the *Apollo* programme, so these books are not as out of date as you might expect. Most importantly, very little of the lunar science detailed in them has proved to be wrong or in need of significant modification.

Perhaps I can also mention the first edition of the book you are currently reading! It was published in 2000 by Cambridge University

Press. It has gone through several reprints but the stock of the final one will likely be exhausted by the time you are reading these words. There are a three main topics in the first edition that are not covered in this edition: the accurate timing of lunar occultations, film-based photography of the Moon, and methods for determining the relative heights of lunar features by measuring the lengths of the shadows they cast. These topics had to go to make way for new material, so you will need a copy of the first edition if you are interested in any of them.

It is just possible that you might come across any of these books in a second-hand bookshop. If you have access to a provincial or national astronomical society's library then there is a better chance of obtaining these and any other out-of-print works that take your interest. Failing that, your local library should certainly be able to get hold of copies of them for you via the inter-library loan service.

7.2   BOOKS CURRENTLY IN PRINT

These are a selection of books which, taking this book as a starting point, you may find of use in further advancing your study of the Moon or practical work at the telescope.

The Modern Moon: A Personal View, by Charles A. Wood (Sky Publishing Corp., 2003).

Epic Moon: A History of Lunar Exploration in the Age of the Telescope, by William Sheehan and Thomas Dobbins (Willmann-Bell, 2001).

Advanced Amateur Astronomy, second edition, by Gerald North (Cambridge University Press, 1997).

Introduction to Webcam Astrophotography, by Robert Reeves (Willmann-Bell, 2006).

The Webcam User's Guide to the Moon and Planets, by Martin Mobberley (Springer-Verlag, 2006).

Introduction to Digital Astrophotography, by Robert Reeves (Willmann-Bell, 2004).

How to Photograph the Moon and Planets with your Digital Camera, by Tony Buick (Springer-Verlag, 2005).

Digital Astrophotography – the state of the art, edited by David Ratledge (Springer-Verlag, 2005). This book includes a chapter on webcam imaging written by Damian Peach.

Video Astronomy, Revised Edition, by Steve Massey, Thomas A. Dobbins and Eric J. Douglas (Sky Publishing Corp., 2004).

Handbook of Astronomical Image Processing, second edition by Richard Berry and James Burnell (Willmann-Bell, 2005). Not for beginners! Large, highly detailed and very advanced reference book on the theory and practise of all types of image processing. Packaged with a CD-ROM including tutorials and *AIP4WIN* software.

Photoshop Astronomy, by R. Scott Ireland (Willman-Bell, 2005).

Practical Amateur Spectroscopy, edited by Stephen F. Tonkin (Springer-Verlag, 2002).

7.3   PRINTED MAPS, CHARTS AND ATLASES

You can often find Moon maps presented in general astronomy books, though they often tend to be limited in size and the details they carry. One fairly good example, containing simple geological maps of the Moon as well as larger-scale topographic maps, is *The NASA Atlas of the Solar System*, by Ronald Greeley and Raymond Batson, published by Cambridge University Press in 1997. *Norton's Star Atlas*, edited by Ian Ridpath, published in its nineteenth edition by Longman in 1998, contains a reasonably good photomosaic Moon chart. However, the serious lunar observer will want something better.

Premier amongst Moon atlases is the superb *Atlas of the Moon* by Antonin Rükl. The latest edition (2004) is published by Sky Publishing Corp. (49, Bay State Road, Cambridge, MA, 02138-1200, USA – accessed through SkyTonight.com). The bulk of its 224 pages is given over to a map of the Moon's near side in 76 sections. These are airbrushed surface-relief drawings of exquisite quality. On the facing pages of each section are lists of feature names, along with their sizes, etc. One thing to beware of: the maps are presented north uppermost. Nonetheless I highly recommend this atlas.

The Clementine Atlas of the Moon, by Ben Bussey and Paul D. Spudis was published by Cambridge University Press in 2004. This atlas uses the visible-light imagery from the *Clementine* space-probe as its source material. The whole of the Moon's surface (including the farside) is mapped in 144 slices. Each section is presented as it would be observed by you looking down from directly above it. Each *Clementine* image is paired with an airbrushed relief map on the facing page on which the principal features are named. The fifty-plus page introduction includes a potted history of lunar exploration and an in-depth overview of the 'geology' of the Moon. It also includes a fairly detailed account of the *Clementine* mission and the main scientific results obtained by it. Highly recommended for the serious student of the Moon.

I have (in Chapter 4) already mentioned *The Hatfield Photographic Lunar Atlas*, edited by Jeremy Cook and published by Springer-Verlag in 1998. The atlas covers the Moon's nearside in 16 overlapping sections to the scale of 64 cm to the Moon's full diameter. Each opens with a detailed hand-drawn map which is extensively labelled, facing a photograph of the same area. The following pages show the same area under differing lighting conditions. In some cases the selection also includes close-ups of specific lunar areas/features at the scale of about 90 cm to the Moon's full diameter. In total there are 88 photographic plates making up this excellent atlas. It ends with an extensive index of all the named features, with the plates on which they appear, their selenographic co-ordinates and their sizes. The plates are Commander Hatfield's originals taken in the 1960s and the resolution is not as high as someone might now obtain with a webcam on even a fairly small telescope. However, the scale and the whole arrangement of the atlas, including having south uppermost, makes it especially useful in the field. I am always referring to my copy.

To be really useful at the telescope in the Earth's northern hemisphere a Moon atlas should either show the inverted view one gets through a normal telescope – south at top and lunar east to the left (Tycho at the top and Mare Crisium to the lower-left), or it should show the view one gets using a normal telescope used with a star-diagonal – most times north at the top but with lunar east still to the left (Tycho at the bottom and Mare Crisium to the upper-left).

Jeremy Cook had virtually completed the work on another version of the Hatfield Atlas that was intended for observers using telescopes with star-diagonals that give mirror-reversed images when he tragically died. His son Tony put the finishing touches to the work and *The Hatfield SCT Atlas* was published by Springer-Verlag in 2005. It also has my recommendation.

Much more of a guidebook for the beginner than the foregoing examples, *Photographic Atlas of the Moon*, by S. M. Chong, Albert C. Lim and P. S. Ang (published by in 2002 by Cambridge University Press), is nonetheless quite detailed. The Moon maps are based on photographs taken by the authors. There is one main photograph of the Moon for each day of the lunar month, plus supplementary photographs, including a few close-up photographs. The main photographs are reproduced at a scale of about 19 cm to the Moon's full diameter (the whole of the Moon being shown in each image). These are labelled with all new features that appear (in the waxing phase) and are about to disappear (in the waning phase) in a narrow band along the terminator. The appendices include a very full index for all named formations, giving the best lunar days to see the formation, its selenographic longitude and latitude, diameter, the year the current name was approved by the IAU, and the corresponding map number in Rükl's atlas in which to find the formation. This atlas also has the benefit of showing the Moon south uppermost.

Here are some other atlases and maps:

Lunar Orbiter Photographic Atlas of the Near Side of the Moon, by Charles J. Byrne (Springer-Verlag, 2005). North uppermost.
Atlas of the Lunar Terminator, by John E. Westfall (Cambridge University Press, 2000). North uppermost.
Sky & Telescope's Field Map of the Moon (Sky Publishing Corp.). This map is north at the top and lunar east to the right.
Sky & Telescope's Mirror-Image Field Map of the Moon (Sky Publishing Corp.). This map is orientated with north at the top and lunar east to the left.
Lunar Quadrant Maps (Sky Publishing Corp.). This set of maps is orientated with north at the top and lunar east to the right.

7.4 SOME USEFUL WEBSITE ADDRESSES CONCERNING EQUIPMENT AND TECHNIQUES

There is a plentiful supply of images of and information about telescopes, ancillary equipment and observing techniques on the Internet. Website addresses tend to be ephemeral, so it is a good idea to make a start by looking in publications such as the latest issues of the *Journal of the British Astronomical Association*, the *Handbook of the British Astronomical Association*, and in *Sky & Telescope* and other magazines for website addresses. Of course, you can also make use of your favourite search engine. Many websites have links to other websites, so it should not take much surfing to uncover plenty of up-to-date information. Here is a limited selection of specific website addresses, active at the time of writing, that you might like to check out to see if they could be useful to you.

Equipment suppliers

Adirondack Video Astronomy (manufacturer of video cameras and the Atmospheric Dispersion Corrector, also suppliers of accessories and the *Philips* ToUcam webcams) — www.astrovid.com

Astro-Physics (manufacturer of very high quality but expensive telescopes and mountings) — www.astro-physics.com

Baader Planetarium (accessories including filters for reducing a refractor's secondary spectrum) — www.alpineastro.com

Celestron (telescopes and accessories, also the NextImage camera) — www.celestron.com

Jims Mobile Inc. (telescope accessories including motorised focusers)   www.jimsmobile.com

Meade (telescopes and accessories)   www.meade.com

Orion Telescopes and Binoculars (USA company – telescopes and accessories, including filters to reduce a refractor's secondary spectrum)   www.telescope.com

Orion Optics Ltd. (UK, unrelated to the USA company – telescopes and accessories)   www.orionoptics.co.uk

ScopeTronix (various, including *Philips* ToUcam webcams)   www.scoptronix.com

Sky-Watcher (telescopes of all types, and accessories)   www.skywatchertelescope.net

Sirius Optics (various, including filters for reducing a refractor's secondary spectrum)   www.siriusoptics.com

Takahashi (see adverts for US and UK retailers – top quality but expensive telescopes and mounts)

Telescopes & Astronomy (various, including *Philips* ToUcam webcams)   www.telescopes-astronomy.com.au

Tele Vue (high-quality optical accessories, including eyepieces and Powermates, along with small apochromatic refractors)   www.televue.com

Image-processing programs

Adobe Photoshop	www.adobe.com/productionstudio
AIP4WIN	www.willbell.com/aip4win/AIP.htm
AstroArt	www.msb-astroart.com
Image Editor	www.ulead.com
Images Plus	www.mlunsold.com
IRIS	http://astrosurf.com/buil/usiris/iris.htm
MaximDL	www.cyanogen.com
Paint Shop Pro	www.corel.com/paintshop/

CCD cameras
Commercial cameras
Starlight XPress   www.starlight-xpress.co.uk
S-BIG   www.sbig.com
Apogee   www.apogee-ccd.com

Do-it-yourself cameras
Audine   http://astrosurf.com/audine
Cookbook   www.wvi.com/~rberry/cookbook.htm

Webcam imaging
RegiStax   http://aberrator.astronomy.net/registax/
K3CCDTools   www.pk3.org/astro
IRIS   http://astrosurf.com/buil/usiris/iris.htm
AstroVideo   www.ip.pt/coaa/astrovideo.htm
AstroStack   www.astrostack.com
Astro-Snap   www.astrosnap.com/index_uk.html
AVIedit   www.am-soft.ru/aviedit.html

Spectroscopy
Buil   http://astrosurf.com/buil/us/spe1/spectro1.htm
Spectr'aude   http://astrosurf.com/buil/us/spectro8/spaude_us.htm
Maurice Gavin   www.astroman.fsnet.co.uk/spectro.htm

As I said, this is a rather small selection of website addresses but it may serve to get your researches started. You can uncover plenty more websites by using your favourite search engine.

7.5   *CONSOLIDATED LUNAR ATLAS, LUNAR ORBITER*
       PHOTOGRAPHIC ATLAS, APOLLO IMAGE ATLAS AND
       RANGER PHOTOGRAPHS ONLINE

To access these wonderful resources all you have to do is visit the website of the Lunar and Planetary Laboratory at:

   www.lpi.usra.edu/resources/cla/

This web site opens with a page of links to enable you to navigate through the *Consolidated Lunar Atlas*, providing you with a number of options of the ways you can view the material.

The original *Consolidated Lunar Atlas, Supplement Number 3 and 4 to the USAF Photographic Lunar Atlas* was created by Gerard Kuiper, Robert Strom, Ewen Whitaker, John Fountain and Stephen Larson and published by the Lunar and Planetary Laboratory of the University of Arizona in 1967. The online digital version was created by Eric Douglas in 2000. By clicking on a link on the page you can even order a copy of this atlas on CD-ROM.

At the bottom of the same web page are further links which will take you to the opening pages of *The Digital Orbiter Photographic Atlas of the Moon* (including a link you can click on to order a copy of it on DVD-R), the *Apollo Image Atlas* (comprising some 25000 images!) and the *Ranger Photographs of the Moon*. Each of these pages has links to click on for the purposes of navigation around the sites.

Another useful website address is NASA's home page:

http//nssdc.gsfc.gov/

This is the National Space Science Data Center (NSSDC) at NASA Goddard Space Flight Center, Greenbelt, MD 20771, USA.

7.6 *CLEMENTINE, LUNAR PROSPECTOR* AND *SMART-1* IMAGES AND DATA ONLINE

There are many websites you can go to. I recommend starting with 'Clementine Mission to the Moon – Online Resources' at:

www.pdsimage.jpl.nasa.gov/PDS/public/clementine/clementine.html

There you will find virtually all the products of the *Clementine* mission. The page includes a link to the 'Clementine Lunar Image Browser', available through the Naval Research Laboratory which provides access to over 170 000 of the *Clementine* images.

Another good website is 'Clementine Moon Maps' at:

http://simkin.asu.edu/clem/

If you wish to you can purchase the set of *Clementine* images on a set of CD-ROMS, as well as atlases derived from the images. The NASA website given at the beginning of Section 7.5 will be of help here, as will some others.

At the time of writing, *Lunar Prospector* and *SMART-1* data and images are less easy to come by. However, that is bound to change, maybe even by the time you are reading these words.

7.7 VIRTUAL MOON ATLAS

There is so much material on the Internet that all I can do in this chapter is to mention a few websites that you might like to have a look at as a starting point, and hope that the majority of them are still active by the time you are reading these words. One I must make special mention of is 'Virtual Moon Atlas (VMA)'. This is an excellent piece of freeware you will find at:

www.astrosurf.com/avl/UK_download.html

Created by Christian Legrand and Patrick Chavalley, you only need a very basic computer package to download and run this program. You get a fairly detailed representation of the Moon created from the US Geological Survey maps. The image is correct for lunar phase, and libration. What you see on the screen can be for the current date and time but you can also 'date zoom' forwards and backwards in time and watch things change. You can change the orientation of the image or even mirror-reverse it to suit the equipment you are using.

A side panel shows lots of relevant data. For instance it can display a list of all the lunar features very close to the terminator at any given time. You can look at the whole Moon or zoom in to specific areas. You can place the cursor over a given feature and read off lots of information about that feature. There is even the option of adding your own notes about that feature. You can use the cursor to measure distances on the lunar surface. I could say more but instead I will just urge you to download a copy of it for yourself!

7.8   LUNAR EPHEMERIDES

Publications such as the *Astronomical Ephemeris* and the *Handbook of the British Astronomical Association* provide useful ephemerides for the lunar observer and many societies and groups also provide hard-copy ephemerides, while the ubiquitous Internet probably has innumerable sites where this information can be found (though it is wise to check on the accuracy of any data given).

There are also plenty of programs available commercially, and some available simply by downloading them from the Internet. Some, such as 'Lunar Calculator' (see the January 1999 issue of *Sky & Telescope* for a review), and 'LunarPhase Pro' (reviewed in the June 2003 issue of *Sky & Telescope*) are highly sophisticated, have vast databases, and can do things such as simulate the views of the Moon that you would get from an orbiting spacecraft. Others present you with Earth-based views taking libration into account, as well as providing you with much useful information such as lighting angles, etc. New and improved software is continuously coming on the market, and ever increasing amounts of freeware are appearing on the Internet, so I will leave you to search out for yourself the up-to-date advertisements and reviews and obtain the product that best suits your own needs.

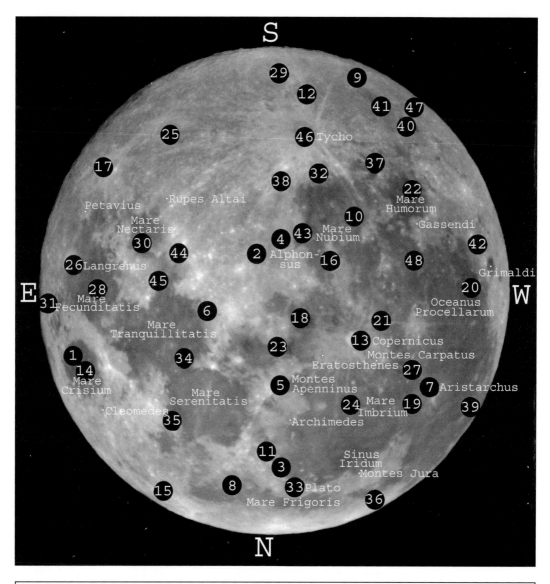

S

29
12
9
41 47
40
25
46 Tycho
17
37
38 32
22
·Rupes Altai   Mare
Petavius   Humorum
        Mare   10   ·Gassendi
        Nectaris   Mare
30   4 43 Nubium   42
26 Langrenus 44 2 Alphon- 16   48   Grimaldi
28   45 sus   20
E 31 Mare   Oceanus   W
   Fecunditatis   Procellarum
        Mare   18   21
        Tranquillitatis   13 Copernicus
1   23   Montes Carpatus
14   34   Eratosthenes 27
Mare   Montes   7 Aristarchus
Crisium   Mare   5 Apenninus   19   39
·Cleomedes Serenitatis   24 Mare   39
        35   ·Archimedes Imbrium
                11   Sinus
                3   Iridum
        8   33 Plato Montes Jura
15   Mare Frigoris 36

N

KEY:

1. Agarum, Promontorium	**9.** Bailly	**19.** Harbinger, Montes	**29.** Moretus	**40.** Schickard
2. Albategnius	**10.** Bullialdus	**20.** Hevelius	**30.** Nectaris, Mare	**41.** Schiller
3. Alpes, Vallis	**11.** Cassini	**21.** Hortensius	**31.** Neper	**42.** Sirsalis, Rima
4. Alphonsus	**12.** Clavius	**22.** Humorum, Mare	**32.** Pitatus	**43.** 'The Straight Wall' (Rupes Recta)
5. Apenninus, Montes	**13.** Copernicus	**23.** Hyginus, Rima	**33.** Plato	**44.** Theophilus
6. Ariadaeus, Rima	**14.** Crisium, Mare	**24.** Imbrium, Mare	**34.** Plinius	**45.** Torricelli
7. Aristarchus	**15.** Endymion	**25.** Janssen	**35.** Posidonius	**46.** Tycho
8. Aristoteles	**16.** Fra Mauro	**26.** Langrenus	**36.** Pythagoras	**47.** Wargentin
	17. Furnerius	**27.** Maestlin R	**37.** Ramsden	**48.** Wichmann
	18. 'Gruithuisen's Lunar City'	**28.** Messier	**38.** Regiomontanus	**39.** Russell

Figure 7.1 Key map of
the Moon, for use with
Chapter 8.

7.9   KEY MAP FOR CHAPTER 8

Chapter 8, occupying about half the length of this book, is given over to a
detailed study of a representative selection of 48 principal areas/features
(many others detailed along with each of these) on the Moon's nearside.
The treatment is slanted to the interests of the telescopist. Herewith
(Figure 7.1) is a key map intended to help you locate each of the numbered features.

The numbers on the map, and the key to it, follow the same sequence
as the numbers and titles of each of the sections in the next chapter. For
instance, number 4 on the map corresponds to number 4 in the key and is
named as 'Alphonsus'. Look up Section 8.4 in the next chapter and you
will find it is headed 'Alphonsus'.

In practice, you will normally use the system the other way round.
You might be reading Section 8.15 in the next chapter, about Endymion.
To find out where Endymion is you can look up number 15 on the map.
Of course, it is also named in the key presented with the map.

Of course you will need a proper Moon map or atlas to help you with
your lunar observing but I hope this key map will prove useful to refer to
while you are reading the next chapter.

CHAPTER 8

'A to Z' of selected lunar landscapes

In the earlier chapters of this book, alongside the descriptions of observational hardware and techniques, I have tried to provide a picture of the Moon and of lunar science past and present. Necessarily this 'picture' has been painted with a rather broad brush. To really get to know the Moon, one must be prepared to examine it in finer detail. To that end, this chapter presents a selection of 48 specific features/areas of the Moon. Taken together, these provide a representative selection of the types of lunar formation one may encounter at the eyepiece of one's telescope.

Why the particular selection that follows? I can only say that this has been my personal choice. I have tried to cover the fullest possible range of lunar features. You will find descriptions of craters both grand and small, conventional and unusual in their profile. Mountains, valleys, domes, rilles (both sinuous and linear), mare flood-plains and other types of terrain are also described. In some of the sections broad areas are described. A few sections concentrate on particular features of special interest. Altogether, over two hundred named formations are examined. Many are the 'old favourites' of novice and experienced observers. Others are more 'off the beaten track'. Sometimes I have provided detailed descriptions. Other times I have only provided sketchy details and leave you, the reader – and I hope the observer – to find out things for yourself.

In some cases the features described provide an object lesson in particular observational techniques and/or pitfalls for the unwary. Others exemplify particular points of lunar science and geology. All are also of interest in their own right. Everything is described from the viewpoint of the observer with his/her backyard telescope. I have tried to provide a range of targets for the full range of observers, from novice to advanced. My aim is that you will be able to go to the telescope and

interpret what you see/photograph/image in terms of lunar science (and particularly lunar evolution).

Look at page 154 and you will see a simple key map which may be of use to you in locating the areas/features on the Moon's face described in this chapter. As previously explained, simply use the section number to locate the item on the map. As an example, the principal feature described in Section 8.26 (Langrenus) will be located at the point labelled 26 on the map (also it is named against the number 26 in the key presented with the map). At the head of each section I give the selenographic latitude and longitude of the principal named formation. This immediately tells you in what quadrant you will find the formation on the key map. Going beyond that, even the roughest estimate of position based on the latitude and longitude figures will easily and quickly enable you to find the numbered formation on the key map.

Of course, the latitude and longitude figures will also be of use in locating the formation/area on other Moon maps and atlases – the key map is only intended as an aid to locating the main features discussed here. You will undoubtedly want a much more detailed map/atlas for general use (see the previous chapter).

Let me entice you to take the first steps of a journey of exploration. After taking those first steps with me, I hope that you will then want to continue on your own. The Moon you will discover is both a thrilling and an eerie place of spectacle and wonder . . .

8.1   AGARUM, PROMONTORIUM [14°N, 66°E]

An impressive cape, projecting into the Mare Crisium, the highest peaks of which reach up to several thousand metres above the mare. From time to time there have been claims of apparent mistiness around the cape, especially to the south, and most frequently soon after local sunrise. The visibilities of some of the tiny craters on the mare in the vicinity also seem variable. I think that these appearances are probably not true Transient Lunar Phenomena (TLP, see Chapter 9), but local variations of albedo with Sun-angle. Why not make a long-term study of this area in order to establish its true behaviour as the Sun rises over it? As explained more fully in Chapter 9, in TLP research it is particularly valuable to establish the true apparent behaviour of lunar surface features as lunations progress and under varying conditions.

Figure 8.1 is centred on the cape. It was taken using the 1.5 m reflector at the Catalina Observatory (of the Lunar and Planetary Laboratory, University of Arizona) on 1966 April 6^d 7^h 18^m UT. At that time the selenographic colongitude was 96°.9.

Figure 8.1 Promontorium Agarum. Details in text. (Catalina Observatory photograph – courtesy Lunar and Planetary Laboratory.)

Figure 8.2 Albategnius (centre) and Hipparchus (lower). Details in text. (Catalina Observatory photograph – courtesy Lunar and Planetary Laboratory.)

8.2   ALBATEGNIUS [11°S, 4°E] (WITH KLEIN AND HIPPARCHUS)

This 136 km diameter crater is very old, as witness its heavily degraded walls and the intrusion of other craters into it, notably the 44 km diameter Klein on its western (right in Figure 8.2) side. Moreover, the floors of both these craters have been flooded with mare-type lavas. Notice how the lofty (and unusually massive, for this size of crater) central peak of Albategnius pokes up through the lava, as does the smaller central peak of Klein. Also notice that Klein has degraded Albategnius and not the other way round. So, we can conclude that Klein is younger than Albategnius. What is your opinion about the age of the small crater that has intruded into the north-east rim of Klein? Yes, I know the answer is fairly obvious. I deliberately chose this as an easy example. The point is that you have made a start in unfolding the dynamic history of the lunar surface.

North of Albategnius (near the bottom of Figure 8.2) lies the even mightier and more complex and more ancient Hipparchus. Hipparchus's heavily degraded (almost destroyed along its western section) rim spans 151 km. Notice the almost parallel set of great scars cutting through the terrain in this region of the Moon. Each channel is orientated, roughly speaking, from south-south-east to north-north-west. Other examples exist in the area further west than is covered by Figure 8.2. If you relish a challenge try to deduce the history of this tortured area of the Moon's surface, using spacecraft (*Orbiter*, *Clementine*, etc.) images and, perhaps, your own telescopic observations. I guarantee that you will be kept busy for a great many hours! To get you started, backtrack the scars northwards and you will find them to be radial to a particular major feature on the surface of the Moon. Enough said?

The photograph was taken with the 1.5 m Catalina Observatory telescope on 1966 September 6^d 11^h 7^m UT, when the selenographic colongitude was 167°.5.

8.3   ALPES, VALLIS [CENTRED AT 49°N, 3°E]

Even when seen through a small telescope, the 'Alpine Valley' (properly called Vallis Alpes) is a truly striking spectacle around the times of first and last quarter Moon. This tremendous gorge, nearly 180 km long, seems to slice straight through the Montes Alpes, linking the Mare Imbrium with the Mare Frigoris.

Certainly it is not simply a channel cut through the mountains by a river of lava. I think that there is probably a connection between this feature and the heavy linear scars that cross the highlands in the vicinity of Albategnius and Alphonsus (see Sections 8.2 and 8.4). Perhaps all of these great valleys were really formed by slumping of the crust along stress fractures as a result of a very slight horizontal expansion of the lunar mantle (or at least the

Figure 8.3 Vallis Alpes (centre) cuts through Montes Alpes in this Catalina Observatory photograph. Details in text. (Courtesy Lunar and Planetary Laboratory.)

deeper layers of the crust) after the regolith had first solidified? The most likely explanation, though, is that the crust has shrunk very slightly after its initial solidification and stress fractures developed as a result, with the ground slumping into each fracture.

This type of formation is known as a *graben*. Perhaps the colossal impact event that created the Imbrium basin was responsible for the faults that ultimately produced the graben? Search through the Montes Alpes and you will find other linear features that are at least approximately radial to the Mare Imbrium, lending support to this idea. However none of these other linear features are anything like as strikingly obvious as Vallis Alpes.

Certainly, though, the floor of the Vallis Alpes has been flooded with lava. Also, a sinuous rille meanders along the length of the floor of this great lunar valley. Perhaps a further, minor, episode of lava flowing after the main formation and flooding processes were over? Under appropriate illumination and excellent seeing conditions the rille can be seen in a 13 cm refractor of first-class optical quality. However, it is elusive and you need not doubt your abilities if you fail to see it even when using a more powerful telescope.

The view of the Vallis Alpes shown in Figure 8.3 was taken using the Catalina Observatory 1.5 m reflector on 1967 January 20^d 01^h 45^m UT, when the Sun's selenographic colongitude was $18°.4$.

8.4   ALPHONSUS [13°S, 357°E] (WITH ARZACHEL, PTOLEMAEUS, ALPETRAGIUS AND HERSCHEL)

The three adjacent craters Arzachel (southernmost), Alphonsus and Ptolemaeus (northernmost) are very distinctive around the times of first and last quarter Moon. The area is shown in Figure 8.4(a), the details of which are the same as for Figure 8.2.

The 97 km diameter Arzachel is obviously the youngest of the three. Even a small telescope is enough to show its richly complex structure. The walls are heavily terraced and rise to a greater height (being about 4.5 km above the immediate surrounds) on the eastern side than on the west (height about 3.4 km – the surrounds being so rough and hummocky, these figures are only very approximate).

The crater floor, itself lying nearly 1 km below the level outside the formation, has obviously been partially flooded with lavas. Yet it is very far from smooth. There are a number of small craters, several hills, and at least one rille on the crater floor which are visible to the users of amateur-sized telescopes. Note how much the 'central' mountain mass is offset from the centre of the crater. All these features are well shown on the incredible image obtained by Terry Platt (and shown in Figure 8.4(b)) using his 318 mm tri-schiefspiegler reflector and *Starlight Xpress* CCD camera (other details not available). I used *Hauppauge Image Editor* software on my own computer to further sharpen Terry Platt's already outstandingly fine image.

Alphonsus is, arguably, one of the most interesting craters on the Moon. This 119 km diameter ring-plain has highly complex walls and many fascinating details can be made out on its flooded floor, especially by users of large telescopes.

Look carefully at Figure 8.4(a) and you will see several small dark patches on the floor of Alphonsus. At the centre of the patches are small craters. These formations are known as *dark halo craters*. At one time these were taken to be volcanic cinder cones, or fumaroles, and were used by the supporters of the endogenic theory to bolster their views on the moulding of the lunar surface. Modern studies do indicate that the Alphonsus dark halo craters really are fumaroles. However, the *Apollo 17* astronauts visited an example of this type of formation (a small dark halo crater named 'Shorty', on the southeastern border of the Mare Serenitatis) and found it to be a conventional impact crater where the impact explosion had excavated dark mare material from beneath a thin layer of lighter regolith. It is likely that most of the other dark halo craters of the Moon have the same explanation.

Several rilles and faults cross the floor of Alphonsus. These are particularly well shown in the superb image Terry Platt made of the crater and which is presented in Figure 8.4(c) (subsequent image sharpening and

(a)

Figure 8.4 (a) Arzachel (upper), Alphonsus (centre) and Ptolemaeus (lower) are the craters that dominate this Catalina Observatory photograph. (Courtesy Lunar and Planetary Laboratory.) Details given in text.

Figure 8.4 (*cont.*)
(b) Arzachel – CCD image
by Terry Platt. Details in
text. (c) Alphonsus – CCD
image by Terry Platt.
Details in text.

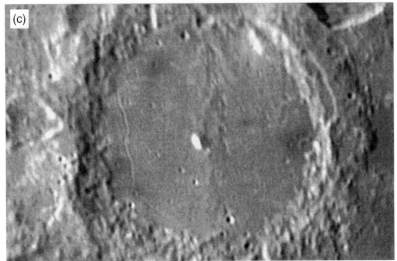

other details as for (b)). The apparent changes in the dark halos are now
known to be variations in relative albedo with illumination angle.

As far as the controversial subject of TLP is concerned, Alphonsus
provided the best 'hard-copy' (as opposed to anecdotal) evidence that at
least a small minority of the reported instances of TLP are real events at
the Moon's surface and not simply illusions or mistakes on the part of the
observers concerned. More on this in Chapter 9.

The northern section of the rim of Alphonsus merges with that of the
magnificent 153 km diameter 'walled-plain' Ptolemaeus. As one might
expect the terrain is highly chaotic and broken down at the merger,
Ptolemaeus pre-dating the formation of Alphonsus. The ancient flooded

floor of Ptolemaeus is covered in small craters and several crater-chains. These are mostly rather delicate objects for those using moderate telescopes under typical backyard conditions. The apparent brightness of the floor of Ptolemaeus changes considerably during a lunation, as does the appearance of the local 'mottlings' on it. The floor brightens considerably under a high Sun, the Moon then being near full, but appears quite dark at times close to first and last quarter Moon.

The area covered in Figure 8.4 overlaps that shown in Figure 8.2, which lies to the east. Notice the same straight scars in the terrain, each hundreds of kilometres long. What story do they tell? (There is no mystery but you might like to deduce the answer for yourself. I offered a clue in Section 8.2.)

The prominent 40 km diameter crater to the north-west of Arzachel and south-west of Alphonsus is called Alpetragius. Notice its prominent central peak, rather 'mound-like' in profile. The walls of this crater are rather finely terraced.

Of similar size to Alpetragius, and positioned just north of Ptolemaeus, is Herschel. You might like to consider making a detailed comparison between Alpetragius and Herschel. Given that the craters are similar in many ways, why are they rather different in others?

8.5 APENNINUS, MONTES [CENTRED AT 20°N, 357°E] (WITH CONON, ERATOSTHENES, PALUS PUTREDINIS, SINUS AESTUUM, WALLACE)

If any feature on the Moon can take the prize for being the most striking when seen through even the smallest telescope at the appropriate time, then surely it has to be the magnificent Montes Apenninus. The 'appropriate time' for this formation occurs twice every lunation: near first and last quarter Moon.

Spanning about 600 km along the south-eastern 'shore' of the Mare Imbrium, this stunningly rugged mountain range strikes a breath-taking spectacle when seen under a low Sun. I admit that I find the fine details quite confusing when seen under good conditions with a powerful telescope. I tend to use a higher magnification than I would normally do under the ambient conditions just to enjoy the sheer awesome effect of the view. This also reduces the confusion greatly. The slight softening of the image, due to over-magnification, hardly matters when one gets the thrilling impression of apparently flying over the complex array of mountain peaks, ridges and valleys!

Its origin dates back about 3850 million years, with the creation of the Imbrium Basin. The impacting projectile caused the highland crust along the south-eastern border to be violently uplifted, so forming the range.

Figures 8.5(a) and (b) are sections of a photograph which was taken with the Catalina Observatory 1.5 m reflector on 1967 January 20^d 01^h 46^m

Figure 8.5 (a) The northernmost extent of the Montes Apenninus. Details in text. (Catalina Observatory photograph – courtesy Lunar and Planetary Laboratory.)

Figure 8.5 *(cont.)* (b) The southernmost termination of the Montes Apenninus with the crater Eratosthenes (upper right). The ruined crater Wallace can be seen near the bottom of this Catalina Observatory Photograph. Details in text. (Courtesy Lunar and Planetary Laboratory.)

UT, when the Sun's selenographic colongitude was 18°.5. Figure 8.5(a) shows the mountain range to its northernmost extent. The conspicuous crater in the midst of the mountain peaks (just to the right of centre of the photograph) is the 22 km diameter Conon. The area of mare extending from the bay a little north of Conon is called the Palus Putredinis ('Marsh of Decay'). Notice the extensive network of fine sinuous rilles in the area.

The *Apollo 15* astronauts visited the north-eastern border of the Palus, close to the foot hills of the Montes Apenninus and examined one of the rilles (Hadley Rille) close-up.

Figure 8.5(b) shows the southernmost extent of the Montes Apenninus, where it terminates with the prominent 58 km crater Eratosthenes. Under a low angle of illumination Eratosthenes is particularly striking and, largely shadow-filled, then appears to be very deep. The rim of the crater actually rises to just over 2.4 km above the surrounding terrain level, while the floor is depressed by about the same distance below the level of the outside surrounds. A lunarnaut standing on the rim of the crater and looking towards the central mountain complex would see the ground sloping away from him with an average gradient of about 1 in 3 (actually

Figure 8.5 (*cont.*)
(c) Eratosthenes and the Sinus Aestuum, photographed by Tony Pacey. Details in text.

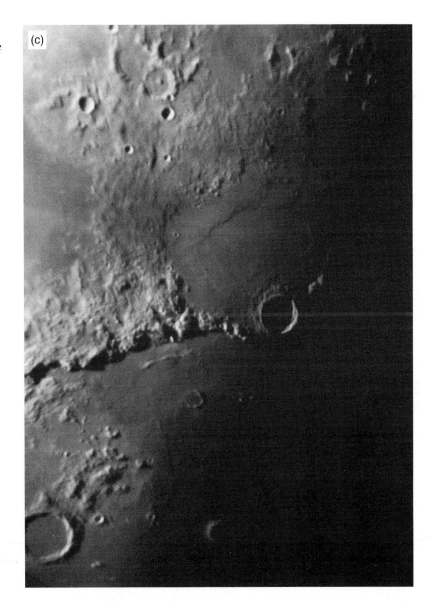

the walls are terraced) down towards the crater floor (at a depth of nearly 5 km vertically below that of the rim). The giant amphitheatre would be harshly sunlit, while the sky above would be inky black. Imagine what a spectacle that would be!

Eratosthenes is somewhat of a lunar chameleon. Despite its stark magnificence when seen at low Sun-angles (see also Figure 8.13(a), further on in this chapter, where it appears alongside Copernicus), the crater takes on a very 'washed-out' appearance when seen under a higher Sun. In fact the crater can even give the completely illusory impression of then being filled with a white, pall-like, mist. Some other craters also show the same optical behaviour and undoubtedly this has led to many a false claim of TLP.

Some of the observers of yesteryear were convinced that changes regularly happened within Eratosthenes (even the growth and dying of vegetation and the migrations of animal/insect life-forms!) but they were certainly mistaken. Near full Moon Eratosthenes becomes really quite difficult to detect, especially as the ray system from the nearby major crater Copernicus then tends to help with the camouflage.

Another notable feature, pictured in Figure 8.5(b), is the ruined crater Wallace. Shown near the bottom of the photograph, the surviving remnants of Wallace's broken walls poke above the level of the Mare Imbrium. This 26 km crater has been almost entirely flooded by the lavas which filled the old Imbrium Basin to form the Mare Imbrium about 3.3 billion years ago.

Figure 8.5(c) provides another view of Eratosthenes, this time almost entirely filled with black shadow. It also shows off one of the more obscure basaltic flood-plains to very good effect. This is the Sinus Aestuum, situated to the immediate south-east of Eratosthenes. This 'bay' spans 230 km. It is quite easy to trace the outline of the original basin that was to be later filled with lava to form the sinus on this excellent photograph by Tony Pacey. He used eyepiece projection to enlarge the image at the f/5.5 Newtonian focus of his 10-inch (254 mm) Newtonian reflector onto FP4 film for this 0.5 second exposure on 1990 February 3^d 19^h 35^m UT, when the Sun's selenographic colongitude was $0°.4$.

At higher Sun-angles the sinus becomes very hard to see. This is particularly so because the rays from the nearby Copernicus, splattered across this region, then dominate.

Notice how the impact cut through, and obliterated, the southernmost extent of the Montes Apenninus. Clearly the Aestuum impactor did its work **after** the much more massive projectile that created the Imbrium Basin had hit the Moon. As such, the Aestuum Basin must be one of the youngest on the Moon, the Imbrium Basin itself being quite youthful. What about the ages of Eratosthenes and the Imbrium and Aestuum flood-plains relative to each other and the basins? The answer is well established, and is quite easy to fathom but I will leave this as an exercise for you.

8.6 ARIADAEUS, RIMA [CENTRED AT 7°N, 13°E] (WITH ARIADAEUS, SILBERSCHLAG, JULIUS CAESAR AND AGRIPPA)

Whenever the terminator lies in the vicinity of the Rima Ariadaeus this rille becomes very easy to see even when using quite small telescopes. Roughly 220 km long, it spans much of the rough terrain between the

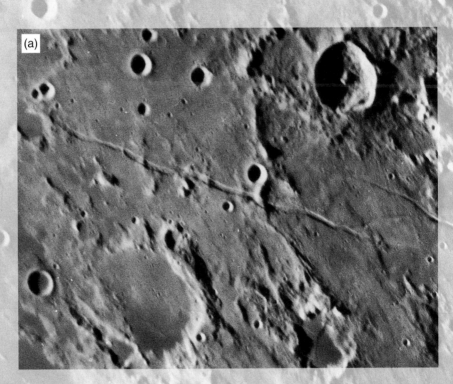

Figure 8.6 (a) Rima Ariadaeus. Details given in the text. (Catalina Observatory photograph – courtesy Lunar and Planetary Laboratory.)

(b)

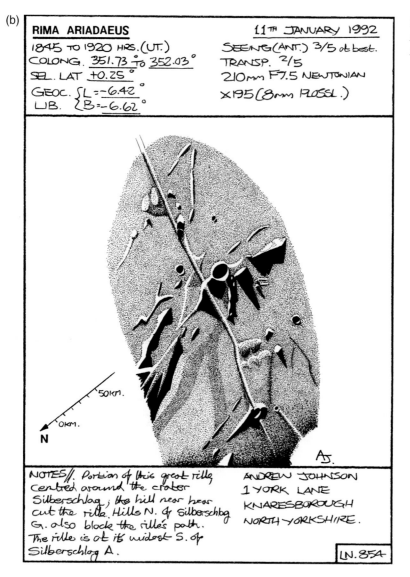

RIMA ARIADAEUS

1845 TO 1920 HRS. (UT.)
COLONG. 351.73 to 352.03°
SEL. LAT +0.25°
GEOC. { L = -6.42°
LIB. { B = -6.62°

11TH JANUARY 1992
SEEING (ANT.) 3/5 at best.
TRANSP. 2/5
210mm F7.5 NEWTONIAN
X195 (8mm PLOSSL.)

50km.

0km.

N

A.J.

NOTES// Portion of this great rille
centred around the crater
Silberschlag; the hill near hear
cut the rille. Hills N. of Silberschbg
G. also block the rille's path.
The rille is at its widest S. of
Silberschlag A.

ANDREW JOHNSON
1 YORK LANE
KNARESBOROUGH
NORTH YORKSHIRE.

LN.854

Figure 8.6 (*cont.*) (b) Rima Ariadaeus drawn by Andrew Johnson.

Mare Tranquillitatis (to its east) and the junction between the Mare Vaporum (to the north-west) and Sinus Medii (to the south-west).

Figure 8.6(a) is a splendid view obtained using the Catalina Observatory 1.5 metre reflector, taken on 1966 May 27^d 03^h 56^m UT when the Sun's selenographic colongitude was 356°.8. The whole area, especially to the west, is rich in rilles and the right-hand side of Figure 8.6(a) also shows the easternmost part of the famous 'Hyginus Rille', Rima Hyginus. Rilles make a fascinating subject for study at the telescope eyepiece. Figure 8.6(b) shows one such study made by Andrew Johnson. Details as written on the drawing, but do note the orientation when comparing it to Figure 8.6(a). Figure 8.6(a) is, in common with most of the other images in this book, reproduced with south uppermost.

It may superficially look like a sinuous rille, except that it is rather bigger and straighter, but Rima Ariadaeus is actually a graben – the vertical slumping of ground along a stress fracture. Look carefully at Figure 8.6(a) and you will see topographical features, such as mounds, along it that clearly match the 'high and dry' features to either side. Notice that Andrew Johnson has recorded some of these delicate features in his drawing which shows the rille in the vicinity of the 13 km crater Silberschlag (which is also shown near the centre of Figure 8.6(a)).

The easternmost extent of Rima Ariadaeus (far left on Figure 8.6(a)) is marked by the small (11 km) bright crater Ariadaeus, notable for the intrusion of a smaller crater into it.

If you experience any trouble in locating Rima Ariadaeus, then first find your way to the imposing 44 km diameter crater Agrippa (upper right on Figure 8.6(a)) and the ancient ruined formation Julius Caesar (roughly 91 km in diameter and shown at the lower left of Figure 8.6(a)). Apart from near full Moon you should easily be able to identify the rille passing between these two craters and close to Silberschlag, which itself is conveniently half-way along a line between these two craters.

8.7 ARISTARCHUS [24°N, 313°E] (WITH HERODOTUS AND VALLIS SCHRÖTERI)

Even in the lowliest binoculars the crater Aristarchus stands out like a brilliant diamond against the grey expanse of the Moon's Oceanus Procellarum. Aristarchus is the brightest of the large formations on the lunar surface. It is quite easy to identify even when illuminated only by earthshine. Indeed, the great eighteenth-century astronomer William Herschel mistakenly believed that Aristarchus was an erupting volcano! The crater is reckoned to be very approximately 300–500 million years old. This is very young for a Moon crater of its size. Its youthfulness is the reason for its high albedo. The solar-wind bombardment has not had time enough to do its work of darkening the materials excavated from below the regolith.

Aristarchus spans about 40 km from rim to rim and close inspection reveals that it has a decidedly polygonal outline. It stands on an extensive plateau, with the rim of the crater rising to over 600 metres above its immediate surrounds. The interior terraced walls slope down to the crater floor at a depth of some 2.1 km below the rim. As Figure 8.7(b) shows, the appearance of the formation is somewhat confusing when seen under a high Sun. With the terminator somewhat nearer the crater, details then stand out readily: contrast Figure 8.7(a) with Figure 8.7(b).

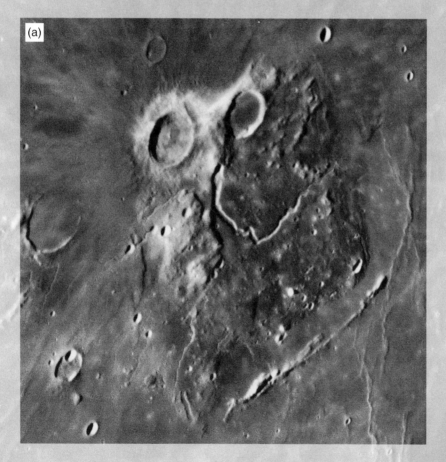

(a)

Figure 8.7 (a) Aristarchus (the largest crater shown), Herodotus (right of Aristarchus) and Vallis Schröteri (below Herodotus) at colongitude 63°.7. Details in text. (Catalina Observatory photograph – courtesy Lunar and Planetary Laboratory.)

Figure 8.7 (*cont.*)
(b) Aristarchus and
environs at colongitude
79°.3. Details in text.
(Catalina Observatory
photograph – courtesy
Lunar and Planetary
Laboratory.)

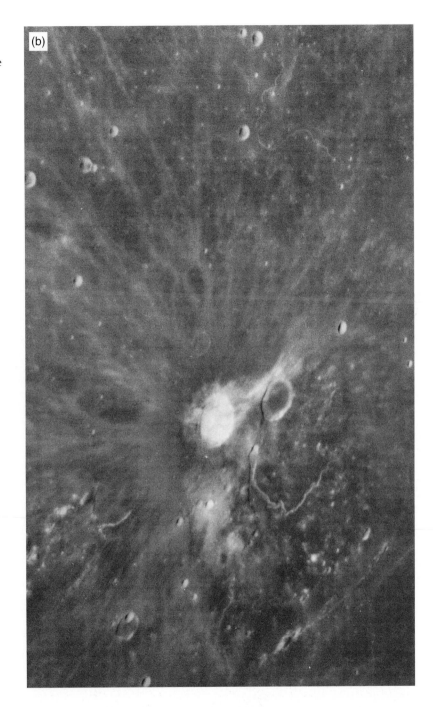

Both photographs were obtained using the Catalina Observatory 1.5 m reflector: (a) was taken on 1965 December 06^d 05^h 14^m UT and (b) was taken on 1966 October 28^d 06^h 28^m UT.

The youthfulness of Aristarchus has already been referred to in connection with its albedo. This characteristic also gives it a degree of thermal lag, when it comes to the diurnal cycle. In the lunar mornings bright features such as Aristarchus show up as cold spots on thermal maps, since they reflect away more of the solar radiation. However, the reverse is the case after sunset. Good reflectors make poor emitters. Then they show up as being warmer than their surrounds. During the local lunar night Aristarchus can remain up to 30 degrees Celsius warmer than its surrounds.

Another indicator of the youthfulness of Aristarchus is the complexity and crispness of the terracing of its walls. Further evidence is provided by its relatively complex floor and central mountain (features on the Moon lose their ruggedness as time passes).

Figure 8.7 (*cont.*) (c) *Orbiter IV* photograph of Aristarchus, Herodotus and Vallis Schröteri. (Courtesy NASA and E. A. Whitaker.)

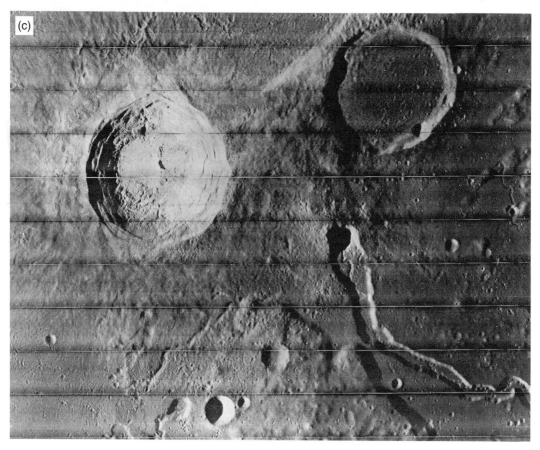

(c)

One intriguing feature of Aristarchus is readily apparent in Figure 8.7(a): the system of radial dark bands that extend up the interior terraces of the crater. They were first drawn by Lord Rosse in 1863 but were not recorded, nor even as much as mentioned, by the earlier observers of the Moon. Why was it that the likes of Mädler, Schmidt, or Neison failed to discover the bands when these great selenographers paid considerable attention to Aristarchus? How could they possibly miss a striking feature that even I as a novice observer could so readily see in my 76 mm reflector at the beginning of the 1970s? Could the bands have gone from being very hard to see to very obvious in the intervening century? I, for one, find this very hard to believe. There is a real mystery here.

Take a look for yourself. You should easily be able to see the two most prominent bands in a small telescope and you might count up to nine of them if you use a large telescope under suitable lighting and good atmospheric conditions. The bands do vary in intensity throughout the lunation, being hardest to see when the terminator is close by. You might like to make a study of their changing appearance.

The striking ray system emanating from Aristarchus tells a story about the impactor that created the crater. As is usual for ray systems on the lunar surface, the Aristarchus rays are most obvious when illuminated by a high Sun. Figure 8.7(b) shows the ray system well. Rays radiate in all directions from Aristarchus but note how the majority of the crater ejecta stream off to the south-west. Clearly the projectile hit the Moon at a fairly low angle, and came from the north-east. Look at the shape and offset of the interior 'central' mountain as shown in Figure 8.7(a) and you will find confirmatory evidence for this hypothesis.

The Aristarchus plateau, already referred to, is an approximately square area of rough and hummocky terrain, extending about 200 km × 200 km. Altimetry data obtained by the *Clementine* space-probe reveals that the southern edge of the plateau is about 2 km higher than the general level of the Oceanus Procellarum and that it gently slopes downwards to the north and north-west (the average slope being about one degree).

To me, the plateau seems to have a rich 'coffee-brown' tint that contrasts strongly with the white Aristarchus and the greenish-grey mare. As noted in Chapter 2, though, perceived colours are not accurate (and not everybody's eyes are colour-sensitive enough to show them) but the colour contrasts are at the least instructive. Proper colorimetric studies do, indeed, reveal that the plateau is much redder than the average hue of the Moon. The multi-waveband images obtained by *Clementine* indicate that the colour arises due to a layer of reddish pyroclastic glasses.

Other interesting features highlighted by *Clementine* include the presence of the mineral olivine distributed along the southern part of the rim of Aristarchus and the presence of anorthosite on the crater's central (or near-central!) peak.

Sitting on the plateau, alongside Aristarchus, is the slightly smaller and much shallower crater Herodotus. Both craters in effect at least approximately define the southernmost boundary of the Aristarchus plateau. The differences between the two craters could hardly be greater. Herodotus is obviously an ancient crater whose floor has been lava-flooded. Superficially the floor looks smooth but some craterpits are revealed by large-aperture telescopes used under good conditions. Space-probe images show the floor of Herodotus to be covered in tiny craters and fissures.

Perhaps the most remarkable feature in this very remarkable region of the Moon is Schröter's Valley, or more properly Vallis Schröteri. This is the Moon's largest sinuous rille, originating at its southernmost end at a deep crater known as the Cobra's Head and winding on for over 160 km to the western corner of the Aristarchus plateau. The impression is that it was created by a river of lava erupting from the Cobra's Head and cutting its way along a winding path to lower ground. Most lunar experts think that this was, indeed, what happened. A finer sinuous rille runs along the length of the floor of the valley, indicating at least one subsequent lava flow. This is best seen in the *Orbiter IV* photograph presented in Figure 8.7(c).

The whole area abounds with finer sinuous rilles. Clearly the geological (I would prefer to say 'selenological' but that term is not in fashion) history of this region is very complex.

The area is especially interesting for those involved in the controversial study of Transient Lunar Phenomena. About a third of all the catalogued reports of TLP involve Aristarchus or its surrounds. It is the single most 'event prone' of the areas if one is to believe all the reports. However, it must be borne in mind that the brilliance of Aristarchus may well be responsible for many illusory reports. Especially so as *spurious colour*, the prismatic splitting of colours along light–dark boundaries caused by the Earth's atmosphere, is especially evident with this crater. Often the southernmost part of the rim, and extending to the southernmost boundary of the ejecta blanket, shows a yellow or even orange-red glow due to this cause (the complementary colour showing up chiefly along the northern rim of the crater). Also there is the problem of observational selection. Being so apparently 'event prone' observers tend to concentrate their efforts to studying this area, so distorting the statistical evidence. However, it is true that there is some evidence for real TLP in this area and it may be significant that the *Apollo 15* particle spectrometer indicated a higher than average emission of radon gas when it flew over Aristarchus.

Aside from the various effects which have been reported involving Aristarchus itself, various reports of mistiness and coloured effects issuing from the Cobra's Head are on record. The transient event that I have most faith in as being something genuine, of any of the very few that I have witnessed myself, involved the crater Aristarchus. More about this and the whole subject of Transient Lunar Phenomena in the next chapter.

8.8 ARISTOTELES [50°N, 17°E] (WITH EUDOXUS AND EGEDE)

The crater Aristoteles stands proudly just south of part of the 'shore' of the Mare Frigoris and a little way east of the Montes Alpes. It is 87 km from rim to rim and possesses very finely terraced walls, rising to over 3.3 km above the floor. The floor, itself, is far from smooth. A low Sun-angle reveals that it is rippled with small hills. A little to the south lies the 67 km diameter Eudoxus, itself positioned at the northern termination of the Montes Caucasus. Its walls rise to a similar height as Aristoteles.

Aristoteles and Eudoxus are a magnificent pair of craters, easily identified at almost all lunar phases. Figure 8.8(a) shows the formations under morning illumination, while Figure 8.8(b) shows them late in the lunar day. Both photographs were obtained using the Catalina Observatory 1.5 m reflector: (a) was taken on 1967 January 20^{d} 01^{h} 45^{m} UT and (b) was taken on 1966 September 04^{d} 10^{h} 03^{m} UT. However, the terminator was

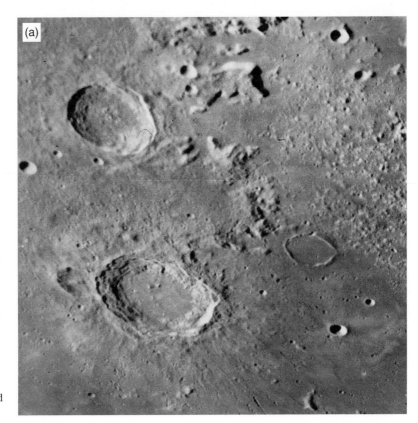

(a)

Figure 8.8 (a) Aristoteles (lower crater) and Eudoxus are the largest craters on this Catalina Observatory photograph taken at selenographic colongitude 18°.4. The small, lava-flooded, crater to the right is Egede. Other details in text. (Courtesy Lunar and Planetary Laboratory.)

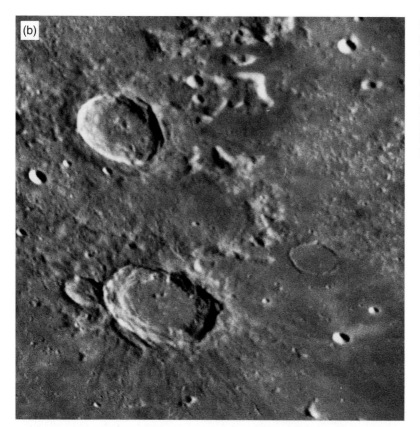

Figure 8.8 (*cont.*) (b) Aristoteles and environs at selenographic colongitude 142°.6. Other details in text. (Catalina Observatory photograph – courtesy Lunar and Planetary Laboratory).

not particularly close for either photograph. I will leave you to study the detailed topography of the area (and I especially commend to you the arc of mountain peaks which span Aristoteles and Eudoxus to the south) and to work out the chronologies.

However, there is something I will draw to your attention: the highly polygonal outlines of the craters. Notice how the small (37 km) flooded and ruined crater Egede (situated just to the west of the arc of mountain peaks just referred to) also shares the polygonal outline. Ask a casual observer what are the shapes of the outlines of the lunar craters and the answer you will probably get is that they are circular. Actually, many of them are decidedly polygonal. There is also some evidence that many of the faults on the lunar surface are aligned in the same general directions as the distortions to the crater outlines. This has been termed *the lunar grid system*, though not everybody accepts its validity. Has some global twisting of the Moon occurred in recent times (even some of the youngest craters have polygonal outlines) to cause the distortion? This certainly seems highly implausible. The mystery remains.

8.9   BAILLY [67°S, 291°E]

Under old terminology the largest craters were called 'walled plains', or 'ring plains'. Bailly, at 303 km diameter, qualified as the largest 'walled plain' on the Moon's Earth-facing hemisphere. Now, though, Bailly is regarded as one of the smallest *basins*. Basins are spread over both hemispheres of the Moon in approximately equal proportions but those on the Earth-facing side are almost all flooded with mare basalts. By contrast, there is almost no mare-type lava flooding on the Moon's farside and so the basins there are still 'raw', as is Bailly. The huge projectiles that created the basins belong to the early history of the Moon. In all probability Bailly is more than 3 billion years old.

There are two probable, and connected, reasons why Bailly has remained free of lava flooding. One is that the impacting projectile was not massive enough (and so did not convey enough kinetic energy) to cause sufficient fracturing of the Moon's crust to drive fissures deep enough to reach through to the upper mantle. Even near the centre of the Moon's

Figure 8.9 (a) Bailly spans this Catalina Observatory photograph taken at a selenographic colongitude of 80°.9. Details in text. (Courtesy Lunar and Planetary Laboratory.)

(a)

(b)

Earth-facing disk (where the crust is thinnest) the flooded plains are all of greater diameter than Bailly. The second reason is that Bailly, like the great basins of the Moon's farside, is positioned over a region where the crust is thicker than the average near the sub-Earth point.

Being so close to the south-western limb, Bailly is best observed just a little before full Moon, when the crater experiences lunar morning. Even then it can be badly affected by libration. The views of it at lunar evening are often unsatisfactory since the Moon is then a very thin crescent, and so is rather too close to the Sun in the sky to allow for good observing conditions. It is probably because of this that the earliest selenographers missed discovering Bailly. Cassini was the first to record it in his map of 1680.

Figure 8.9(a) provides a magnificent view of this huge formation. This photograph was taken on 1966 January 06$^{\rm d}$ 05$^{\rm h}$ 45$^{\rm m}$ UT with the 1.5 m reflector of the Catalina Observatory. The formation's great age is evident by its general state of ruin. Note how the ramparts have been smoothed

Figure 8.9 (*cont.*)
(b) Bailly is largely in darkness but part of the rim of Bailly B can be seen emerging into the morning sunlight in this view taken at a selenographic colongitude of 61°.0. Other details in text. (Catalina Observatory photograph – courtesy Lunar and Planetary Laboratory.)

and eroded, particularly by aeons of impacts. Even so, in places they still soar upwards to over 4 km above the mean floor level!

When Bailly is on view and the conditions are suitable you might like to examine it for yourself. You will find the floor of this formation littered with myriads of craters and ridges. However, the foreshortening will always prove a challenge to successful examination. Of the two large and overlapping craters at the south-western rim of Bailly, the smaller is known as Bailly A. It is 38 km in diameter and actually crosses the rim of Bailly. The larger of the two was, for a time, known as Hare but has now reverted to its original designation of Bailly B. Bailly B is very deep (over 4 km from rim to floor) for its 65 km diameter.

Watching/drawing/photographing lunar formations as the terminator passes over them is highly instructive. It is in this sort of activity that the very few (and dwindling) possibilities for useful and original topographic study lie. Take a look at Figure 8.9(b). This is another Catalina Observatory photograph but this one was taken on 1967 February 27^d 03^h 43^m UT at a slightly earlier lunar phase. This time the floor of Bailly is **almost** entirely in black shadow. Notice, though, the rim of Bailly B catching the morning sunlight ...

An ancient lunar formation full of fascination but a degree of dedication is needed in order to pursue its study at the telescope eyepiece.

8.10   BULLIALDUS [21°S, 338°E] (WITH KÖNIG, LUBINIEZKY AND WOLF)

Measuring just 61 km from rim to rim, Bullialdus may not be one of the largest craters on the Moon but it makes up for that in being one of the most beautifully formed.

Despite its small size, it is very easy to locate, sitting proudly and prominently on the Mare Nubium. Figure 8.10(a) is a Catalina Observatory photograph, taken with the 1.5 m reflector on 1966 December 23^d 04^h 54^m UT. The terraced walls (about 2.4 km in vertical height) are evident in the photograph, as is the somewhat polygonal outline of the crater and its slightly convex floor.

Bullialdus contains a wealth of detail to entertain and interest the observer equipped with a moderate telescope. The central mountain mass is complex and it changes its appearance quite considerably over the lunation as the various shadows develop with changing Sun-angle. Another shadow effect concerns the crater floor. Under a low Sun in the local lunar morning it tends to be dark and fairly evenly shaded (the effect of the convexity accepted) but it brightens considerably as the Sun angle increases and various dark markings appear, change shape and prominence, and then fade as local sunset approaches. This effect is

Figure 8.10 (a) Bullialdus is the largest crater shown on this Catalina Observatory photograph, taken at a selenographic colongitude of 39°.9. Bullialdus A nearly joins Bullialdus and Bullialdus B is just above this. At the top, and over to the right, is König, whilst below König and down at the bottom is Lubiniezky. Other details in text. (Courtesy Lunar and Planetary Laboratory.) (b) Bullialdus is over to the right and is filled with shadow and Wolf is the prominent formation in the upper left of this Catalina Observatory photograph, taken at a colongitude of 22°.6. Other details in text. (Courtesy Lunar and Planetary Laboratory.)

caused by the rough and slightly lumpy nature of the crater floor, the individual shadows generated within the surface relief being too small to appreciate individually but the combined effect producing the behaviour that is noticeable through the telescope.

Near the time of full Moon, the central mountain complex becomes very bright as does much of the crater rim and the interior develops many bright spots.

Away from the time of the full Moon, the careful observer may discern some landslips and even some small craters in Bullialdus's inner terraces. Notice the lines of thin black shadow defining the terraces in the southwest of the interior. Clearly here the steps slump backwards by several degrees. The black spot at the southern end of this section can also become very prominent, indicating a hollow. Under a higher Sun, a ridge, running a short way radially down the terraces, becomes visible in the same location. Look closely at Figure 8.10(a) and you should be able to discern an intriguing raised ridge crossing south-east from the central mountain complex to the foot of the wall terraces.

The outer surrounds of Bullialdus are also of particular interest. The complex array of ridges, most of which are radial to the crater, and chains of secondary craters and some blanket ejecta are visible in Figure 8.10(a). Figure 8.10(b), another Catalina Observatory photograph – this one taken on 1966 May 29^d 04^h 41^m UT – shows some of these features to better effect, here the interior of Bullialdus itself being filled with black shadow.

Almost abutting onto the southern rim of Bullialdus is the 26 km Bullialdus A. The rough ground of their shared outer flanks is interesting. Which crater do you think was formed first? The answer to that ought to be fairly obvious but I will leave that as an exercise for you.

With a gap between their rims of very roughly 20 km, Bullialdus B lies a little further south-south-west of Bullialdus A. Bullialdus B is 20 km in diameter. It is instructive to compare the outlines of these smaller craters with that of Bullialdus itself.

Roughly 80 km, or so, to the west-south-west of Bullialdus B we find the similarly sized (23 km diameter) and even more polygonal König. It has a slightly more 'broken down' appearance than Bullialdus A or B. All three share slight central elevations and generally mound-ridden and pock-marked interiors.

North-north-west of Bullialdus is the broken down and lava-flooded crater Lubiniezky. The 44 km diameter walls are completely demolished in the direction of Bullialdus, as is evident in Figure 8.10(a). Also an intriguing bright streak crosses the floor of Lubiniezky, exactly aligned with a similar streak running tangentially to the exterior of Bullialdus. Careful examination reveals this to be the brightest member of a whole pattern of closely spaced parallel light streaks in the locale, but most prominent in the interior of Lubiniezky. As might be expected, the ancient basalt lava covering of Lubiniezky is ridden with craters but most of these pose a challenge to the eyesight of an observer using even a large telescope in excellent conditions.

The upper-left section of Figure 8.10(b) shows a very peculiar formation, known as Wolf. I will leave this as a challenge for you to investigate, along with the ghost ring which is visible near the middle top of the same photograph. A thoroughly fascinating region of the Moon!

8.11   CASSINI [40°N, 5°E] (WITH THEAETETUS)

Cassini is a rather striking crater situated on the Palus Nebularum and at the southern head of the Montes Alpes. The northernmost extent of the Montes Caucasus lies a short distance to the east. The crater is quite conspicuous at all but the highest Sun-angles and so it is surprising that it was not recorded by the earliest selenographers. Cassini was the first to record it on his 1692 map. Just in case there should be any doubt, do let me add that there is no suspicion, whatever, of it being a crater only just over three centuries old! Dating the original impact precisely may well be problematical but we are certainly reckoning in billions of years, not mere centuries.

Figure 8.11(a) is a photograph centred on Cassini taken with the 1.5 m reflector of the Catalina Observatory on 1966 September 06^d 10^h 44^m UT, when the Sun's selenographic colongitude was 167°.3. Cassini, itself, is 57 km in diameter and has complex and rather broad outer ramparts. As is obvious from the photograph, the crater has been partially filled with mare-type lavas, presumably during the period of major flooding that filled the Imbrium Basin about 3.3 billion years ago. However, the crater-forming impact might conceivably have come a little after, perhaps resulting in fissuring of the still thin crust under the crater allowing a subsequent upwelling of lavas.

Certainly, though, the floor of the crater is old, and consequently saturated with small craters, together with a few larger examples. The largest of these, Cassini A, is 15 km in diameter and is situated somewhat north of the centre of Cassini. Near the south-western flank of the interior is the 9 km diameter Cassini B. Other floor details include various hummocks and ridges, as well as more, rather smaller, craters. The most delicate of these are a test for an observer with a powerful telescope working under good conditions. An astoundingly good image of Cassini is shown in Figure 8.11(b). It was taken by Damian Peach on 2005 September 22^d, using his *Celestron* C-14 telescope and *Lumenera* Lu075M camera.

The nearest major crater to Cassini, of the order of a hundred kilometres to the south-east and nestling close to the western foothills of the

Figure 8.11 (a) Cassini (centre) and Theaetetus (upper left of Cassini). Details in text. (Catalina Observatory photograph – courtesy Lunar and Planetary Laboratory.)

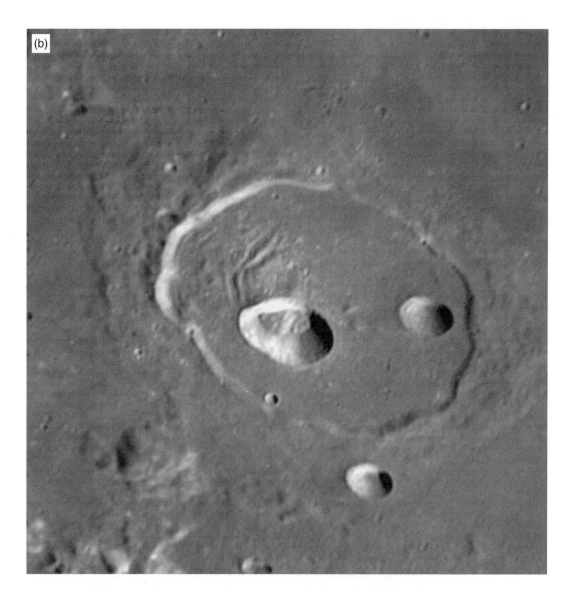

(b)

Montes Caucasus, is Theaetetus. It is about 25 km in diameter but its outline is obviously rather distorted. The rim of Theaetetus rises some 600 metres above the level of the outer surrounds but the floor of this crater is at a depth of over 2 km below the rim. It possesses a small, rather low, central mound. The terrain to the north and to the east of this crater is highly complex.

The occasional odd appearance has been reported in the vicinity of Theaetetus, by W. H. Pickering and by other observers. In 1902 the French astronomer Charbonneaux, using the 830 mm refractor of the Meudon

Figure 8.11 (cont.) (b) Cassini – image by Damian Peach on 2005 September 22^d when the Sun's selenographic colongitude was circa 144°. See text for more details.

Observatory, recorded the formation of a temporary 'white cloud' near the crater and in 1952 Patrick Moore, using his 12½-inch (318 mm) Newtonian reflector, saw 'a hazy line of light' crossing the otherwise shadow-filled interior of the crater.

8.12 CLAVIUS [58°S, 345°E] (WITH PORTER, RUTHERFURD, CLAVIUS C, D, J, K, L AND N)

A huge formation in the Moon's southern highlands, Clavius is surely one of the best known and easiest lunar formations to identify. It is a great crater, of the type that used to be called a 'walled plain', some 225 km in diameter.

At a lunar phase of about 8–9 days (selenographic colongitude about 17°) it is entirely filled with black shadow. At these times you only need good eyesight, and no optical aid, to see it as a distinct notch in the terminator. Sunrise over the formation is spectacular, with the central regions coming into view first. Andrew Johnson captures something of the grandiosity of the scene in his drawing, which is shown in Figure 8.12(a). This demonstrates that the floor of the crater, though rough and cluttered with detail, does at least follow the general curvature of the Moon's surface. In fact, it would not be at all obvious to a hypothetical observer stationed within Clavius that he was inside a crater at all. From the centre he could not see the walls and if he was in sight of one wall he could not see its continuation round to the other side of the crater.

Figure 8.12(b) is a Catalina Observatory photograph which shows the formation under a slightly higher Sun angle than the view shown in (a). It was taken with the 1.5 m reflector on 1967 January $20^d\,01^h\,52^m$ UT, when the Sun's selenographic colongitude was 18°.5.

Clavius seems to be of Nectarian age. In other words, it is somewhere around 4 billion years old. Hence it slightly pre-dates the Mare Imbrium and the global flooding of the large basins which formed the lunar maria. However, I would bet that it is slightly younger than Bailly. I will leave you to make a detailed comparison of Bailly and Clavius for yourself.

Although the official classification of a basin is reserved for formations larger than 300 km across, there can be little real doubt that Clavius is just a smaller example of the same. The walls of Clavius rise but little above the outer surrounds. Indeed, to the south they are heavily broken down and here the rim is rather ill defined. This formation is really a great trough sunk over 3.5 km below the outer surface. The nature of the walls vary somewhat going round the crater. From the north and going round to the west the terracing is coarse and broad, becoming narrower and more cliff-like round to the south. The very complicated and hummocky nature of walls bordering the rest of the crater is evident in Figure 8.12(b).

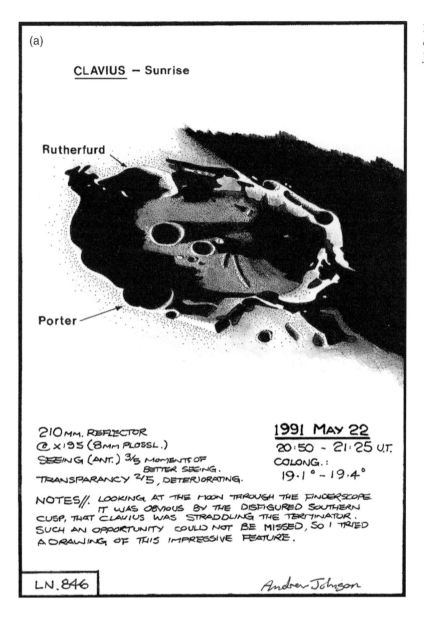

CLAVIUS – Sunrise

Rutherfurd

Porter

210MM. REFLECTOR
@ x195 (8MM PLOSSL.)
SEEING (ANT.) 3/5 MOMENTS OF
                    BETTER SEEING.
TRANSPARANCY 2/5, DETERIORATING.

1991 MAY 22
20:50 – 21:25 U.T.
COLONG. :
    19.1° – 19.4°

NOTES//. LOOKING AT THE MOON THROUGH THE FINDERSCOPE
       IT WAS OBVIOUS BY THE DISFIGURED SOUTHERN
CUSP, THAT CLAVIUS WAS STRADDLING THE TERMINATOR.
SUCH AN OPPORTUNITY COULD NOT BE MISSED, SO I TRIED
A DRAWING OF THIS IMPRESSIVE FEATURE.

LN.846

Andrew Johnson

Figure 8.12 (a) Clavius drawn by Andrew Johnson.

Apart from its size, the most striking feature of Clavius must be its interior craters. Starting with the 48 km Rutherfurd (shown on the oldest maps as Clavius A) an arc of craters of decreasing size curves across the floor of Clavius. Clavius D is next, diameter 28 km, then C (21 km), N (13 km), and J (12 km). These are the main ones forming the arc of craters but smaller examples abound, tending to form a closed loop which doesn't quite reach the south-western ramparts of Clavius but

Figure 8.12 (*cont.*)
(b) Clavius. (Catalina
Observatory photograph –
courtesy Lunar and
Planetary Laboratory.)

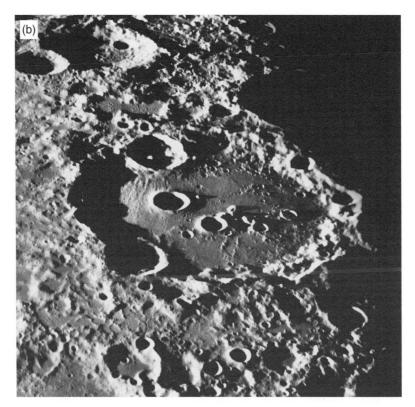

curves back towards Rutherfurd. Of course, this closed loop is only
approximately circular and is very roughly defined when we are considering the smallest craters. The whole effect is, nonetheless, striking and
supporters of the endogenic theories of crater formation made much of
Clavius and its interior craters as evidence to support their views. It is not
hard to see why.

I find it very difficult to accept that the formation of Clavius's interior
craters is the product of purely random impacts. However, I am certainly
not advocating the abandonment of the impact theory of crater formation! It is just that I think the situation involving Clavius is a little more
involved. Could it be that the impactors arrived together and in some sort
of formation? That is perhaps not as fantastic as it sounds. One might
envisage a partially fragmented asteroidal or cometary body slamming
into the already formed great ring of Clavius itself. Having said that,
there is one difficulty. The craters show signs of **not** being the same
age – at least as regards their ejecta patterns. Clavius D looks to be the
youngest. However, the morphology of the area is highly complicated
and so the age differences might be illusory. I must here emphasise that

the official view is that the arrangement of craters within Clavius is purely a chance one. Whatever the truth, Clavius is certainly not one of the easiest of lunar formations to investigate!

Even if the foregoing idea of a single fragmented body creating the curve of interior craters of Clavius is correct, then Rutherfurd was probably not included in this great event. The reason I say this is that Rutherfurd, unlike the others so far mentioned, shows definite signs of being created by an oblique-angle impact. It has the offset central peak, other interior details, and ejecta pattern (admittedly faint and hard to discern) that suggest a low-angle impact from the south-east. In addition to the details mentioned previously, the interior of Rutherfurd is complex and rather untypical of that of craters of its size. Sets of ridges radiate outwards from the rim of Rutherfurd across part of the floor of Clavius.

While Rutherfurd spans the rim of Clavius to the south-east, Porter does the same in the north-east. At 52 km diameter it is slightly larger than Rutherfurd and is noticeably non-circular. It, too, has a rather complex and hummocky interior and hummocky outer ramparts. On the oldest maps Porter is referred to as Clavius B. Other large craters 'splashed' into the rim of Clavius are Clavius L (diameter 24 km) to the west and Clavius K (diameter 20 km) to the south-east of Clavius L.

One oddity is that the floor of Clavius has some smoother areas, mostly in the east. Perhaps the impact events that created Rutherfurd and Porter are responsible for this. However, could a degree of volcanism have played a part in the early history of Clavius? Spectacular as it is when seen even through the smallest of telescopes, there is much about Clavius that we do not yet fully understand.

8.13   COPERNICUS [10°N, 340°E]

One early example of satire, and surely the only one to involve a heavenly body, must be Riccioli's naming of the lunar crater Copernicus in the seventeenth century. He detested the very idea of the Earth being in orbit around the Sun, as championed by Nicolaus Copernicus a century earlier. So he, in his own words, "flung Copernicus into the Ocean of Storms".

Sited prominently on the expanse of the Oceanus Procellarum as the crater is, Riccioli could hardly be accused of trying to 'bury the opposition'. In fact, now we know that the Solar System really is Sun-centred, it is fitting that one of the most prominent of the lunar formations bears the name of Copernicus. I wonder what Riccioli's reaction would have been if he could have known how his raillery was to backfire so spectacularly!

Figure 8.13 (a) Sunrise over Copernicus (the right-hand crater). Webcam image by John Gionis, Mike Butcher and the author on 2005 May 17^d 19^h 08^m UT when the Sun's selenographic colongitude was 20°.7. *Philips* ToUcam Pro webcam and the 19½-inch (0.5m) reflector of the Breckland Astronomical Society stopped off-axis to 20 cm. The AVI (10 frames per second for 12 seconds) was aligned and stacked using *RegiStax 3*. Image processed using 'Wavelets' in *RegiStax 3*, then using *Image Editor* by the author.

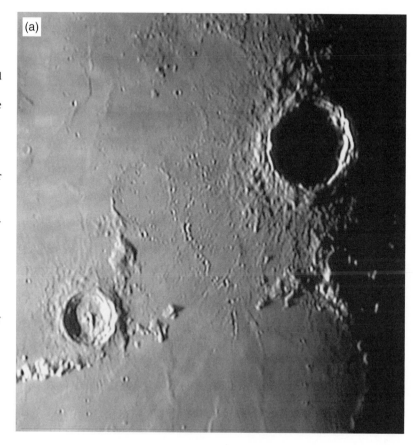

In their 1874 book *The Moon*, Nasmyth and Carpenter wrote of the crater Copernicus:

This may deservedly be considered as one of the grandest and most instructive of lunar craters. Although its vast diameter (46 miles) is exceeded by others, yet, taken as a whole, it forms one of the most impressive and interesting objects of its class. Its situation, near the centre of the lunar disc, renders all its wonderful details, as well as those of its immediate surrounding objects, so conspicuous as to establish it as a very favourite object.

Few can disagree with their opinion on the matter. T. G. Elger, the noted observer of the Moon, author of a Moon book, and first Director of the Lunar Section of the British Astronomical Association, christened the crater Copernicus "the Monarch of the Moon".

The modern value for the diameter of Copernicus is 93 km. Just before first and last quarter Moons the crater is entirely filled with black

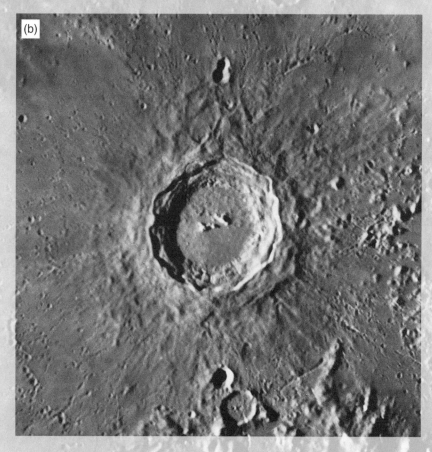

Figure 8.13 (*cont.*)
(b) Copernicus
photographed using
the 1.5 m Catalina
Observatory reflector
on 1967 January 21^d
02^h 44^m UT, when the
Sun's selenographic
colon-gitude was 31°.4.

Figure 8.13 *(cont.)*
(c) Copernicus drawn by
Andrew Johnson.

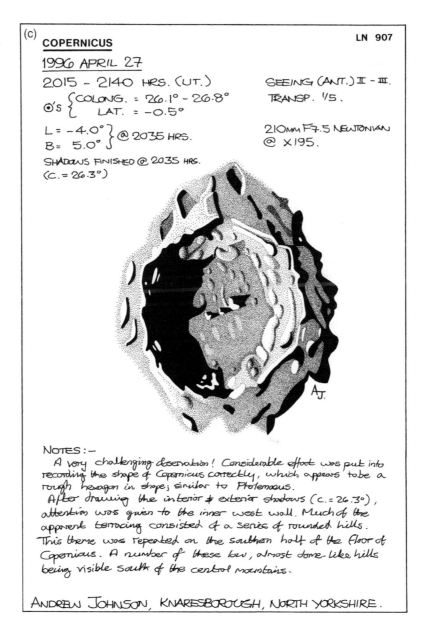

(c)

COPERNICUS   LN 907

1996 APRIL 27

2015 - 2140 HRS. (UT.)   SEEING (ANT.) II - III.

⊙'s { COLONG. = 26.1° - 26.8°   TRANSP. 1/5.
     {   LAT. = -0.5°

L = -4.0° } @ 2035 HRS.   210mm F7.5 NEWTONIAN
B= 5.0° }   @ X195.

SHADOWS FINISHED @ 2035 HRS.
(C. = 26.3°)

A.J.

NOTES :-
    A very challenging observation! Considerable effort was put into
recording the shape of Copernicus correctly, which appears to be a
rough hexagon in shape; similar to Ptolemaeus.
    After drawing the interior & exterior shadows (c.=26.3°),
attention was given to the inner west wall. Much of the
apparent terracing consisted of a series of rounded hills.
This theme was repeated on the southern half of the floor of
Copernicus. A number of these low, almost dome-like hills
being visible south of the central mountains.

ANDREW JOHNSON, KNARESBOROUGH, NORTH YORKSHIRE.

shadow. Figure 8.13(a) shows the Sun beginning to rise over the formation. More and more of the spectacular interior of this crater is revealed
as the Sun rises higher and the shadows retreat to the eastern flanks until
views such as that shown in Figure 8.13(b) are displayed.

The first thing to notice is that the outline of the crater is far from being
perfectly circular. It consists of roughly linear sections of varying lengths,

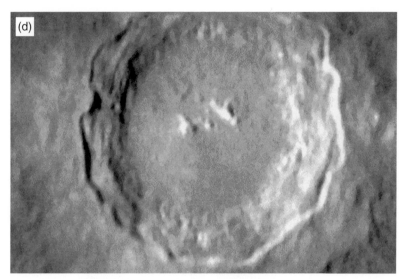

(d)

Figure 8.13 (*cont.*)
(d) Copernicus imaged
by Terry Platt, using
his 318 mm tri-
schiefspiegler reflector
and *Starlight Xpress* CCD
camera (other details not
available).

broken again by irregularities on the smaller scale. This polygonal outline
is at least approximately carried down towards the crater floor by the
complex system of terraces. These were created by the rebounding and
interfering shock waves during, and in the immediate aftermath, of the
colossal explosion that created the crater (which has been estimated to be
of magnitude equivalent to about 20 trillion tons of TNT). The terraces
themselves are not the sharply cut steps they appear to be in small telescopes but are somewhat rounded and softened into mounds and ridges.
Andrew Johnson represents this appearance in a drawing he made using
his 210 mm reflector, and which is shown in Figure 8.13(c).

The terracing of Copernicus also shows evidence of some slumping in
places. This can best be seen in Figure 8.13(d) which shows one of Terry
Platt's CCD images. Note that the Sun-angle is higher for this view,
revealing more details of the eastern part of the crater interior.

The floor of the crater is an almost circular plain about 62 km in
diameter and lies 3.8 km below the crater rim and about 2.9 km below
the general level of the outer surrounds. It has a central mountain complex, comprising several peaks in an arrangement which extends in a
roughly east–west direction. The highest of the peaks reaches up about
1.2 km above the crater floor. The southern part of the crater floor is
noticeably rougher than the northern section.

The outer slopes of Copernicus are complex, with a chaotic terrain of
mounds and radial ridges and crater ejecta. In fact, Copernicus holds a
special trophy in that it was the first crater around which secondary
craters were depicted (by Cassini in his map of 1680). Look carefully at
Figure 8.13(b) and you will see many of the small craters and chains of
small craters surrounding Copernicus. These must have been produced

by the debris ejected explosively from the impact site during the crater's formation. Figure 8.13(e) provides a wider-angle view. Imagine what it must have been like in the region in the aftermath of that great explosion – blocks of rock and other debris raining down as the seismic shock waves still rumbled through the ground!

It is reckoned that Copernicus was created about 0.8 billion years ago, the time from then to the present day now being known as the *Copernican era* of lunar chronology. The *Apollo 17* astronauts brought back rock samples, including some which are taken to be part of the Copernican ejecta blanket. If these have been correctly identified then we can be

Figure 8.13 (*cont.*) (e) Wider-angle view of (b), showing the extensive pattern of secondary craters which surround Copernicus.

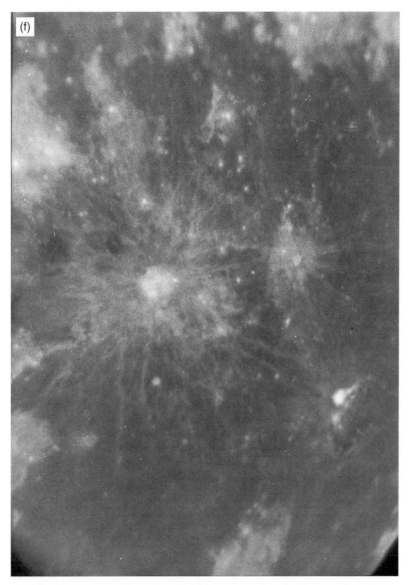

Figure 8.13 (*cont.*) (f) The ray systems of Copernicus (just left of centre) and Kepler (to the right) show up well in this near full-phase photograph by Tony Pacey, taken on 1991 November 23[d] using his 10-inch (254 mm) Newtonian reflector and eyepiece projection on *Ilford* FP4 film. The bright feature with a comet-like tail, below and a little to the right of Kepler, is Aristarchus.

even more precise with our dating of Copernicus. Laboratory analysis of these rocks suggest an age of 810 million years.

Large craters of similar ages as Copernicus, and younger, tend to have bright interiors and possess ray systems. The ray system associated with Copernicus is the second most prominent and extensive on the Moon.

While traces of the rays can be faintly seen at all but the lowest Sun angles, the rays really only become very prominent around the time of full Moon. Figure 8.13(f) shows a wide-angle view of the Copernican ray system under a high Sun. The photograph, obtained by Tony Pacey using

Figure 8.13 (*cont.*)
(g) *Orbiter IV* photograph
of Copernicus and its
environs to the north-
east, showing the inner
part of the ray system
and the ejecta pattern of
secondary craters.
(Courtesy NASA and E. A.
Whitaker.)

his 254 mm reflector, shows that the interior of the crater is brighter than the rays. These are wispy and plume-like, unlike the longer and straighter rays of Tycho (the premier rayed crater, discussed in Section 8.46). The whole ray system is very confused with many ray components not being exactly radial to the centre of the crater. Some rays even meet the crater rim tangentially. The confusion is compounded by the Copernican ray system merging with the rays emanating from the crater Kepler to the west of Copernicus.

Figure 8.13(g) shows an *Orbiter IV* photograph of Copernicus with the inner part of its ray system and the secondary craters in the quadrant to the north-east of the crater.

8.14 CRISIUM, MARE [CENTRED AT 17°N, 59°E] (WITH CLEOMEDES, LICK, PEIRCE, PEIRCE B, PICARD, PROCLUS, YERKES)

Easily visible to the naked eye as a dark patch near the Moon's north-east limb, the Mare Crisium also tends to grab the attention of the telescope user. This is especially the case when the terminator begins to cross it a little after full Moon (see Figure 8.14(a)). In part, this is because it is completely detached from the main system of lunar maria. It is reckoned that the Crisium Basin was formed about 3.9 billion years ago and the main episode of lava flooding occurred a few hundred million years after

Figure 8.14 (a) Evening falls over the Mare Crisium. Below the mare is the prominent large crater Cleomedes. Just to the right of the mare is the brilliant small crater Proclus, with its asymmetric ray system. Photograph by Tony Pacey, taken using his 10-inch (254 mm) f/5.5 Newtonian reflector on 1992 January 22^d 00^h 05^m UT, when the Sun's selenographic colongitude was 115°.7. The image was projected by eyepiece to approximately f/50 onto *Ilford* FP4 film for this 0.5 second exposure.

Figure 8.14 (*cont.*) (b) The western section of the Mare Crisium. Same details as for Figure 8.1 (see text, page 159). The largest crater on the mare, here on the left, is Picard. To the upper right of Picard is the incomplete ring of the flooded crater Lick (with a small, well-formed crater immediately below it). The even less complete Yerkes lies on the border of the mare to the right of Picard. To the lower right of Picard are Peirce (the larger crater) and Peirce B. To the west of the mare (and to the middle right on this photograph) is the brilliant Proclus, with its prominent and highly asymmetric ray system. (Catalina Observatory photograph – courtesy Lunar and Planetary Laboratory.)

(c)

Figure 8.14 (*cont.*)
(c) The northern to the western regions of the Mare Crisium photographed with the 1.9 m reflector of the Helwan Observatory at Kottamia on 1965 August 1^d 20^h 39^m UT (Sun's selenographic colongitude 311°.1). Lower right of the mare is the prominent crater Cleomedes. The 'flying eagle' effect of Yerkes is particularly apparent here. (Reproduced with the kind permission of Dr T. W. Rackham.)
(d) Peirce and Pierce B drawn by Roy Bridge.

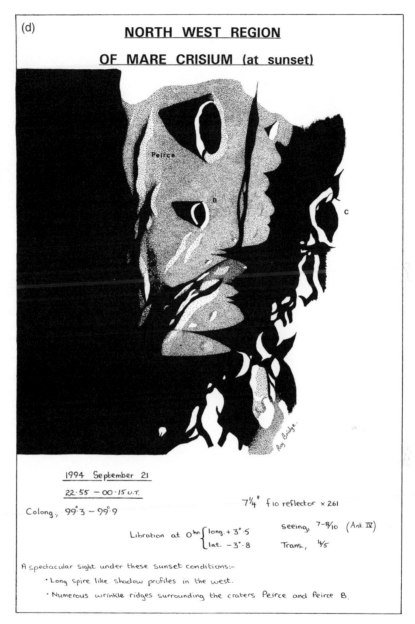

(d)

NORTH WEST REGION
OF MARE CRISIUM (at sunset)

1994 September 21

22.55 — 00.15 U.T.

Colong., 99.3 – 99.9

7¼" f10 reflector × 261

Libration at 0^{hr} { long. + 3°.5
{ lat. − 3°.8

Seeing, 7-8/10 (Ant. IV)

Trans., 4/5

A spectacular sight under these sunset conditions:-

• Long spire like shadow profiles in the west.

• Numerous wrinkle ridges surrounding the craters Peirce and Peirce B.

that. The basaltic covering probably extends to about 1 km deep at the middle of the mare, this being the deepest point of the Crisium Basin.

Though the Mare Crisium looks very elliptical and longest in the north-south direction, the appearance is deceptive because of the fore-shortening near the limb. Actually it is rather hexagonal (more than truly elliptical) in outline – and it is extended in the east-west direction

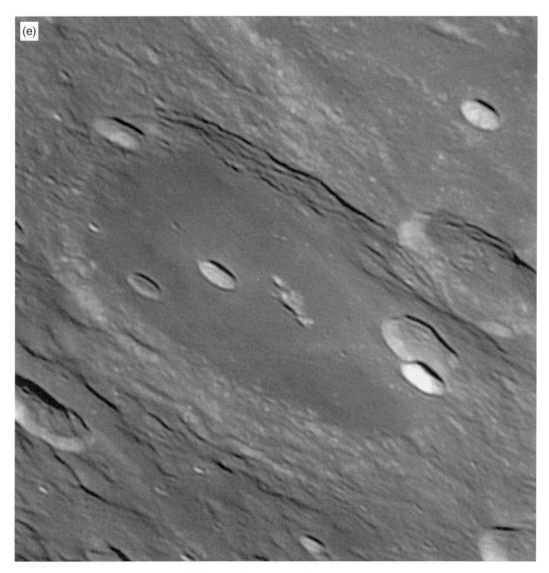

(570 km, against its north–south diameter of 450 km)! Clearly the impactor that created the Crisium Basin hit the surface at a very low trajectory.

The south-eastern sector of the Mare Crisium is notable for the Promontorium Agarum, and this is described in Section 8.1 (and shown in Figure 8.1), earlier in this chapter. Figure 8.14(b), reproduced here, shows much of the western half of the mare. The details are the same as for the photograph in Figure 8.1 (see text, page 159). The northern half of the Mare Crisium is well shown in Figure 8.14(c), which is a photograph taken with the 1.9 m reflector of the Helwan Observatory at Kottamia in Egypt.

Figure 8.14 (*cont.*)
(e) This incredible image of Cleomedes was made by Damian Peach using his 9.25-inch *Celestron* Schmidt–Cassegrain telescope and *Lumenera* Lu075M camera, from Barbados in April–May 2005. The Sun's selenographic colongitude was circa 104° at the time.

Numerous odd appearances have been reported at various locations in the mare, especially near the Promontorium Agarum (see Section 8.1). These have usually been instances of apparent mistiness obscuring details. However, local changes of albedo with Sun-angle are probably the real cause of these. It also used to be said that Mare Crisium shows a very strong green tint to observers, more so than the other lunar seas. However, I have never seen the greenish tint to be any stronger than elsewhere. In my opinion the strongest mare colouration is that of Mare Tranquillitatis, which often looks an inky blue to my gaze. Do let me repeat, though, that the real colours of the Moon are various shades of brown. The observer's eye tends to normalise the overall colour as white, so producing the range of apparent tints actually observed.

Various mottled patches, light spots and streaks abound on the mare and some wrinkle ridges show up under the lowest angles of illumination. The largest crater on the mare is the 23 km diameter Picard. It has a sharp rim and a small central mound on the deepest part of its floor, which lies about 2.4 km below the rim.

South-west of Picard, against the coastline of the mare, is the remnant of an old flooded crater named Lick. Further along the shoreline, and to the west of Picard, lays another broken partial ring, Yerkes. With the raised ridge that joins the surviving wall of Yerkes to a small crater, the whole forms an effect that has been aptly nicknamed 'the flying eagle'. Figure 8.14(c) shows this appearance particularly well.

Going approximately northwards from Picard, Peirce is the next largest of the well-formed craters on the Mare Crisium. It is almost as deep as Picard, despite its much smaller diameter (19 km). Again of similar depth, though even smaller, is Peirce B, just to the north of Peirce. On older maps Peirce B is referred to as Graham, or sometimes Peirce A. The IAU-approved designation is Peirce B. Peirce and Peirce B can look extremely dark under early morning illumination, as is well shown in Figure 8.14(c). Compare their appearance with the views shown in Figure 8.14(a) and (b). An impressive drawing of their appearance under late evening illumination is shown in Figure 8.14(d).

About 70 km west of the western shore of the Mare Crisium resides the 28 km diameter, but rather polygonal, crater Proclus. The early morning view in Figure 8.14(c) shows its shape well. However, under a high Sun the crater becomes one of the most brilliant on the Moon and its structure is then much harder to make out; see Figure 8.14(b) for comparison. At these times it also possesses a very bright system of rays, as can be seen on Figures 8.14(a) and (b). The ray pattern is very asymmetric. Faint rays do cross onto the Mare Crisium but mostly they extend towards the north-west.

Proclus, and particularly its rays, often takes on a distinctly yellowish colour but this is mainly due to spurious colour (the prismatic effect due to the Earth's atmosphere that often produces false colours along brightness boundaries in the images of celestial bodies as seen through telescopes).

Lunar north of the Mare Crisium (north-west as far as the orientation as seen in a telescope goes) lies the magnificent crater Cleomedes. About 100 km of rough ground separates the peculiarly straight section of the border of the mare from the rim of Cleomedes. Cleomedes is very deep. The roughly terraced walls in places plunge more than 2.7 km down to the convex floor of the crater. It contains several interior craters and other features of interest to the telescope user. The crater is well shown in Figure 8.14(c). Figure 8.14(e) shows it under the opposite lighting effect, and in incredible detail.

8.15 ENDYMION [54°N, 57°E] (WITH ATLAS, ATLAS A, BELKOVICH, CHEVALLIER, HERCULES, MARE HUMBOLDTIANUM)

The limb regions of the Moon offer a challenge to the observer because all details are distorted by foreshortening. The dimensions are only what they seem to be along arcs concentric with the limb of the Moon. Maximum contraction occurs along lines which are radial to the centre of the Moon's disk and the effect increases rapidly with proximity to the limb.

To compound matters, libration often conspires to move formations even closer to the limb just when the sky is clear and the lighting would be suitable over the chosen formation for studying it. On the other hand, libration can sometimes help by moving limb features further on to the Moon's Earth-facing side. One just has to make the best of the opportunities that arise.

The crater Endymion serves as a fairly prominent marker to some limb-hugging formations of particular interest. It is an old ring, 125 km in diameter, with a rather smooth and dark floor, it having been flooded with mare-type basalts. Endymion is shown at the centre of Figure 8.15(a). Before its flooding the crater must have been quite deep. As it is, the walls rise to over 4.5 km above the present flood-plain. Various spots and streaks are visible on the crater floor but any surface relief (craters, etc.) is very difficult for the backyard observer to detect. Figure 8.15(b) is a superb CCD image of Endymion, by Gordon Rogers, in which the shadows cast by the rugged peaks of the crater rim are seen thrown across the crater floor.

A spectacular pairing of craters, Atlas and Hercules, lie to the south-west of Endymion. They are well shown upper right in Figure 8.15(a) and

Figure 8.15 (a) Endymion lies near the centre of this Catalina Observatory photograph. It was taken with the 1.5 m reflector but I can't give the date, time and exact colongitude (though I estimate this to be about 38°) as the data are not available. The craters Atlas and Hercules are to the upper right (with Atlas A and Chevallier to their left). Part of the Mare Humboldtianum can be seen near the lower left and the crater Belkovich can (with difficulty) be identified in the lower middle of the frame. (Courtesy Lunar and Planetary Laboratory.)

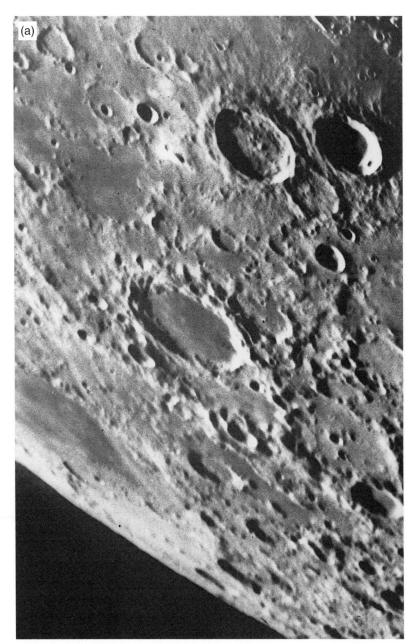

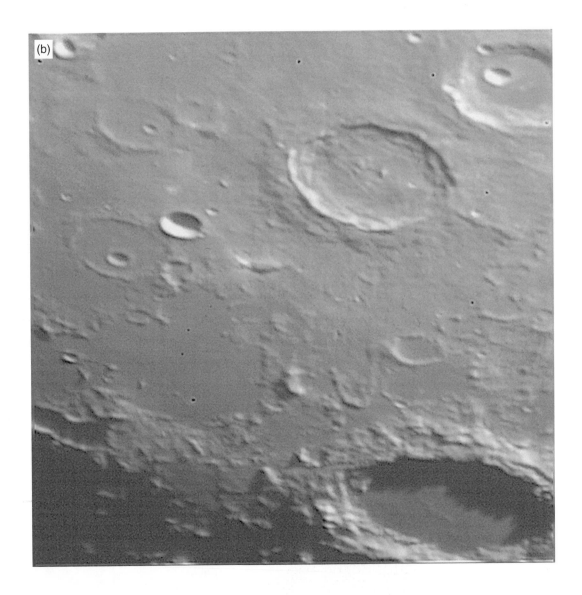

in the upper part of Figure 8.15(b). The larger of the two is Atlas, at 87 km diameter. Its rim averages about 3 km in height above the deepest part of the crater, near its centre. As is well shown in Figure 8.15(b), the floor is very rough and hummocky and it has a ring of mountains at its centre, rather than a central peak. Numerous fissures and small craters can be seen under particularly good conditions if one is using a powerful telescope.

Hercules is 69 km in diameter and it is obviously older than Atlas, having walls which are clearly broken down to a greater degree. Various

Figure 8.15 (*cont.*)
(b) Endymion (lower-right corner), Chevallier, Atlas A, Atlas and Hercules (extending from middle left to upper right) are shown on this image made by Gordon Rogers using his 16-inch (406 mm) *Meade* LX200 telescope and *Starlight Xpress* CCD camera on 1996 November 28^d 00^h 50^m UT, when the Sun's selenographic colongitude was 107°.7.

landslips in the terraces also become evident when the lighting is correct and I wonder how much of the damage we see today was caused by the impact which gave rise to the nearby Atlas? The floor of Hercules is more heavily covered with larger craters than that of Atlas, another indicator of its greater age.

If you take a look at Figure 8.15(a) and project a line from the upper part of the rim of Hercules, head it towards the upper part of Atlas and continue it on for a distance roughly equal to the diameter of Atlas you will come to a small but prominent well-formed crater. This is Atlas A,

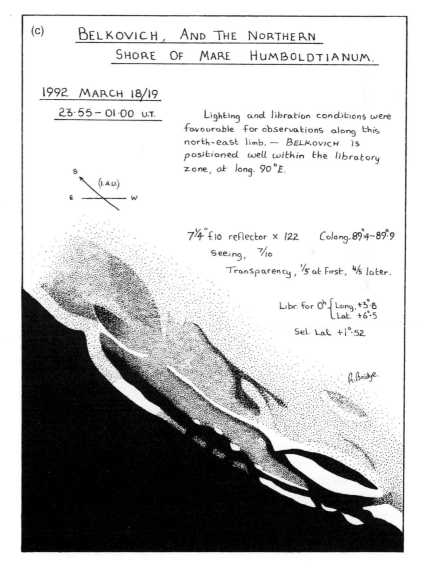

Figure 8.15 (*cont.*)
(c) Belkovich and northern Mare Humboldtianum drawn by Roy Bridge.

which is 22 km in diameter. Immediately to the left of Atlas A is a 'ghost ring' crater – one that has been flooded almost up to its rim. This is the 52 km diameter Chevallier. Note the small crater within Chevallier, obviously post-dating the episode of flooding.

In fact, it probably strikes you that this is an area of the Moon which shows extensive evidence of mare-type flooding. This characteristic is very evident in Figure 8.15(b). Yet this area in not actually on a lunar sea. The Mare Frigoris and Lacus Mortis are nearby, though, and the demarcation between lunar maria and lunar terrae is here rather less definite than in most other places.

The extreme limb features, first referred to, are the Mare Humboldtianum and the crater Belkovich. Mare Humboldtianum was named by Mädler, with a sense of appropriateness, after the German explorer Alexander von Humboldt. Humboldt's discoveries spanned, and you could say linked, the eastern and western hemispheres of the Earth.

Mare Humboldtianum is one of the smaller of the lunar 'seas'. Its diameter averages about 260 km but its shape is decidedly irregular when seen in plan view. Belkovich attaches to it on the north-western side. It is an old crater (of what used to be called the 'walled plain' variety), 198 km diameter, with two large craters intruding into its walls on the west and a further flooded ring on the eastern flank (and spanning the connection to the mare). The northern part of the mare is shown on the lower left of Figure 8.15(a) and Belkovich can be made out almost attaching to the mare's northernmost section. Neither are well shown, despite the very favourable libration when this Catalina Observatory photograph was taken. Roy Bridge has made a valiant effort to draw these features and the result is shown in Figure 8.15(c). If you relish a challenge, you might like to try recording this very difficult area yourself.

8.16   FRA MAURO [6°S, 343°E] (WITH BONPLAND, GUERICKE, PARRY)

This area of the Moon lies a little beyond the north-west border of the Oceanus Procellarum and on older maps occupied part of the Mare Nubium. However in 1964, after the successful flight of *Ranger 7*, the area was re-named Mare Cognitum (the Known Sea). *Ranger 7* transmitted the first close-range photographs of a lunar mare before it was deliberately crashed in this section of the mare. Officially the Mare Cognitum extends out to an average radius of 170 km centred on a position 10°S, 337°E.

Several spacecraft have either been deliberately crashed or have softlanded in this area. In particular, the *Apollo 14* astronauts Alan Shepard and Ed Mitchell landed in the foothills less than 30 km to the north of the northern rim of the great crater Fra Mauro (see Chapter 6 for

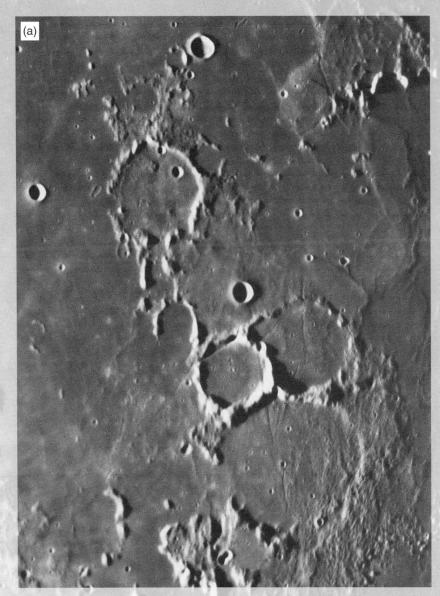

Figure 8.16 (a) Fra Mauro is the largest crater in the lower right section of this Catalina Observatory photograph, taken on 1966 May 29^d 03^h 51^m UT (at a colongitude of 22°.2) with the 1.5 m reflector. The two craters attached to the upper part of Fra Mauro are Parry (on the left) and Bonpland. The largest crater in the upper-left section of this photograph is Guericke. (Courtesy Lunar and Planetary Laboratory.)

(a)

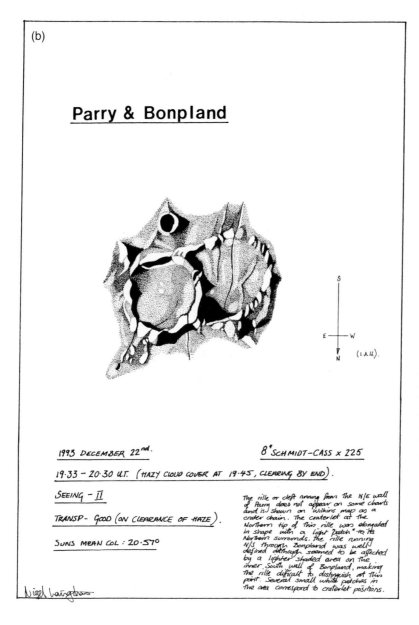

Parry & Bonpland

1993 DECEMBER 22nd.   8" SCHMIDT-CASS × 225

19.33 – 20.30 U.T. (HAZY CLOUD COVER AT 19.45, CLEARING BY END).

SEEING – II

TRANSP – GOOD (ON CLEARANCE OF HAZE).

SUNS MEAN COL : 20.57°

The rille or cleft running from the N/E wall of Parry does not appear on some charts and is shown on Wilkins' map as a crater chain. The craterlet at the Northern tip of this rille was elongated in shape with a light "patch" to its Northern surrounds. the rille running N/S through Bonpland was well defined although seemed to be affected by a lighter shaded area on the inner South wall of Bonpland, making the rille dificult to distinguish at this point. Several small white patches in the area correspond to craterlet positions.

Nigel Longshaw

Figure 8.16 (*cont.*)
(b) Parry and Bonpland, drawn by Nigel Longshaw. Necessarily being reduced in size for reproduction in this book, some of the hand-written notes are too small to read easily. The block of descriptive text reads (with no editing by me): *The rille or cleft running from the N/E wall of Parry does not appear on some charts and is shown on Wilkins' map as a crater chain. The craterlet at the northern tip of this rille was elongated in shape with a light 'patch' to its northern surrounds. The rille running N/S through Bonpland was well defined although seemed to be affected by a lighter shaded area on the inner south wall of Bonpland, making the rille difficult to distinguish at this point. Several small white patches in the area correspond to craterlet positions.*

more details). Whenever I look at the area north of Fra Mauro through a telescope I can never help but imagine them scuttling about 'down there', wheeling their Modular Equipment Transporter ('handcart', to you and me) all those years ago.

The Imbrium ejecta blanket is very evident in this region of the Moon and the patterns of ridges and scarring, radial to the Imbrium Basin, are

Figure 8.16 (*cont.*)
(c) Parry, drawn by Nigel Longshaw.

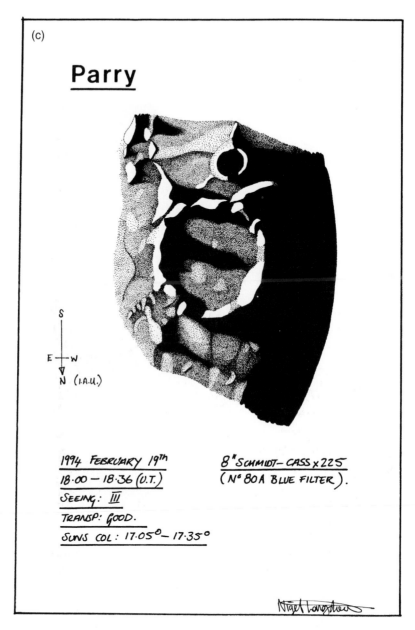

(c)

Parry

S
E — W
N (I.A.U.)

1994 FEBRUARY 19th
18·00 — 18·36 (U.T.)
SEEING: III
TRANSP: GOOD.
SUNS COL: 17·05° — 17·35°

8" SCHMIDT- CASS x 225
(N° 80A BLUE FILTER).

obvious even through small telescopes. Spacecraft results show this area to be very rich in KREEP (see Chapter 6), being one of the main 'hot spots' in gamma-ray maps.

Fra Mauro, itself, is 95 km in diameter. As can be seen from Figure 8.16(a), it is an old, ruined and flooded, 'ring-plain'. A section of the wall to the east is completely missing. The floor is extensively

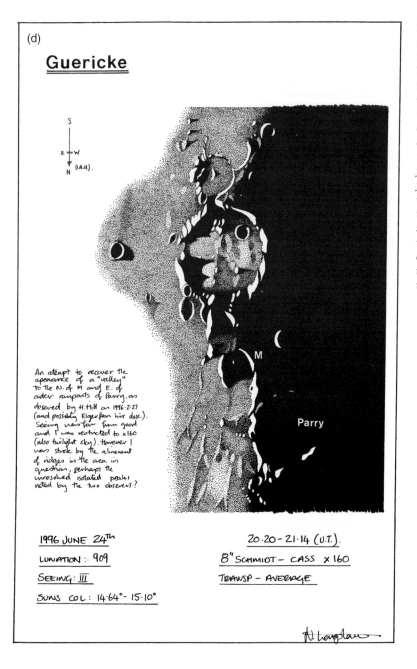

(d)

Guericke

An attempt to recover the appearance of a "valley" to the N. of M and E. of outer ramparts of Parry, as observed by H. Hill on 1996.2.27 (and possibly Elger from his desc.). Seeing was far from good and I was restricted to x160 (also twilight sky). However I was struck by the alinement of ridges in the area in question, perhaps the unresolved isolated peaks noted by the two observers?

1996 JUNE 24ᵀʰ

LUNATION :· 909

SEEING: III

SUNS COL: 14·64° - 15·10°

20·20 - 21·14 (U.T.)

8" SCHMIDT - CASS x160

TRANSP - AVERAGE

Figure 8.16 (cont.)
(d) Guericke, drawn by Nigel Longshaw. The block of descriptive text reads: *An attempt to recover the appearance of a 'valley' to the N. of M and E. of outer ramparts of Parry as observed by H. Hill on 1996.2.27 (and possibly Elger from his desc.). Seeing was far from good and I was restricted to ×160 (also twilight sky). However, I was struck by the alignment of ridges in the area in question, perhaps the unresolved isolated peaks noted by the two observers?*

cratered and ridged, mostly in the roughly north–south direction in common with the radial ejecta pattern from the Imbrium Basin. Another two 'ring-plain' craters attach to Fra Mauro: Bonpland and Parry.

Bonpland is the larger of the two, at 60 km diameter, and shares its rim with that of Fra Mauro along the southern section of the latter. Although Bonpland superficially looks older (more degraded) than Fra Mauro, a close examination of the shape of the augmented rim at the intersection suggests the opposite. In turn both Fra Mauro and Bonpland have been intruded upon by the still younger Parry. Parry is 48 km in diameter. The grouping is well shown in Figure 8.16(a) and in the special studies by Nigel Longshaw in Figures 8.16(b) and 8.16(c).

To the south-south-east of the 'Fra Mauro trio', and separated from them by about 100 km of very interesting terrain, is another ancient relic of a crater: Guericke. It has a diameter of 64 km. It, too, is well shown in Figure 8.16(a). Figure 8.16(d) is another of Nigel Longshaw's splendid drawings.

Rather than me providing all the intricate details of this very interesting area of the Moon, I thought I would leave this as a project for you. With the features generated by the Imbrium Basin event 3.85 billion years ago as a 'time marker', you might like to have a go at 'untying the temporal knot' and reconstruct the sequence of events which generated the features we see today.

The Fra Mauro region of the Moon may not be the most attention grabbing when one is looking through the telescope eyepiece, but it more than makes up for that if one is prepared to really study the small details.

8.17 FURNERIUS [36°S, 60°E] (WITH FRAUNHOFER, FURNERIUS B, FURNERIUS J, PETAVIUS)

Furnerius is the southernmost member of what used to be called 'the great western chain of craters'. This was before the IAU reversed east and west on the Moon, so we might now call the arrangement 'the great eastern chain'. All lying on virtually the same meridian, the 'chain' comprises the craters Furnerius, Petavius, Vendelinus and Langrenus, with the Mare Crisium also included and the crater Endymion to the north of the Mare Crisium (Cleomedes is a little further from the shared meridian than the other members). This 'chain' was used as evidence that craters were distributed in definite patterns, rather than randomly, and so must be of endogenic rather than impact origin.

The evidence for the impact scenario is so overwhelming that it surely must be the correct one. However, 'the great eastern chain', as I propose to call it, certainly makes one pause for thought when the ambient

(a)

Figure 8.17 (a) The two prominent craters, largely filled with black shadow, on the Moon's terminator are Furnerius (upper) and Petavius (lower, with central mountain poking up into the sunlight). Photograph by Tony Pacey. He used his 10-inch (254 mm) Newtonian reflector, with eyepiece projection to obtain this view on 1991 November 23[d]. Others details not available.

lighting throws it into prominence (see Figures 8.17(a) and (b)). It really does seem, though, to be a coincidence. Certainly the members of the 'chain' are all of different ages.

Of the other 'chain' members, Vendelinus and Langrenus are discussed in Section 8.26 further on in this chapter, while the Mare Crisium is discussed in Section 8.14 and Endymion in Section 8.15, earlier in this chapter.

The 125 km diameter Furnerius is obviously very old, as witness its somewhat degraded appearance and the number of large craters strewn over its interior. The crater walls climb to about 3.5 km above the level of its interior. Figure 8.17(c) is a magnificent drawing of this feature by Nigel

Figure 8.17 (*cont.*)
(b) Furnerius (upper
crater), Petavius (middle)
and Vendelinus (lower)
photographed by
Tony Pacey on 1989
October 16[d], same details
as for (a).

Longshaw. The largest crater in the interior of Furnerius is Furnerius B. This 22 km diameter crater is situated near the western rim of Furnerius. The slightly larger (24 km) crater that spans the north-east rim is Furnerius J. The floor of Furnerius abounds in interesting details, including the prominent rille that snakes from the north-west wall through the centre of the crater.

One month earlier, Nigel Longshaw observed Furnerius at a slightly later colongitude. His drawing is shown in Figure 8.17(d). The crater to

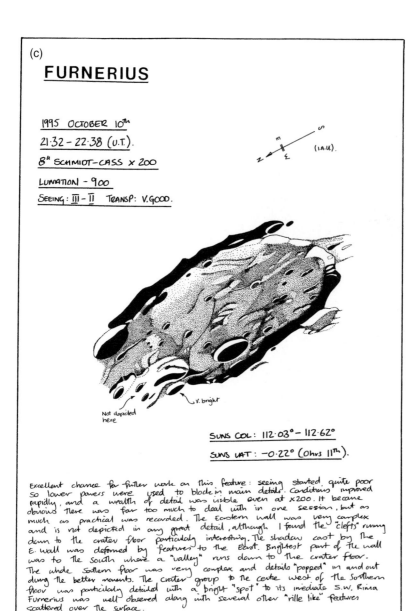

(c)

FURNERIUS

1995 OCTOBER 10TH
21.32 – 22.38 (U.T.).
8" SCHMIDT–CASS × 200
LUMATION – 9.00
SEEING: III – II TRANSP: V.GOOD.

(I.A.U).

SUNS COL: 112.03° – 112.62°

SUNS LAT: –0.22° (Ohrs 11TH).

Excellent chance for further work on this feature: seeing started quite poor so lower powers were used to block in main details. Conditions improved rapidly, and a wealth of detail was visible even at ×200. It became obvious there was far too much to deal with in one session, but as much as practical was recorded. The Eastern wall was very complex and is not depicted in any great detail, although I found the "clefts" running down to the crater floor particularly interesting. The shadow cast by the E. wall was deformed by features to the East. Brightest part of the wall was to the south where a "valley" runs down to the crater floor. The whole Southern floor was very complex and details "popped" in and out during the better moments. The crater group to the centre west of the Southern floor was particularly detailed with a bright "spot" to its immediate S.W. Rima Furnerius was well observed along with several other "rille like" features scattered over the surface.

N. Longshaw.

Not depicted here

V. bright

Figure 8.17 (cont.)
(c) Furnerius, drawn by Nigel Longshaw. The block of descriptive text reads (with no editing by me): *Excellent chance for further work on this feature: seeing started quite poor so lower powers were used to block in the main details. Conditions improved rapidly, and a wealth of detail was visible even at ×200. It became obvious there was far too much to deal with in one session, but as much as practical was recorded. The eastern wall was very complex and is not depicted in any great detail, although I found the 'clefts' running down to the crater floor particularly interesting. The shadow cast by the E. wall was deformed by features to the east. Brightest part of the wall was to the south where a 'valley' runs down to the crater floor. The whole southern floor was very complex and details 'popped' in and out during the better moments. The crater group to the centre west of the southern floor was particularly detailed with a bright 'spot' to its immediate S. W. Rima Furnerius was well observed along with several other 'rille like' features scattered over the surface.*

the south of Furnerius, and almost attached to it, is the 57 km diameter Fraunhofer. Sequence drawings of the Sun rising and/or setting over a lunar feature are particularly instructive as the detailed height relationships are then revealed. Note which parts of Furnerius are last to sink into the darkness of lunar night on Nigel's drawing. An *Orbiter IV* photograph of Furnerius is presented in Figure 8.17(e).

Figure 8.17 *(cont.)*
(d) Sunset over Furnerius, drawn by Nigel Longshaw. The block of descriptive text reads:
Seeing conditions rather poor at commencement, but as Moon rose seeing improved greatly. A wealth of detail was visible along the eastern wall of Furnerius, during the steady moments too much to depict. It was interesting to follow the retreat of the visible surface to the southern floor and details 1 and 2 to the right depict the 'shrinking' of this feature (extent shown on main drawing detailed at commencement of sketch.)

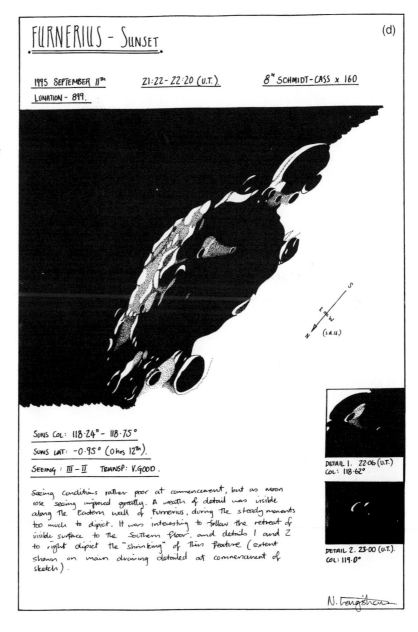

North of Furnerius, the crater Petavius must rank as one of the most beautifully sculpted objects on the Moon. It is the centre one of the 'chain' of craters shown in Figure 8.17(b). The outer walls of this great 'dinner plate' of a crater span 177 km but notice the unusually wide inner terraces, even tending to a double-ring type of structure along the western periphery. The inner ramparts extend upwards to approximately 2.1 to 3.3 km above the floor, the height varying around the crater.

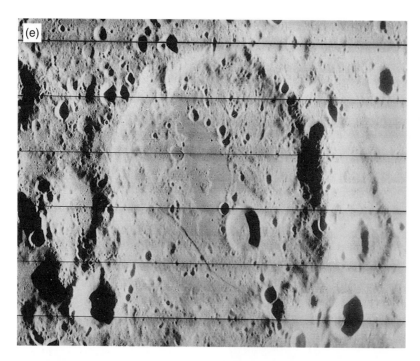

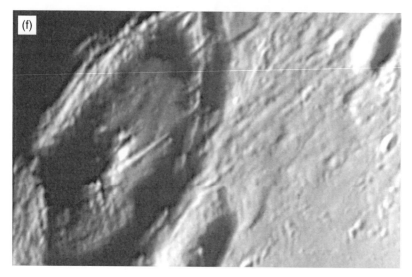

Figure 8.17 (*cont.*)
(e) *Orbiter IV* photograph of Furnerius. (Courtesy NASA and E. A. Whitaker.)
(f) Petavius. CCD image by Terry Platt, obtained using his 318 mm tri-schiefspiegler reflector and *Starlight Xpress* CCD camera. No other details available. The author has applied slight image sharpening and brightness and contrast re-scaling.

The mighty central mountain complex soars up to 1.7 km above the crater floor but is raised another 0.5 km, or so, above the level at the periphery because the crater floor is highly convex.

It is fascinating to watch the progression of shadows when Petavius is close to the terminator. The area around the central mountains, and then the central mountains themselves, are always last to disappear at sunset and first to appear at dawn, with the rest of the interior of the crater largely filled with deep-black shadow at these times. Something of the effect is shown in Figure 8.17(a).

Figure 8.17(f) shows one of Terry Platt's CCD images. The fineness of the detail shown is readily apparent by considering that only part of the formation fits into the frame!

The convexity of the floor gives a clue to the origin of one of the most outstanding of Petavius's features: the remarkable fissure that crosses the floor from the central mountains to the south-west wall. This used to be called 'the Great Cleft of Petavius' but the term 'cleft' is now obsolete. I suppose that it should now be known as 'the Great Rille of Petavius'. It seems to be a graben, a deep-seated stress fracture caused when the ground is pulled apart to either side and the ground along it slumps downwards into the crack. Other rilles are visible, mostly at least approximately radial to the central mountains. However, these are all very much harder to see than 'the Great Rille'. When the Sun-angle is low over the area 'the Great Rille' is easy to see even through a 60 mm refractor. Of course when the Sun is high it becomes a difficult object to view even through a large telescope.

As I said, the convexity of the floor provides the clue to its formation. It seems that enormous forces have built up under the floor of the crater, raising its floor and causing the stress fractures. As to the cause of the forces ...

8.18 'GRUITHUISEN'S LUNAR CITY' [5°N, 352°E]

Baron Franz von Gruithuisen was born in Bavaria in 1774. He took a medical degree but turned to astronomy as a profession, becoming Professor of Astronomy at Munich in 1826. He was an energetic seleno-grapher and generally a good observer. However he did tend to bring ridicule upon himself by making some extraordinary claims – the fruits of a vivid imagination.

Most famously, in 1824 he announced his 'discovery of many distinct traces of lunar inhabitants, especially of one of their colossal buildings'. He further described 'a lunar city' with 'dark gigantic ramparts'. The site of this edifice is quite near the centre of the Moon's disk, less than a hundred kilometres north of the ruined 35 km diameter crater Schröter.

(a)

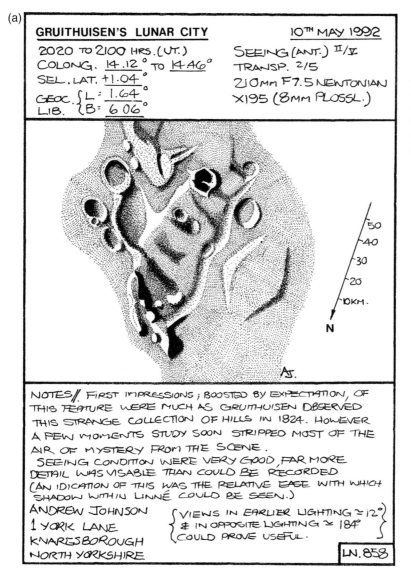

GRUITHUISEN'S LUNAR CITY   10ᵀᴴ MAY 1992

2020 TO 2100 HRS. (UT.)   SEEING (ANT.) II/V
COLONG. 14.12° TO 14.46°   TRANSP. 2/5
SEL. LAT. +1.04°   210MM F7.5 NEWTONIAN
GEOC. { L = 1.64°   X195 (8MM PLOSSL.)
LIB. { B = 6.06°

NOTES// FIRST IMPRESSIONS; BOOSTED BY EXPECTATION, OF
THIS FEATURE WERE MUCH AS GRUITHUISEN OBSERVED
THIS STRANGE COLLECTION OF HILLS IN 1824. HOWEVER
A FEW MOMENTS STUDY SOON STRIPPED MOST OF THE
AIR OF MYSTERY FROM THE SCENE.
SEEING CONDITION WERE VERY GOOD, FAR MORE
DETAIL WAS VISABLE THAN COULD BE RECORDED
(AN IDICATION OF THIS WAS THE RELATIVE EASE WITH WHICH
SHADOW WITHIN LINNÉ COULD BE SEEN.)
ANDREW JOHNSON   { VIEWS IN EARLIER LIGHTING ≈ 12°}
1 YORK LANE   { & IN OPPOSITE LIGHTING ≈ 184°}
KNARESBOROUGH   (COULD PROVE USEFUL.
NORTH YORKSHIRE   LN.858

Figure 8.18 (a) What a pity there are no selenites busying themselves in the morning within the confines of the 'lunar city', as Gruithuisen had interpreted this structure! Drawing by Andrew Johnson.

Actually, the southern point of the 'lunar city' is the 10 km crater Schröter W.

Of course, there is nothing but a rough arrangement of hills to be seen in that location. Figures 8.18(a) and 8.18(b) show two splendid studies of Gruithuisen's fabled 'city' made by Andrew Johnson.

Not everything one does at the eyepiece of the telescope has to be for serious study. Just for the fun of it you might like to take a look at the area yourself, particularly around the times of first and last quarter Moon, and

Figure 8.18 *(cont.)*
(b) Gruithuisen's 'lunar city' in the late afternoon, as drawn by Andrew Johnson.

(b)

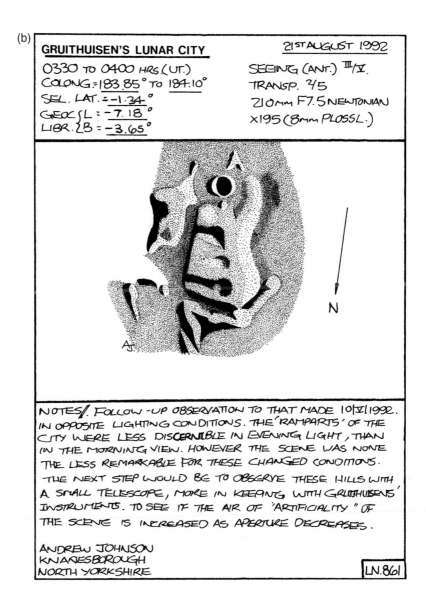

see if you can force your imagination to create a lunar city out of the jumbled topographic features in the area just north of Schröter.

Before we laugh too loudly at the memory of Gruithuisen, we should remember that he made many good contributions to the study of the Moon in his day and has been commemorated with a 15 km crater named after him positioned at 33°N, 320°E on the lunar surface, at the junction of the Mare Imbrium and Oceanus Procellarum. He can even be said to be the originator of the impact theory of the formation of lunar craters – after many years of dispute among experts, now the accepted scenario!

8.19   HARBINGER, MONTES [27°N, 319°E] (WITH PRINZ)

The Harbinger Mountains, more properly Montes Harbinger, are a small but interesting cluster of hills situated in a fairly barren part of the Oceanus Procellarum just a little north-east of the prominent crater Aristarchus (see Section 8.7). The proximity of Aristarchus is evident in Figure 8.19(a), which is a Catalina Observatory (1.5 m reflector) photograph taken on 1965 December 6^d 05^h 14^m UT, when the Sun's selenographic colongitude was 63°.7. You might find this view useful in helping to locate the mountain group through the telescope eyepiece.

The individual peaks are like islands rising up from the Ocean of Storms. A ruined crater, Prinz, attaches to the south-west of the range. Prinz forms an incomplete ring with a diameter of 47 km. Clearly the Procellarum lavas have partially buried this crater, along with the lower parts of the Harbinger Mountains and doubtless other low-lying features. Prinz is completely open to its south-west.

Numerous rilles and even some *domes* (volcanic swellings) are evident in the area and become evident at different stages of illumination. Figure 8.19(b) is a drawing by Roy Bridge showing the group under the

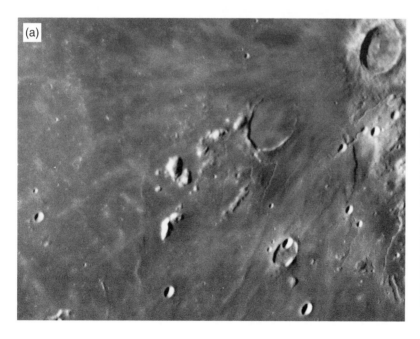

Figure 8.19 (a) Montes Harbinger is at the centre of this photograph, with the incomplete crater Prinz abutting the mountain group to the upper right. The crater at the top-right corner of this Catalina Observatory photograph is Aristarchus. (Courtesy Lunar and Planetary Laboratory.)

Figure 8.19 (*cont.*)
(b) Sunrise over Montes
Harbinger, drawn by Roy
Bridge.

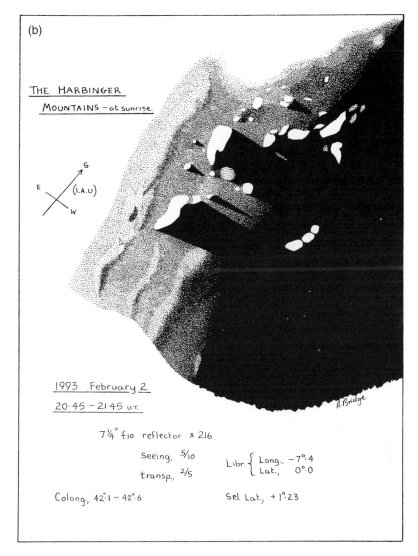

(b)

THE HARBINGER
MOUNTAINS — at sunrise.

S
E (I.A.U)
W

1993 February 2
20·45 – 21·45 U.T.

7¼" f10 reflector × 216

Seeing, 5/10

transp., 2/5

Libr. { Long., −7°·4
Lat., 0°·0

Colong, 42°·1 – 42°·6

Sel. Lat., +1°·23

R.Bridge

first light of lunar dawn. Figure 8.19(c) shows the area under a much higher Sun. Figure 8.19(a) shows it under a slightly higher Sun-angle, still.

I would unhesitatingly recommend the aspiring draughtsman of the lunar scene to gain practice at drawing mountains by observing and recording the Harbinger–Prinz complex under as many different illuminations as possible. The details are delicate enough to make it a real challenge but not so overwhelmingly complex as to be off-putting. After all, even the most practised observer–draughtsman would balk at

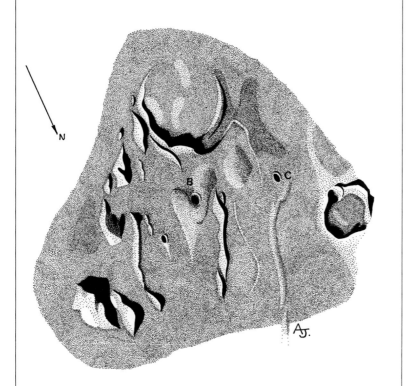

(c) **PRINZ & ENVIRONS** — morning illumination   LN. 880

N

B

C

AJ.

1994 FEBRUARY 22
1940 – 2020 HRS. (U.T.)
☉: { COL. 54.4° – 54.7°
      LAT. 1.6°

GEOC. LIBRATION { L = – 6.9°
@ 2000 HRS.   { B = 5.3°

210 MM F7.5 NEWTONIAN
×195
SEEING (ANT.) III.
TRANSP. 2–3/5
(SOME CLOUD INTERUPTIONS.)

NOTES: Observation under good libration conditions. Tried to detect
what I could of the Prinz rille system. Only definitely saw Prinz I, a
fine white line further N. may have been Prinz II. Other rillas appear
to have been beyond my scope. Though only average seeing, plus illumination a little too advanced. H. Hill's 18/1/1989 observation better on this
score. (C = 47.5°)

Figure 8.19 (*cont.*)
(c) Prinz and Montes
Harbinger, drawn by
Andrew Johnson.

the prospect of trying to record even part of extensive ranges, such as Montes Apenninus!

8.20 HEVELIUS [2°N, 293°E] (WITH CAVALERIUS, GRIMALDI, LOHRMANN)

The impression of the apparent non-accidental alignments of some craters on the Moon is, at least in part, generated by the presence of the terminator. How many prominent chains of large craters can you find that extend in even a roughly east–west direction? Yet time and time again crater chains along the north–south meridians spring into prominence as the morning and evening terminator looms close. Such an example is provided by the craters Grimaldi, the southernmost member, and extending northwards, Lohrmann, Hevelius and Cavalerius. Seen just before full Moon this line of large craters appears strikingly prominent even in the smallest telescope.

Figure 8.20(a) shows a 'snapshot' of the spectacle I took myself. Figure 8.20(b) is another example, though this time with the terminator slightly less advanced and the largest crater (Grimaldi) here entirely filled with shadow and forming a distinct 'notch' in the Moon. This 'notch' is obvious even when seen through low-power binoculars.

Figure 8.20 (a) The line of craters Grimaldi to Cavalerius is very apparent along the terminator in this photograph taken by the author on 1983 October 19[d], using his 0.46 m Newtonian reflector. The camera, fitted with a 58 mm lens, was handheld to a 44 mm Plössyl eyepiece (EFR = f/7.4) for a 1/500 second exposure on Ektachrome 200.

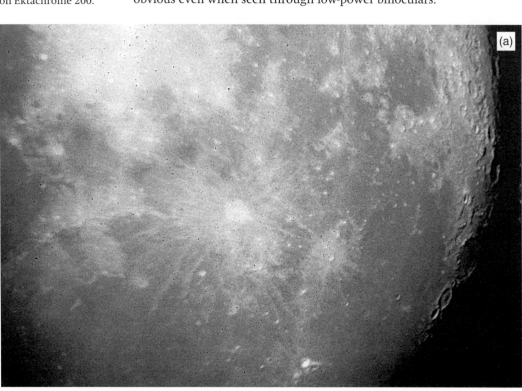

(a)

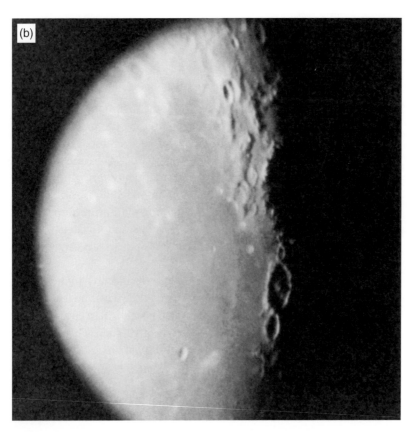

Figure 8.20 (*cont.*)
(b) Grimaldi is entirely shadow-filled, creating a 'notch' in the Moon, while Lohrmann, Hevelius, and Cavalerius are prominently displayed below Grimaldi on this photograph taken by the author using his 0.46 m reflector on 1985 January 5^d 20^h 23^m UT (when the Sun's selenographic colongitude was $67°.8$). The camera, fitted with a 58 mm lens, was hand-held to a 9 mm Orthoscopic eyepiece (EFR = f/36) for the 1/125 second exposure on *Fuji* HR1600 film.

Figure 8.20(c) shows a detailed study of Hevelius, together with Lohrmann and Cavalerius, made by Andrew Johnson. Hevelius is a magnificent crater, 120 km in diameter, with some terracing and much fine detail visible in its somewhat irregular walls. It can be said to represent an example of a formation intermediate in type between the 'saucer-shaped' and 'walled-plain' craters. It is, though, rather closer to being of the 'walled-plain' variety than the other.

The floor is convex as is evident in Figure 8.20(b). The walls vary in height around the crater but typically soar up to about 1.8 km above the floor. The crater has a central mountain, as well as other floor details such as rilles and small craters. Hevelius is named after Johann Hewelcke, a seventeenth century Danzig astronomer and selenographer, and the crater is called 'Hevel' on old maps.

Lohrmann is a 34 km diameter crater which seems to sit uneasily between Hevelius and Grimaldi. It has a rather hummocky floor and a central mound.

Figure 8.20 (*cont.*)
(c) Hevelius (largest
crater), Lohrmann (above
Hevelius) and Cavalerius,
drawn by Andrew
Johnson.

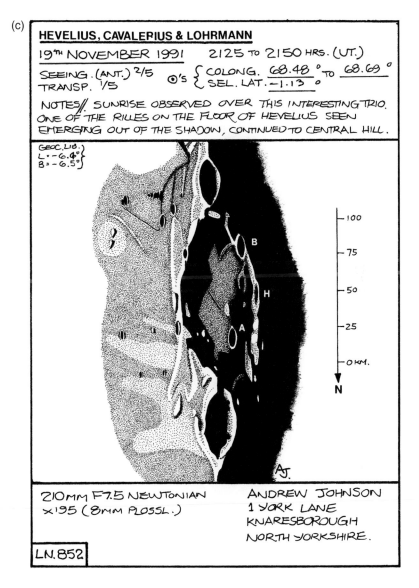

North of Hevelius, and just intruding into it, is the 64 km diameter Cavalerius. That Cavalerius is more youthful than Hevelius is evident both in its crisper appearance and in the fact that it is obviously Cavalerius which has intruded across the rim of Hevelius and not the other way round.

There is, though, some slumping of the rims of both at their intersection. In the case of Cavalerius the slumping is slight. In the case of

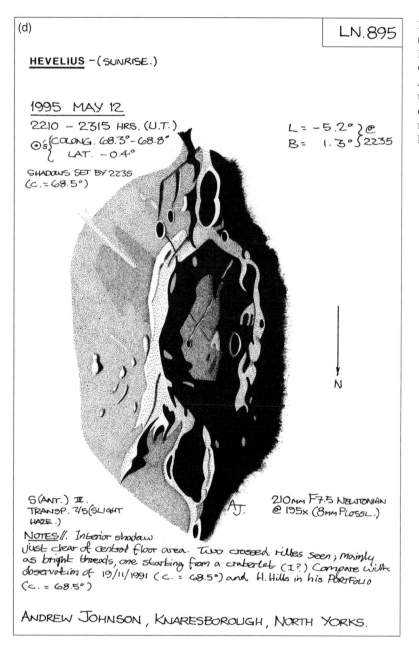

(d)   LN.895

HEVELIUS — (SUNRISE.)

1995 MAY 12
2210 - 2315 HRS. (U.T.)
☉'s {COLONG. 68.3° - 68.8°   L = -5.2° } ℂ
        LAT. -0.4°   B = 1.3° } 2235
SHADOWS SET BY 2235
(C. = 68.5°)

N ↓

S (ANT.) III.   210mm F7.5 NEWTONIAN
TRANSP. 2/5 (SLIGHT   @ 195x (8mm PLOSSL.)
HAZE.)   A.J.

NOTES//. Interior shadow
just clear of central floor area. Two crossed rilles seen; mainly
as bright threads, one starting from a craterlet (I?) Compare with
observation of 19/11/1991 (C. = 68.5°) and H. Hills in his PORTFOLIO
(C. = 68.5°)

ANDREW JOHNSON, KNARESBOROUGH, NORTH YORKS.

Figure 8.20 (cont.)
(d) Hevelius, with
Lohrmann and
Cavalerius, drawn by
Andrew Johnson. Note
the differences to the
drawing in (c), largely the
result of the different
libration.

Hevelius it is considerable. Also the ground outside Hevelius in which Cavalerius was formed is lower, anyway, than the rim of Hevelius. This creates the illusion of a deep channel apparently joining Hevelius to Cavalerius when the formations are close to the terminator. Even the low-resolution view of the arrangement in Figure 8.20(b) shows this effect well. The exaggerated effect persists to quite a high Sun-angle because the rim of Hevelius just east of the point of intersection casts a long shadow down into Cavalerius.

Similar is the so-called 'Miyamori Valley'. This is an apparent chasm extending from Lohrmann south-westwards to the major crater Riccioli. Again it is an exaggerated effect for the most part generated by the shadows cast by the rather linear north-north-east section of the outer ramparts of Grimaldi. Any real valley is much less deep and well defined, being at most some low ground that threads between assorted hummocks and craters. Figure 8.20(e) shows a study of it by Roy Bridge. Notice how the shadow fades away, like the grin of a Cheshire cat, at the eastern end of the 'valley' as the terminator moves westwards. This sort of sequence drawing is highly instructive.

Figure 8.20 (*cont.*) (e) The 'Miyamori Valley', drawn by Roy Bridge.

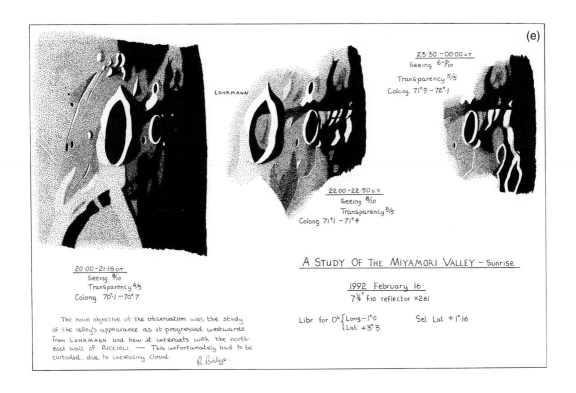

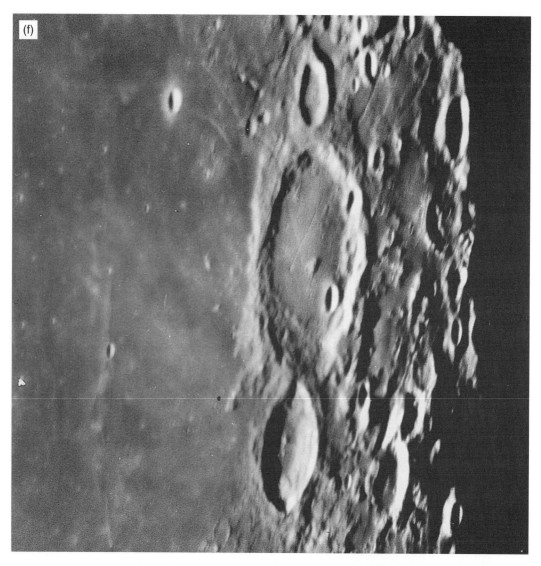

(f)

Figure 8.20(f) shows the area under higher Sun. The 'valley' has all but disappeared. However, some arcuate rilles are now in evidence curving westwards away from Hevelius.

Being so close to the western limb of the Moon, the appearance of the formations can alter quite significantly because of the effect of libration in longitude. Compare Figure 8.20(c) with Figure 8.20(d), which is another drawing by Andrew Johnson. Here the values of the Sun's selenographic colongitude are very similar and yet there are significant differences in the way Andrew has represented the features – in large part caused by the difference in the librations.

Figure 8.20 (*cont.*)
(f) Portrait of Hevelius, Cavalerius and Lohrmann, made using the 1.5 m reflector of the Catalina Observatory on 1966 February $4^{d} 06^{h} 53^{m}$ UT, when the Sun's selenographic colongitude was $74°.2$. (Courtesy Lunar and Planetary Laboratory.)

Figure 8.20 (*cont.*)
(g) Grimaldi (centre)
dominates this Catalina
Observatory photograph,
taken just 10 minutes
after the one shown in (f).
(Courtesy Lunar and
Planetary Laboratory.)

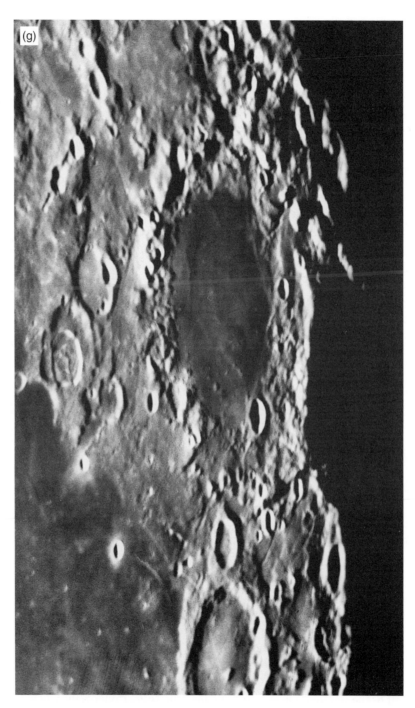

South of Lohrmann, and with a tract of rough and cratered ground separating the two, is the great 'walled-plain' Grimaldi. Completely in shadow in Figure 8.20(b), the floor of this crater is just beginning to receive the rays of the rising Sun in Figure 8.20(a). Figure 8.20(g) shows the formation (together with Lohrmann and the southern part of Hevelius) in full sunlight.

That Grimaldi is much larger than Hevelius is very obvious. However, the exact size of Grimaldi is a little problematical. The very dark, lava-flooded, floor of the formation is roughly 140 km in diameter but, as can be seen from the photographs, is rather irregular in outline. Look carefully and you will see that the rough surrounds of the flood-plain climb upwards and form a vague crater rim (more obvious to the north and west) of diameter exceeding 220 km. There are even, in places on the Moon, faint traces of a secondary concentric ring at nearly twice the radius of the first. Clearly the impactor that created Grimaldi packed quite a wallop!

Occasional bright flashes and patches of colour and apparent mistiness have been reported from time to time on Grimaldi's lava-flooded floor and many small craters, mounds, spots, streaks and wrinkle ridges provide an endless source of study for the telescopist.

8.21   HORTENSIUS [6°N, 332°E] (WITH ASSOCIATED LUNAR DOMES)

Hortensius is a fairly unremarkable crater, 15 km in diameter, situated in the Oceanus Procellarum just west of the great crater Copernicus. It has a sharp rim and is quite deep for its size. Rim to lowest point in the bowl-shaped depression measures nearly 2.9 km. The real point of interest lies on the mare just to the north of the crater: several of the raised mounds known as lunar domes.

Most casual observers of the Moon never get to see lunar domes. They are invariably small and elusive, only showing up well at low Sun-angles. Small-scale maps, and even some of larger sizes, do not show the location of these intriguing formations. The domes near Hortensius are notable in that their location is easily defined. Also they are at their most evident close to the time of first quarter Moon, the most popular time for lunar observing (they are again evident at last quarter Moon, of course, but far fewer amateurs are out with their telescopes in the small hours of the morning to observe the Moon).

As far as advice about locating the domes goes, I can do no better than to refer you to Figure 8.21. On it you will see part of Copernicus to the far left of the photograph, and so can gauge the scale. Upper right you will see Hortensius and just below Hortensius you should be able to make out a cluster of blister-like mounds. They are the lunar domes.

Figure 8.21 The subject of this photograph, Hortensius and its associated lunar domes, has been placed near the upper-right corner in order to show their location with respect to the nearby major crater Copernicus (partly shown on the left). Photograph taken with the 1.5 m reflector of the Catalina Observatory on 1967 January 21^d 02^h 44^m UT, when the Sun's selenographic colongitude was 31°.4. (Courtesy Lunar and Planetary Laboratory.)

Historically, the proponents of the endogenic origin theory of lunar craters used the domes to support their views. The domes, to them, were blisters that had failed to burst (on the mud-bubble theory of crater formation) and so were examples of the crater-formation process 'caught in the act'. We can dismiss this theory in the light of modern evidence but the question still remains: what are they and how do they fit into the scheme of things lunar? Here opinions vary.

As is so often the case about features and processes on the Moon, various authorities tend to give the impression that things are very 'cut-and-dried' and they usually give very definite explanations. The trouble is, you then read what another authority says on the same subject, presented with equal certainty, and yet differing from the explanation of the first one!

Some experts say that lunar domes are just mountains. Others, probably the majority, say that domes are volcanic swellings of the crust caused by a magma build-up from below. Others compare them to earthly cinder-cones and say they are true volcanoes.

I tend to side with the opinion that they are true volcanoes, though I don't agree with the cinder-cone interpretation. Many of them have what appear to be calderas situated at their summits. In fact, most of the Hortensius domes have summit craters; I think too many for chance impacts to be responsible. I cannot help wondering about the outpouring of low-viscosity lavas that flooded the great lunar basins. In particular, I wonder about the sites of the eruptions. There is strong evidence that the major lava flows originated from long fissures near the peripheries of the basins. However could there have been other vents further in? It is certain that the lavas did not switch off suddenly. Evidence for successive lava flows abound on the lunar maria. The vents which ceased eruption early were undoubtedly 'levelled over' with mare lavas but what of those that were last to finish? Maybe the last vestiges of volcanism brought to the surface rather less-fluid lavas – viscous enough to build up some vertical structure around the caldera? Certainly the impact melts that would have pooled inside the basins might provide the source of higher-viscosity lavas.

If you will indulge me while I build on my speculations, perhaps the domes have some connection with the newly formed mare striving to achieve isostatic equilibrium. This is the theoretically expected process of slight sinking of the layers of mare basalts after solidification. The cause for this is the higher density of the basalt compared to that of the bedrock. There is some evidence that this isostatic levelling process actually occurred on the Moon, in the form of faulting and apparent tide-marks at the basin boundaries, etc. As the solidified 'seas' sank they inevitably squeezed down on the layers below. Perhaps the domes resulted as a little of the more viscous sub-mare lavas escaped through fissures?

I must emphasise that this idea is not the slightest bit 'official'. It is merely the result of my musings and may be completely wide of reality – as indeed may be many of the current 'official' ideas on the subject. If only one thing can be taken as certain, it is that everybody cannot be right!

A lot of questions, with no universally agreed answers. Are the domes really primarily up-lift features but with internal fissures creating lava vents that open out at the peak in many cases? Maybe they instead represent the last vestiges of the ancient lunar volcanism? I wonder. I expect the definitive answer will have to wait until lunarnaut geologists conduct seismic sounding experiments (setting off small percussive charges and monitoring the resultant shock waves) at dome sites in future decades.

The lunar domes are as interesting as they are a challenge to observe. There are many more examples elsewhere on the Moon, a number of them in the same general area as Hortensius. You will find domes very hard to see in all but the lowest angles of illumination. Also they are a test for good optics and good observing conditions. Image contrast is of prime importance. If you observe with a large-aperture reflector, perhaps one of the popular 'Dobsonian' telescopes, and can't detect domes when you think they should be visible, you might like to try masking the aperture with an off-axis hole. Despite the reduction in light-grasp you might find that the increase in contrast is enough to reveal the domes you could not see before. This technique is especially useful if the atmospheric conditions and/or the telescope optics are of indifferent quality. Happy dome hunting!

8.22 HUMORUM, MARE [CENTRED AT 24°S, 321°E] (WITH DOPPLEMAYER, GASSENDI, GASSENDI A, VITELLO)

Mare Humorum (Sea of Moisture – a rather inappropriate name, that!) might be one of the smaller of the lunar 'seas', its diameter averaging 400 km, but I think that it is certainly one of the most interesting. It is pictured in Figure 8.22(a), a photograph taken by the 1.5 m reflector of the Catalina Observatory on 1967 February 22^d 03^h 35^m UT, when the Sun's selenographic colongitude was 61°.

The mare basalts fill the very ancient basin. In fact the Humorum Basin is probably one of the earliest of them, at nearly 4.2 billion years old. This is evident by the total absence of any identifiable ejecta pattern, obliterated by the subsequent impacts and other events which have moulded the lunar surface.

By contrast, the surface of the mare appears to be one of the youngest of the great 'seas'. This is implied by the crater-record and is consistent with the results gained by the *Apollo* missions concerning the relative youthfulness of the westernmost of the lunar mare flood-plains. It is probable that the final significant eruptions of mare basalts that covered the Humorum Basin occurred not much more than 3 billion years ago. So, Mare Humorum is both one of the oldest and one of the youngest of the Moon's great 'seas', depending on whether we are talking about the basin or the lava flood-plain!

You might be wondering how the age of the mare can be gauged from the craters on it. To understand that, remember how the meteoritic bombardment of the Moon was severe early in its life and dwindled, both in the sizes of the impactors and the frequency of impacts, with time. From widespread studies of the numbers and sizes of craters on the Moon, taken together with certain well-established age benchmarks obtained by laboratory measurements of lunar samples brought to

Figure 8.22 (a) Mare Humorum. The largest crater (at the bottom of the photograph) is Gassendi. Details in text. (Courtesy Lunar and Planetary Laboratory.)

Earth, planetologists are in a position to gauge the age of a given surface on the Moon. Crudely put, if a given area of surface contains lots of large craters then it is old. If a greater fraction of the area is weighted to smaller craters then it is young. By making painstaking counts of craters and measurements of their sizes scientists can quite reliably determine the age of particular surfaces, such as the maria, on the Moon.

I find it difficult to believe that the Humorum Basin, along with the other western basins, remained completely 'dry' while the basins on the Moon's eastern hemisphere were busily filling with basaltic lavas; the site of the action only to switch to the Moon's western hemisphere when things had quietened down in the east. I think it much more

Figure 8.22 *(cont.)*
(b) A low Sun-angle over
the Mare Humorum
reveals low-relief
features such as the
wrinkle ridges on the
mare and the graben on
the junction between it
and the Palus
Epidemiarum (towards
the upper left). Details in
text. (Courtesy Lunar and
Planetary Laboratory.)

likely that the lava flooding started in the west at more or less the same
time as in the east. However, I think that it **continued for longer** in
the west.

If that is the correct scenario then certain consequences must follow.
For one thing, the basalt lavas of the western 'seas' must have a greater
tendency to overflow the basin rims (given the finite depth of the basins)
and spread over a much greater area of the Moon's surface. Look at any
photograph of the full Moon and you will see that is indeed the case. Not
only is a much greater area of the Moon's surface covered by dark mare
material but the edges of the lunar seas are much less well defined by
their parent basin rims.

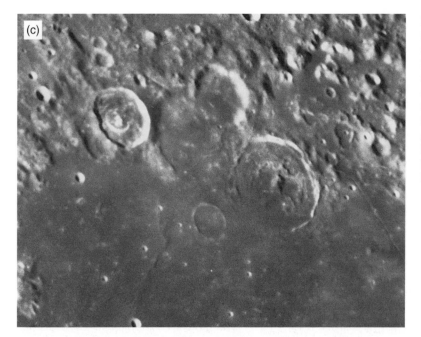

Figure 8.22 (*cont.*)
(c) Southern section of the 'shore' of the Mare Humorum. The well-formed crater on the left of the group is Vitello. To the lower right of Vitello is the largest crater of the group, Dopplemayer. Details in text. (Courtesy Lunar and Planetary Laboratory.)

The westernmost half of the Mare Imbrium spills into the Oceanus Procellarum, which itself merges with the Mare Cognitum and Mare Nubium. Even the Humorum Basin wall is breached to the north-east and here the Mare Humorum spills into the Mare Nubium, and again to the south-east where the flood-plain extends into the Palus Epidemiarum (the ill-defined mare area south of the Mare Nubium), which itself curves round the area outside the south-western periphery of the Mare Humorum.

I would conjecture that future lunarnaut surveyors will find that the western seas are composed of a number of sheet-like strata, and will find fewer strata and generally thinner coverings to the basins in the east of the Moon.

As far as an answer goes as to why the crust should be thinner to the west of the Earth-facing meridian, well, maybe it has to do with the 'Gargantuan Impact' originally proposed by Dr Peter Cadogan. On his theory the Oceanus Procellarum is really the lava in-fill of a 2400 km diameter basin that resulted from a colossal impact in Pre-Nectarian times. Here, though, I cannot claim to be on anything but thin ground, if you will excuse the pun!

As has already been said, there is a paucity of large craters actually on the Mare Humorum but there are some beautiful examples of craters situated around the periphery of the mare that post-date the creation of the basin and yet pre-date the lava flooding.

Again, this provides proof of a substantial interval between the two episodes. If you doubt this, then consider how these craters could survive the event that created the basin. They couldn't. The craters must have been formed after the basin. Then take a close look at the craters and you

Figure 8.22 (*cont.*)
(d) Vitello, drawn by
Andrew Johnson.

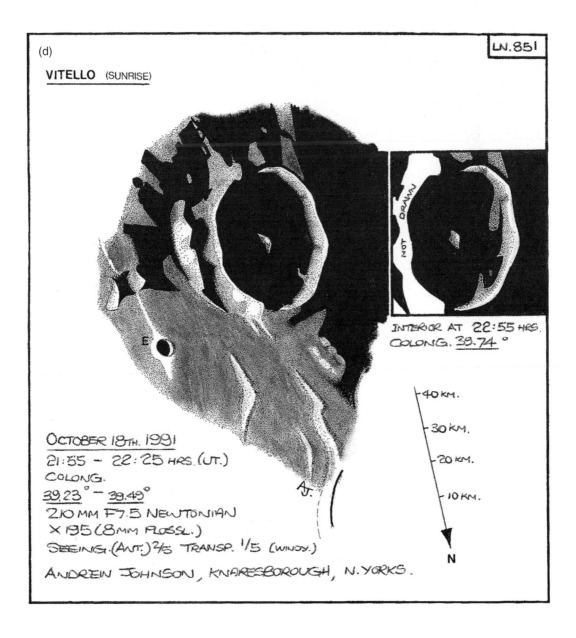

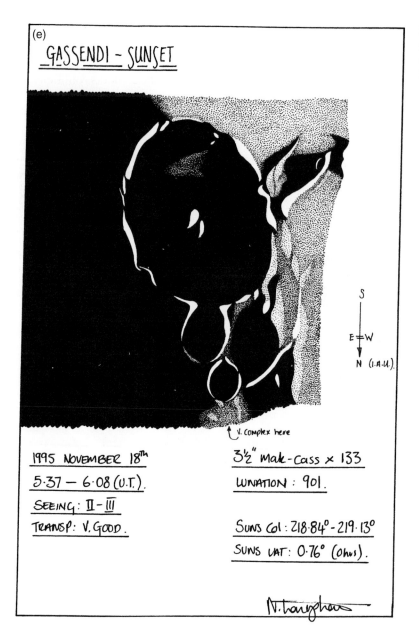

(e)

GASSENDI – SUNSET

↑ V. Complex here

1995 NOVEMBER 18ᵀᴴ
5·37 – 6·08 (U.T.).
SEEING: II – III
TRANSP: V. GOOD.

3½" Mak-Cass × 133
LUNATION : 901.

SUNS Col : 218·84° – 219·13°
SUNS LAT : 0·76° (Ohus).

N. Longshaw

Figure 8.22 (*cont.*)
(e) Sunset over Gassendi,
drawn by Nigel
Longshaw.

will see that in many cases the mare basalt has intruded into them,
typically breaching the walls that face in to the mare, and partially
flooding their interiors. Obviously the flooding came after the craters
were formed – and hence the interval between the basin creation and the
flooding with mare basalts.

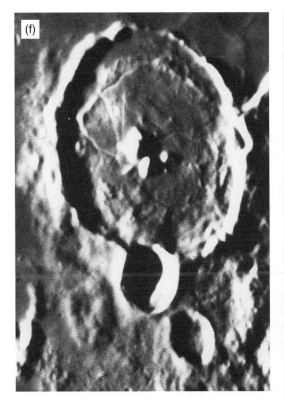 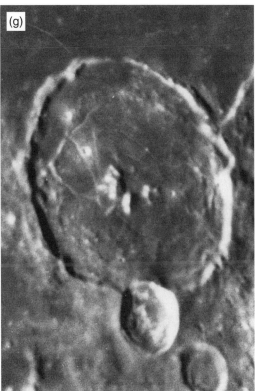

Figure 8.22 (*cont.*)
(f) Gassendi at
colongitude 48°.7. Details
in text. (Courtesy Lunar
and Planetary
Laboratory.) (g) Gassendi
at colongitude 61°.0.
Other details in text.
(Courtesy Lunar and
Planetary Laboratory.)

As always, a low Sun-angle reveals details of small vertical relief. Figure 8.22(b) is another photograph taken with the 1.5 m Catalina Observatory telescope, this time on 1966 December 23^d 04^h 54^m UT when the Sun's selenographic colongitude was 39°.9 and the terminator bisected the Mare Humorum. Several roughly concentric wrinkle ridges show up under this lighting. These are thought to be the result of compressional forces.

Further out, crossing the junctions of the Mare Humorum and Palus Epidemiarum and Mare Nubium, are several, also roughly concentric, graben. Each is of the order of 55 km wide and extends several hundreds of kilometres in length. These slumped features are thought to be the result of crustal stretching. Notice how they even carve their way through the mountainous regions and the oldest of the craters (but not the youngest examples).

The cluster of craters at the southern periphery of Mare Humorum are particularly beautiful. Figure 8.22(c) shows the group. It is an enlarged portion of Figure 8.22(a). The easternmost of the main craters, and the least degraded, is Vitello. Vitello is 45 km across and 1.7 km deep in its complex interior. It has a central hill and the interior shadows are far from regular when the Sun is low, as can be seen in the fine study by Andrew Johnson presented in Figure 8.22(d).

To the west of Vitello is the remnant of a large lava-flooded crater, with another partial ring, again completely flooded, abutting to its south-west. In effect, these form 'bays' in the Sea of Moisture. Just to the north-west of these and, at 64 km diameter, the largest of the formations pictured in Figure 8.22(c) is Dopplemayer. It has been flooded and eroded and yet its lofty central mountain soars about 760 m above the rippled floor of the crater.

It is Gassendi, though, which is the real jewel of the Mare Humorum. Spanning both 'shore' and 'sea' on the northern sector of the mare this 110 km diameter 'ring-plain' formation appears like a black lake at the times when the morning and the evening terminator just reaches it (see Figure 8.22(e)). With the Sun at a higher angle, the interior is seen to be highly complex. Figure 8.22(f) shows a photograph of it taken with

Figure 8.22 (cont.) (h) *Orbiter V* photograph of Gassendi. (Courtesy NASA and E. A. Whitaker.)

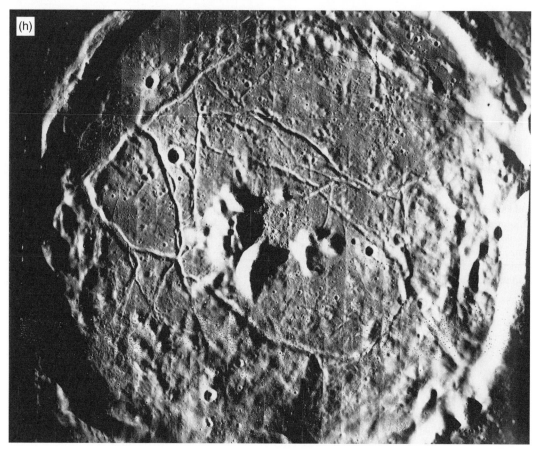

(h)

the 1.5 m Catalina Observatory telescope on 1966 April 2^d 08^h 12^m UT when the Sun's selenographic colongitude was 48°.7. Figure 8.22(g) shows it under a higher Sun (selenographic colongitude 61°.0). In fact, Figure 8.22(g) is an enlarged portion of part of Figure 8.22(a). You might notice how the relative prominence of the interior features has altered quite considerably over what is a relatively small change of lighting angle. This, in addition to the overall complexity of the formation, led to the selenographers of old varying quite considerably in the way they represented the crater. Inevitably, this led to debates about possible physical changes to the crater during the lifetimes of the observers – an idea now long abandoned, I hasten to add.

The walls vary in height around the crater, being highest on the west. An average height for them is about 1.8 km above the crater floor. The crater Gassendi A (33 km diameter, 3.6 km deep) intrudes into Gassendi to the north. Gassendi A is rather hexagonal in outline and has a complex interior. The southern section of the wall of Gassendi is highly eroded and even gives the impression of having been melted down by the lavas of Mare Humorum. The mare materials have clearly entered the crater here. The average hue of the floor of Gassendi is lighter than that of the Mare Humorum, except for the smooth crescent-shaped sector originating at the wall breach, which is of the same shade.

The crater floor is of the order of 600 m higher than the average level of the outer surrounds and is criss-crossed by a remarkable network of rilles. Many of these are visible in quite small telescopes (80–100 mm aperture) under suitable conditions of lighting. For a really detailed view of Gassendi take a look at the *Orbiter V* photograph shown in Figure 8.22(h).

As well as having a network of rilles over it, the floor is also rather rough and hummocky and there is an impressive central mountain complex. This is really the remnants of a central ring, as is evident in Figure 8.22(h). The tallest of the central peaks soars to above 1 km in height.

Gassendi is one of the Moon's 'hot spots' of Transient Lunar Phenomena, with some reliable reports of bright flashes and red glows seen in the crater. Significantly it turns out that it is also one of the sites of enhanced radon emission.

What story does the craggy face of Gassendi have to tell? Has the interior been pushed upwards by forces from below? What are the order and the time-scales of the events that have led to the Gassendi we see today?

8.23   HYGINUS, RIMA [CENTRED AT 8°N, 6°E] (WITH HYGINUS,
        RIMAE TRIESNECKER, TRIESNECKER)

The current thinking about rilles is that they are of two distinct types, each with a different mechanism of formation. *Sinuous rilles* are thought

Figure 8.23
Triesnecker and its
associated rilles (upper
right) and Rima Hyginus
(extending from middle
left to lower right)
photographed using the
1.5 m reflector of the
Catalina Observatory. At
the time of the exposure,
1966 May 27^d 03^h 56^m UT,
the Sun's selenographic
colongitude was 356°.8.
(Courtesy Lunar and
Planetary Laboratory.)

to have been formed by running lava. Possibly the lava cut across the surface of the mare (they are peculiar to maria), or maybe it cut a tunnel just under the surface – the roof of the tunnel collapsing at a later date. They are characteristically about 1–2 kilometres wide (but wider examples exist, such as Vallis Schröteri) and they tend to snake about in the same manner as rivers do here on Earth.

The other type are the *linear rilles*. These tend, as their name suggests, to be somewhat straighter. Any changes of direction are rather more angular than is the case for the sinuous rilles. They also tend to be wider, typically 5–60 km, and can cross mare and highland boundaries. They are evidently lineaments where the ground to either side has been pulled apart slightly, creating slumped channels.

We call any channel of this sort a graben. Rima Ariadaeus, discussed in Section 8.6, seems to be of this type, and the arcuate rilles on the junction between the Mare Humorum and Palus Epidemiarum (see Section 8.22) even more definitely so.

Look at Figure 8.23 and you will find examples of both types of rilles. This Catalina Observatory photograph is of the region of the Moon extending from the Sinus Medii (at upper right) to the Mare Vaporum (lower left). Near the upper right of the frame is the 26 km diameter crater Triesnecker. It is itself an interesting object, 2.7 km deep and with a central mountain complex. However, it is the system of rilles extending from its eastern flank that grab the attention. These are the Rimae Triesnecker. They seem to be of the sinuous variety.

Crossing much of the frame from the lower right to the middle left of Figure 8.23 is the very much more prominent Rima Hyginus. At the extreme left of the photograph you will see the 'tail end' of another rille running parallel to the eastern end of the Rima Hyginus. This is the Rima Ariadaeus already referred to, and discussed more fully in Section 8.6.

Notice how a small crater (it is 10 km across) is situated at the sharp bend in the rille. This is the crater Hyginus. It is deeper than the rille and extends to a depth of about 770 m. Look carefully at the section of the rille to the east (left in the photograph) of Hyginus and you will see further, though much smaller, craters threaded along it like pearls on a necklace. High-resolution images from space-probes reveal even more of the crater-chain appearance of Rima Hyginus. Other linear rilles show a similar structure.

It is not easy to understand exactly how these craters fit into the scheme of things. Are they also collapse features? Could at least some of the 'linear rilles' have been created in the same way as the sinuous rilles, except that the underground channels were much wider? Perhaps there was extensive lava flowing in thin sheets underneath the freshly solidified surfaces of the mare? Perhaps the roof falling in at widely spaced intervals creates crater chains? (Please note, I am here just referring to the small craters associated with rilles. I am not seeking to revive the spent arguments about the creation mechanisms of the Moon's major craters!) If the roof-falls occur closer together do they then form rilles of the Hyginus type? What about the crater Hyginus, itself, though? It is much deeper than the rille of which it seems to be such an important part.

If I seem to have posed a lot of questions to which you think that we ought to have definite answers by now, then I can only say that many experts differ in their interpretations of the rilles – and a few openly say that they are not at all sure of the mechanisms that have created these intriguing features.

The Ariadaeus, Hyginus and Triesnecker rille systems are visible with quite small-aperture telescopes near the times of first and last quarter Moon. I heartily commend you to seek them out and ponder on their significance for yourself.

8.24   IMBRIUM, MARE [CENTRED AT 35°N, 345°E] (WITH ARCHIMEDES,
        ARISTILLUS, AUTOLYCUS, BIANCHINI, HELICON, MONTES JURA,
        PROMONTORIUM HERACLIDES, PROMONTORIUM LAPLACE,
        SINUS IRIDUM, TIMOCHARIS)

The eastern edge of the Mare Imbrium (Sea of Rains) begins to come into sunlight about a day before first quarter Moon. At first quarter the area takes on a grand spectacle, as the photograph by Tony Pacey in Figure 8.24(a) shows. However, the nature of this important feature is probably best appreciated at a lunar age of about 10–11 days. Figure 8.24(b) is a photograph I took under just such lighting and conditions. I used my 18¼-inch (0.46 m) reflector to image the Moon directly onto HP5 film (processed in 'Celer Stellar' developer) in the camera mounted at the telescope's Newtonian focus (no eyepiece). Hence the effective focal ratio (EFR) is simply the focal ratio of the telescope (f/5.6).

If my choice of film seems odd to you, the reason is that I had wanted to use image projection to a high effective focal ratio. Hence the high-speed film I had loaded into my camera. As it was, the night turned out to be totally unsuitable, with poor transparency and very bad atmospheric turbulence, so I contented myself with a couple of low-resolution photographs of the Moon. The 1/500 second exposure was made at 1978 July 15^d 21^h 23^m UT, when the Sun's selenographic colongitude was 25°.3.

Figure 8.24(b) shows the outline of the circular Mare Imbrium very well, only the westernmost extremity of it lying over the dark side of the terminator. This basalt-filled basin is a whopping 1250 km in diameter. Only the Oceanus Procellarum has a greater area of mare-type basalt and the Imbrium Basin is the largest that is clearly identifiable as such on the Moon today. The impactor that created the basin also created ejecta and a pattern of ridges and linear faults radial to it that can be discerned (if you know what to look for) covering a large part of the Moon's visible face. Some of these have been featured in other sections in this chapter.

Several of the *Apollo* missions have obtained samples which included those identified as Imbrium impact ejecta (and *Apollo 15* actually landed on the periphery of the Mare Imbrium). It is from the laboratory analysis of these samples that a fairly precise date has been determined for the basin-creating impact. The results indicate that it happened 3.85 billion years ago, the uncertainty being plus or minus 0.05 billion years. This is one of the primary benchmarks that has been used to build up our picture of the sequence of events which sculpted the surface of the Moon that we see today. Actually, the Imbrium Basin is the youngest of the really large basins on the Moon's Earth-facing hemisphere. The Orientale Basin is the only large one that is younger, but very little of it shows on the Earth-facing hemisphere.

Figure 8.24 (a) Sunrise over the Mare Imbrium photographed by Tony Pacey on 1987 April 6^{d} $20^{\mathrm{h}}\,00^{\mathrm{m}}$ UT (approx.). He used his 10-inch (254 mm) reflector, with eyepiece projection onto *Ilford* FP4 film for this 0.25 second exposure.

The age of the surface covering of the Mare has been determined mainly by the method of counting the numbers of craters of specific sizes, keyed with the determined ages of the *Apollo 15* rock samples. That and spectrophotometric studies (reflectance at specific wavelengths) has shown that episodes of lava flooding occurred from 3.7 to 3.2 billion years ago. The lavas that form the major part of the visible surface of the Mare Imbrium are taken to be about 3.3 billion years old.

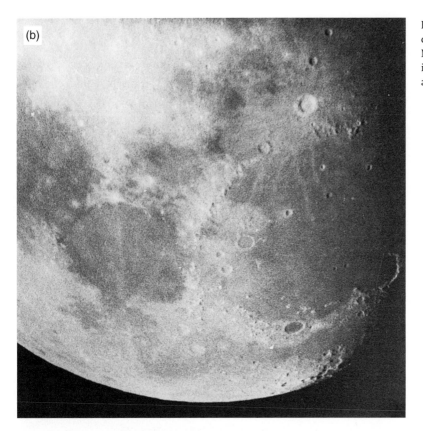

Figure 8.24 (*cont.*) (b) The circular outline of the Mare Imbrium is evident in this photograph by the author. Details in text.

Some of these lava flows are visible as areas of slightly differing hue, even to the observer using simple eyeball-to-eyepiece methods. You might like to try looking for these yourself.

As always with lunar maria, a low Sun-angle shows up many wrinkle ridges and a peppering of small craters also covers the mare, though most are beyond the powers of amateur-sized telescopes and normal backyard observing conditions. See what examples you can find.

Figure 8.24(c) shows the south-eastern sector of the mare, while Figure 8.24(d) shows its north-eastern sector. Both are photographs taken with the 1.5 m reflector of the Catalina Observatory in Arizona. Figure 8.24(c) was taken on 1967 January $20^d\,01^h\,46^m$ UT, when the Sun's selenographic colongitude was $18°.5$ (morning illumination) and Figure 8.24(d) was taken on 1966 September $6^d\,10^h\,44^m$ UT, when the colongitude was $167°.3$ (late afternoon).

The spectacular Montes Apenninus defines the south-eastern boundary of the mare (see Figure 8.24(c)). These are chiefly blocks of lunar crust which have been violently uplifted by the Imbrium impact event. This mountain range is discussed in more detail in Section 8.5, earlier in

Figure 8.24 (*cont.*)
(c) South-eastern Mare Imbrium, bordered on the left by the Montes Apenninus. The largest crater (at the bottom of the photograph) is Archimedes. The crater to its left is Autolycus and that to its right is Timocharis. Further details in text. (Courtesy Lunar and Planetary Laboratory.)

this chapter, together with the crater Eratosthenes, which marks the northernmost point of the Apennine range. To the west of Eratosthenes lies a further, smaller, range of mountains, the Montes Carpatus, with the major crater Copernicus just to their south (see Section 8.13).

The major part of the eastern boundary of the old basin is marked by the Montes Caucasus, another uplift-created mountain range. Between the Montes Apenninus and the Montes Caucasus there is a breach in the mountains where mare lavas have flowed to connect the Mare Imbrium with the adjacent Mare Serenitatis. Figure 8.24(a) shows this particularly well.

Along the north of the Mare Imbrium lies another mountain range, the Montes Alpes (discussed along with the Vallis Alpes, in Section 8.3).

You will notice that the innermost boundary of this mountain range is also concentric with the mare but it lies along an arc of smaller radius than the Montes Carpatus, Apenninus and Caucasus. Actually the Imbrium Basin was a 'multi-ring' basin. The outermost ring is defined by the mountain ranges previously mentioned while the Montes Alpes is the surviving part of the middle one of three original rings of uplifted lunar crust. All that survives of the innermost ring is a few isolated mountains that poke up through the lava flood-plain. I will leave you to obtain a map of the mare (I recommend using a plan-view) and identify, and plot, the vestiges of the rings for yourself.

There is almost no sign of any remaining ring-relics along the western-most boundary of the Mare Imbrium. Here it simply merges with the vast expanse of the Oceanus Procellarum. Examine the area at full Moon and

Figure 8.24 (*cont.*)
(d) North-eastern Mare Imbrium, bordered on the left by the Montes Caucasus, and on the lower left by the Montes Alpes. Further details in text. (Courtesy Lunar and Planetary Laboratory.)

Figure 8.24 *(cont.)*
(e) Archimedes (the
largest crater), Autolycus
(upper-left crater), and
Aristillus (lower-left
crater). Enlarged portion
of Figure 8.24(d).
(Courtesy Lunar and
Planetary Laboratory).

you will see clear indications that the darker Imbrium lavas have over-flowed into the slightly lighter Procellarum lavas, rather than the other way round.

In the bay at the southern extremity of the Montes Alpes and the north-ernmost section of the Montes Caucasus is situated the interesting crater Cassini. This crater and its environs are discussed in detail in Section 8.11.

The prominence of the largest crater on the mare, the 83 km diameter Archimedes, is such as to be a 'focal point' of the mare even though it is positioned considerably off-centre. Figure 8.24(e) (which is an enlarged part of Figure 8.24(d)) shows it very well. Clearly the impact that created the crater happened after the Imbrium Basin was formed (it could not have survived the Basin-forming event) but before the mare flooding episode (the mare lavas remain undisturbed right up to where the crater emerges from the mare). It also indicates the shallowness of the mare flood-plain. Most of this form of crater exists as a depression below the surface of the surrounds (there are plenty of examples all over the Moon to bolster this assertion). Yet the rim still rises up to nearly 1.9 km above the flood-plain in places. Notice also the hummocky tract of ground extending southwards towards the Montes Apenninus and how the mare lavas have interacted with it – a further indication of the shallow-ness of the lunar 'sea'. Even at the centre of the Imbrium Basin, the depth of mare basalt is probably no greater than 1.5 km.

Figure 8.24 (*cont.*)
(f) Sinus Iridum, bordered by the Montes Jura. Details in text. (Courtesy Lunar and Planetary Laboratory.)

The fact that Archimedes has been filled in with mare-type lava almost to the level of the Mare Imbrium surrounds (actually, about 200 m below) is interesting. I estimate that the lava-filling inside the crater must extend to around 2 km depth, deeper than the extreme depths of almost all of the Moon's 'seas'!

Though the crater superficially has a bland appearance, there are some delicate features to discern in Archimedes. The easiest of these is the spire-like shadows cast by the crater rim across the floor when the Sun-angle is low. These are instructive as they are a magnified (though distorted) profile of the rim itself. The next easiest to discern is a faint pattern of east–west bands covering the floor of the crater. They are most obvious under a high Sun. What do you think caused the pattern? Hardest to see are some very tiny craters within Archimedes. They are best under a low Sun but also need a good telescope and fine seeing conditions. The ones I have seen most often are a pair close to the north-west rim. I find I can discern some others but only rarely. How many can you see?

Figure 8.24 *(cont.)* (g) 'The Golden Handle' effect. South is to the right in this drawing by Roy Bridge.

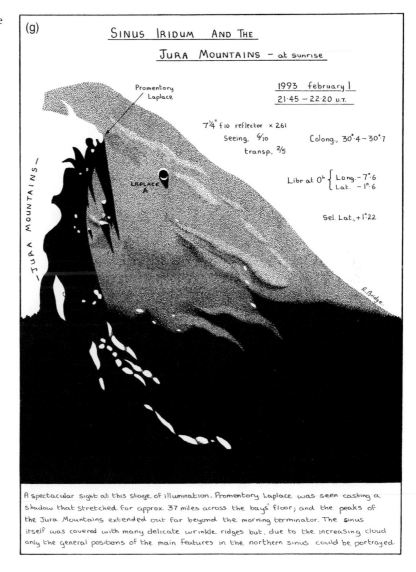

(g)

SINUS IRIDUM AND THE

JURA MOUNTAINS – at sunrise

Promentory / Laplace

1993 february 1
21·45 – 22·20 u.t.

7¼" f10 reflector × 261
Seeing, 6/10
transp, 2/5

Colong., 30°·4 – 30°·7

JURA MOUNTAINS

LAPLACE A

Libr at 0ʰ { Long. – 7°·6
Lat. – 1°·6

Sel. Lat., + 1°·22

R. Bridge

A spectacular sight at this stage of illumination. Promentory Laplace was seen casting a shadow that stretched for approx. 37 miles across the bays' floor; and the peaks of the Jura Mountains extended out far beyond the morning terminator. The sinus itself was covered with many delicate wrinkle ridges but, due to the increasing cloud only the general positions of the main features in the northern sinus could be portrayed.

The next major crater to the west of Archimedes (to the right of Archimedes on Figure 8.24(c), both craters being near the bottom of the photograph), is the 35 km diameter Timocharis. It is 3.1 km deep, with terraced walls rising to a sharply defined rim. It has a faded ejecta pattern which becomes most prominent near full Moon. At these times the crater also takes on a rather 'mist-filled' appearance, the same mirage that afflicts Eratosthenes.

The first major crater to the east of Archimedes is Autolycus. This is 39 km in diameter and 3.4 km deep. It has a ray system which is

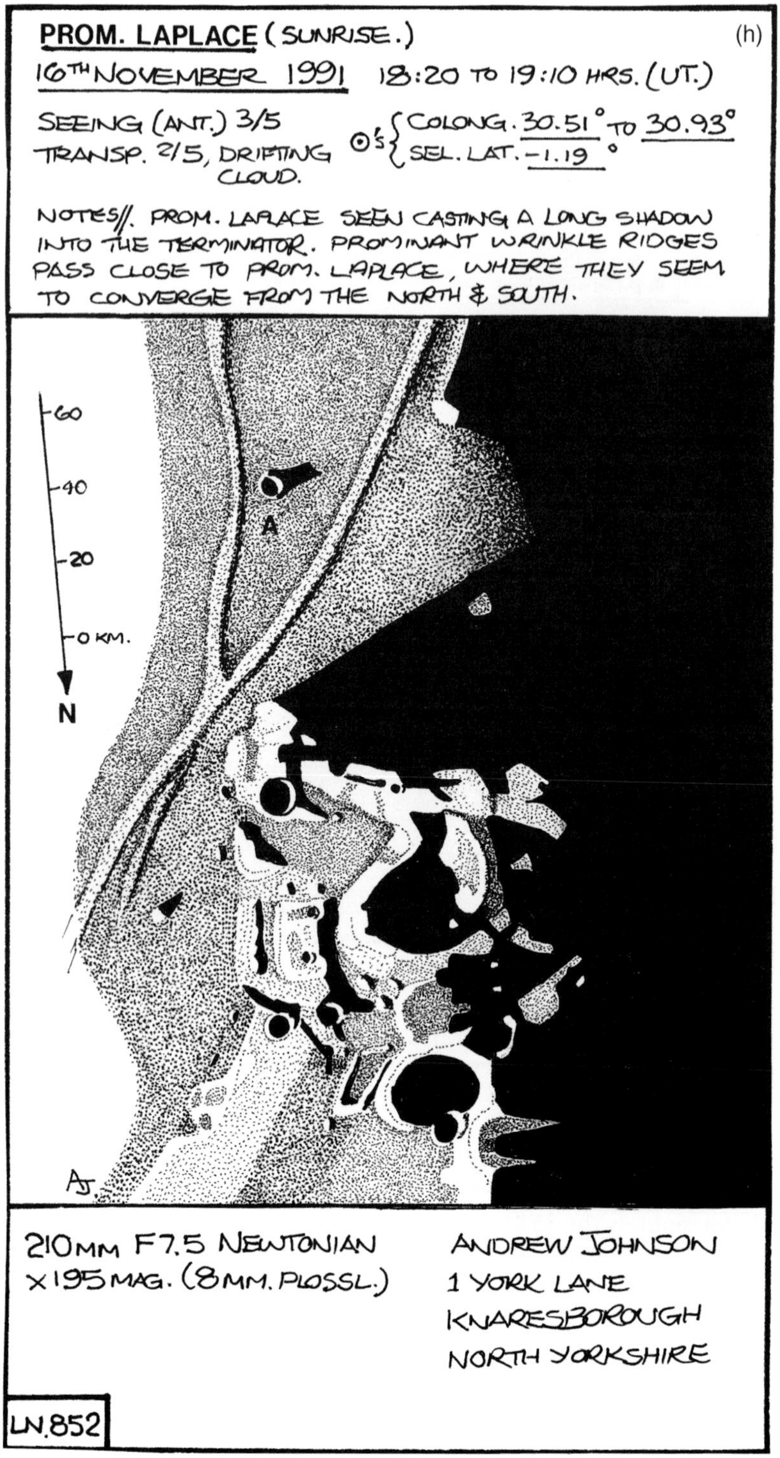
PROM. LAPLACE (SUNRISE.)
(h)
16TH NOVEMBER 1991 18:20 TO 19:10 HRS. (UT.)
SEEING (ANT.) 3/5
TRANSP. 2/5, DRIFTING CLOUD.
⊙'s { COLONG. 30.51° TO 30.93°
SEL. LAT. -1.19 °
NOTES//. PROM. LAPLACE SEEN CASTING A LONG SHADOW INTO THE TERMINATOR. PROMINANT WRINKLE RIDGES PASS CLOSE TO PROM. LAPLACE, WHERE THEY SEEM TO CONVERGE FROM THE NORTH & SOUTH.
60
40
20
0 KM.
N
A
AJ.
210MM F7.5 NEWTONIAN
x195 MAG. (8MM. PLOSSL.)
ANDREW JOHNSON
1 YORK LANE
KNARESBOROUGH
NORTH YORKSHIRE
LN.852

Figure 8.24 (*cont.*) (h) Promontorium Laplace, drawn by Andrew Johnson.

fainter than that of Timocharis, with a gentler interior slope and wider terraces. To the north of Autolycus is the larger, 55 km diameter, Aristillus. It is 3.6 km deep and also has significant interior terracing. Its ray pattern is the most prominent of the three craters. The prominent triple mountain peaks rise to almost 1 km above the crater floor. All three of these craters have raised outer slopes and ejecta patterns splashed across the mare, so they were obviously formed **after** the mare-flooding episode.

At the western end of the Montes Alpes there is a bay, the Sinus Iridum (Bay of Rainbows), the bordering mountain range being named Montes Jura (the Jura Mountains). This bay merges with the Mare Imbrium at the junction between it and the Oceanus Procellarum. Obviously the Jura Mountains are the surviving remnants of the uplifted crust defining another impact basin, of about 250 km diameter, which overlapped the Imbrium Basin. Which, though, came first? The answer is not at all obvious but planetologists consider the wide tract of light-coloured, hummocky, terrain along the northern 'shore' of the Mare Imbrium and extending to the Jura Mountains and a little beyond to have been created contemporaneously with the Imbrium impact. If this is correct then it follows that the Iridum Basin was formed after the Imbrium Basin (can you see why?).

Figure 8.24(f) shows the Sinus Iridum in full morning sunlight. It is another Catalina Observatory photograph, this one taken on 1967 January 22^d 03^h 31^m UT, with the Sun's selenographic colongitude 43°.9. The cape at the eastern end of the Montes Jura has been named Promontorium Laplace (Cape Laplace). That at the western end is Promontorium Heraclides (Cape Heraclides). The large (39 km diameter) crater in the hinterland mid-way between the capes is called Bianchini. The crater at the top left of Figure 8.24(f) is Helicon. It has an unusual structure, 25 km in diameter with a single interior terrace in the 1.9 km high walls above the flooded crater floor. Helicon actually resides beyond the boundary of the Sinus Iridum, on the Mare Imbrium. Many wrinkle ridges cross the floor of the Sinus Iridum. Some of these are roughly concentric to the Iridum Basin, others to the Imbrium Basin. The junction between the two is the most wrinkled part of all.

At the times when the terminator bisects the Sinus Iridum, part of the Montes Jura are in full sunlight, and the rest of the range pokes up into the sunshine, producing an effect called 'The Golden Handle'. I captured this appearance in the photograph presented as Figure 8.24(b). A more detailed view is provided by Roy Bridge in his drawing, which is shown in Figure 8.24(g). Figure 8.24(h) shows a more detailed view of the Promontorium Laplace at local dawn, as drawn by Andrew Johnson.

Lastly, I should mention the 'Great Black Lake' crater Plato which is situated at the north point of the Mare Imbrium. It is important enough to demand a section all of its own. You will find details of it and some of the nearby features on the Mare Imbrium, such as Mons Pico and Piton, in Section 8.33 further on in this chapter.

8.25   JANSSEN [45°S, 42°E] (WITH FABRICIUS, METIUS, RHEITA, VALLIS RHEITA)

The crater-ridden south-eastern highlands of the Moon can be a highly confusing place to find one's way around. Figure 8.25(a) is a wide-field view of this portion of the lunar disk, taken with the 1.5 m reflector of the Catalina Observatory in Arizona. The exposure was made at 1966 June 24^d 03^h 39^m UT, the Sun's selenographic colongitude being 339°.8 at the time.

Look to the left of centre on Figure 8.25(a) and you should be able to discern the outline of one of the most important formations in this area of the Moon. This is the crater Janssen (not to be confused with Jansen – a small crater in the Mare Tranquillitatis!). Figure 8.25(b) is an enlarged portion of the same photograph, in which Janssen nearly fills the frame.

Janssen's hexagonal outline is a mighty 190 km in diameter. It is clearly an extremely old structure, having been extensively modified and covered in craters large and small. The largest of the intruding craters, at 78 km diameter, is Fabricius. This is sited in Janssen's north-east sector. Notice the partial inner ring, concentric with the terraced walls of Fabricius. Also of note is the rille (judging by the look of it, a graben) that curves from the south-south-west rim of Fabricius right across the floor of Janssen to its far wall. The whole interior of Janssen has the decidedly 'tortured' appearance of a lunar formation of the greatest antiquity. You will find endless hours of interest in the study of this one feature alone.

Adjoining Fabricius to its north-east is the somewhat larger (88 km diameter) crater Metius. Though part of it is seen in Figure 8.25(b), it is seen fully in Figure 8.25(c), another enlarged portion of the same Catalina Observatory photograph.

Metius has a much more 'smoothed down' appearance than Fabricius, especially in its floor. Undoubtedly this has much do with the shaking Metius received during the impacts that created the craters around it, especially Fabricius which clearly post-dates it. Figure 8.25(a) shows that many of the old impact-craters share the same eroded characteristics in this region of the Moon. Going further north-east there is another old crater, Rheita. This is 70 km in diameter and its general form seems to be something of a cross between that of Fabricius and that of Metius (but more like that of Metius).

(a)

Figure 8.25 (a) The southeastern highlands of the Moon. Note the hexagonal form of the giant crater Janssen, just to the left of the centre of the photograph, and the gorge-like Vallis Rheita, close to the left. Details in text. (Courtesy Lunar and Planetary Laboratory.)

However, the main interest here is not Rheita but rather the enormous gorge-like valley its western rim intrudes into. This remarkable formation dominates Figure 8.25(c).

Depending on where you define the beginning and end of the valley to be, it is about 180 km long and is about 25 km wide along much of its length. Best seen two or three days after full Moon, it resembles a line of craters, all overlapping and with the walls between them broken down.

Both camps of crater creationists (impact and volcanism) claimed this feature as definite proof of their ideas. To the 'volcanists' the valley

(b)

Figure 8.25 (*cont.*) (b) Close-up of Janssen. This is an enlarged portion of (a). The crater Fabricius is in the lower-left part of Janssen. (Courtesy Lunar and Planetary Laboratory.)

Figure 8.25 (*cont.*)
(c) Close-up of Vallis
Rheita. The crater Rheita
intrudes into the lower
end of the valley. Upper
right of Rheita, and on
the other side of the
valley, the largest crater
is Metius. Adjoining
Metius, and to the upper
right of it, is the crater
Fabricius. Part of Janssen
is also shown in this
enlarged portion of (a),
which also overlaps with
the view shown in (b).
(Courtesy Lunar and
Planetary Laboratory.)

represented a chain of calderas aligned over a massive fault. Meanwhile the 'impactists' thought that the Vallis Rheita, and a few other examples of the same type of formation, were created by blocks of material ejected from the mare basins when they were formed.

It now seems pretty certain that the 'impactists' have got it right. The closest mare is Mare Nectaris and there are a number of less prominent valleys in the area which are radial to it. You might like to examine Figure 8.25(a), or better still use your telescope to observe the region. How many valleys can you find?

The commentators I have read who make reference to the creation of Vallis Rheita all say that the ejecta from the Nectaris Basin event is responsible. However, Vallis Rheita seems to lie at a slightly different angle to the valleys I can identify as being radial to the Mare Nectaris. I, at least, think that the impact event that created the Mare Imbrium is the culprit for this particular 'scar' on the lunar surface. Certainly the alignment seems to better fit the centre of the Mare Imbrium, rather than the centre of the Mare Nectaris.

I think that the Moon-shaking Imbrium impact of 3.85 billion years ago caused some huge blocks of material to be blasted from the site into ballistic trajectories. If I am right, one meteor-like collection of these splattered across the Moon, like a blob of paint from a flicked paintbrush, to create the valley. Hence the valley is really a secondary impact feature!

8.26 LANGRENUS [9°S, 61°E] (WITH ANSGARIUS, HOLDEN, KAPTEYN, KÄSTNER, LAMÉ, LOHSE, LANGRENUS A, LA PÉROUSE, VENDELINUS)

The splendid crater Langrenus is situated on the eastern shore of the Mare Fecunditatis. It is large and looks imposing when the terminator is nearby. It also has a bright interior and so readily stands out under a high Sun. In fact, Langrenus is striking whenever it is sunlit (see Figure 8.26(a)).

It is a prominent member of 'the Great Eastern Chain' of craters, the most southerly member of which is Furnerius. Going northwards from Furnerius, first comes Petavius (Furnerius and Petavius are detailed in Section 8.17), then Vendelinus (see Figure 8.26(b), where it is pictured with Petavius) and then Langrenus (see Figure 8.26(c), where Vendelinus and Langrenus are pictured together). The more northerly members of the 'chain' are the Mare Crisium (see Section 8.14) and Endymion (see Section 8.15).

The details for Figures 8.26(a) and (b) are given in the accompanying captions. Figure 8.26(c) is a Catalina Observatory photograph, taken with the 1.5 m reflector, on 1966 May 6^d 08^h 12^m UT. At the time of the exposure the selenographic colongitude was 103°.4.

Figure 8.26 (a) General area of Langrenus (south is diagonally towards the upper right in this view). Photograph taken by the author, using his 0.46 m reflector, on 1977 August 29^d 22^h 33^m UT, when the Sun's selenographic colongitude was 85°.3. Eyepiece projection was used (EFR = f/17) for the 1/30 second exposure on FP4 film, developed in *Microphen*. (b) Vendelinus (lower formation) photographed, with Petavius, by Tony Pacey on 1992 January 21^d 23^h 55^m UT, when the Sun's selenographic colongitude was 103°.6. Tony used eyepiece projection on his 10-inch (254 mm) Newtonian reflector to produce an EFR of about f/50. A ½ second exposure was given on FP4 film.

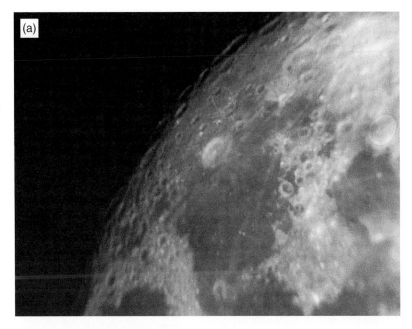

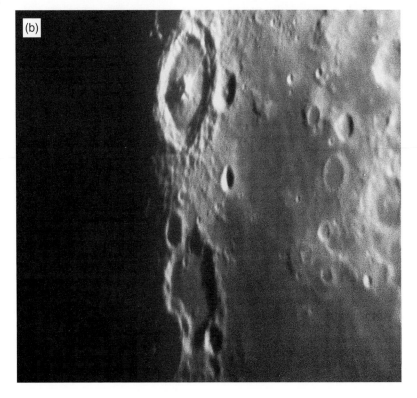

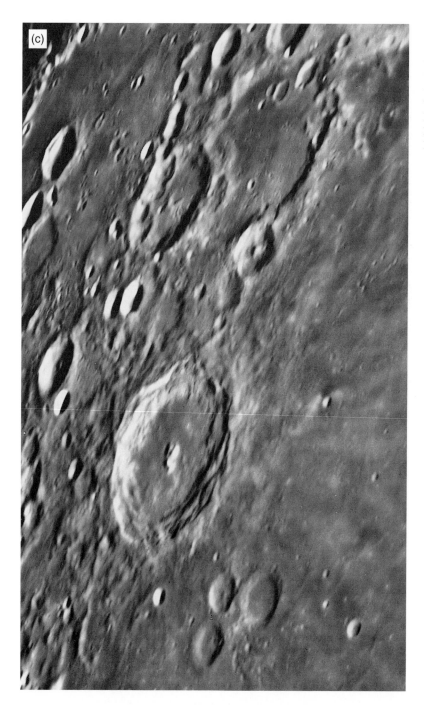

Figure 8.26 (*cont.*)
(c) Langrenus (lower crater) and Vendelinus (the large, less well-defined formation in the upper part of the frame). Details in text. (Courtesy Lunar and Planetary Laboratory.)

Figure 8.26 (*cont.*)
(d) Area extending
eastwards from
Langrenus. Going
increasingly left from the
top of Langrenus are the
major craters: Langrenus
A, Kapteyn and La
Pérouse. To the upper left
of La Pérouse is the larger
crater Ansgarius. To the
lower left of La Pérouse is
the old 'walled-plain'
Kästner. Further details
in text. (Courtesy Lunar
and Planetary
Laboratory.)

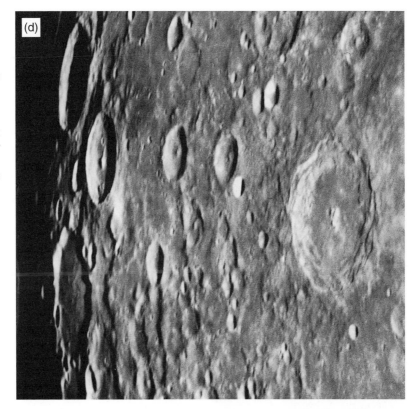

Langrenus's roughly terraced walls rise to about 2.7 km above the hilly floor. Rim-to-rim it spans a diameter of 132 km. The walls may look much higher than 2.7 km, in relation to the crater's diameter, but this is an illusion caused by a combination of our foreshortened view of the crater and the very gentle slope of the inner terraces. Owing to the interior slope, the main floor arena of the crater spans about 87 km, much smaller than its rim-to-rim diameter. The prominent and somewhat oddly shaped central mountain rises up to about 1 km above the crater floor.

There is much fine detail in the interior of Langrenus, and in the complex outer ramparts, to intrigue the observer equipped with even a quite small telescope. This is especially so as the appearance of the crater changes quite dramatically as the Sun rises over it. Figure 8.26(d) shows a view of the crater at a colongitude of only 6°.1 different from that in Figure 8.26(c), and yet the visual differences are obvious.

At high Sun angles I find that the interior of Langrenus takes on a distinctly yellowish-brown tint, compared to its surrounds. To see the colour clearly I use a magnification of no more than ×144 on my

18¼-inch (0.46 m) reflector. Can you see any colour tint inside Langrenus through your telescope?

Vendelinus is a very different type of formation to Langrenus, as is clearly shown in Figure 8.26(c). In fact, you may find it quite difficult to make out Vendelinus from the general confusion of lumps, bumps and smaller craters. This feature really only looks distinctive under a low Sun.

Figure 8.26 (*cont.*) (e) Kästner, drawn by Roy Bridge.

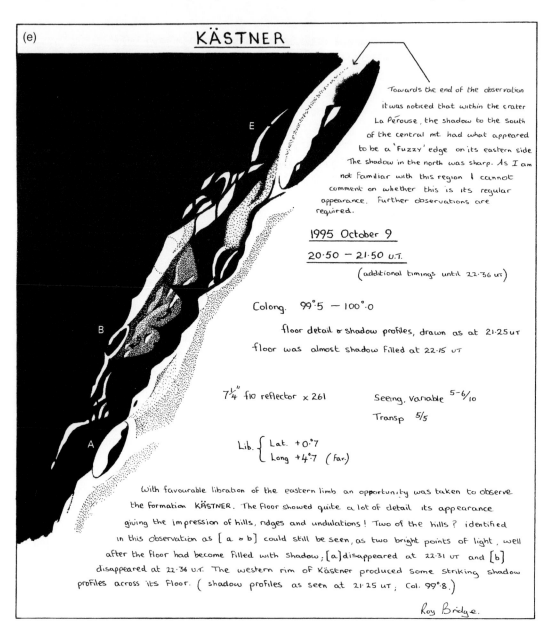

(e) KÄSTNER

Towards the end of the observation it was noticed that within the crater La Pérouse, the shadow to the south of the central mt. had what appeared to be a 'fuzzy' edge on its eastern side. The shadow in the north was sharp. As I am not familiar with this region I cannot comment on whether this is its regular appearance. Further observations are required.

1995 October 9

20·50 − 21·50 U.T.

(additional timings until 22·36 UT)

Colong. 99°·5 − 100°·0

floor detail & shadow profiles, drawn as at 21·25 UT
floor was almost shadow filled at 22·15 UT

7¼" f10 reflector × 261

Seeing. variable 5−6/10
Transp 5/5

Lib. { Lat. +0°·7
Long +4°·7 (far.)

With favourable libration of the eastern limb an opportunity was taken to observe the formation KÄSTNER. The floor showed quite a lot of detail its appearance giving the impression of hills, ridges and undulations! Two of the hills? identified in this observation as [a & b] could still be seen, as two bright points of light, well after the floor had become filled with shadow; [a] disappeared at 22·31 UT and [b] disappeared at 22·36 U.T. The western rim of Kästner produced some striking shadow profiles across its floor. (shadow profiles as seen at 21·25 UT; Col. 99°·8.)

Roy Bridge.

First take a look at the small-scale image of it in Figure 8.26(b) and you should then be able to trace its outline in the much more detailed view presented in Figure 8.26(c). This 147 km diameter 'walled-plain' is clearly very old. Indeed, one could argue that it has just begun to lose its identity as a crater in its own right. Even if you think that is an exaggeration, you must surely agree that Vendelinus is a crater in an advanced state of ruin.

The largest crater to intrude into it (on its north-east portion) is the 84 km diameter Lamé. This creates the large 'notch' in the outline of Vendelinus which is so very apparent in Figure 8.26(b), Lamé here being filled with shadow. The other two sizeable craters which encroach onto the rim of Vendelinus are Holden to the south-east and Lohse to the north-west. These have diameters of 47 and 42 km, respectively. As is the case with Langrenus, there is much fine detail within Vendelinus for the enthusiast to study. The different characters of these lunar neighbours are as instructive as they are interesting.

Figure 8.26(d) is already referred to as providing another view of Langrenus. It is another Catalina Observatory photograph, this time taken on 1966 April 6^d 08^h 00^m UT when the Sun's selenographic colongitude was 97°.3. It shows the area eastwards of Langrenus. Take a line from the top of Langrenus, on the photograph, and extend it to the left and it will cut through three sizeable craters. The first is Langrenus A (diameter 42 km). Then comes Kapteyn (diameter 49 km). The last is La Pérouse (diameter 78 km). Notice the evolution of the forms of these craters going from the smallest. To the south-east (upper left in the photograph) is the even larger (94 km diameter) Ansgarius. To the north-east of La Pérouse is another ancient 'walled-plain' type of crater, the 119 km diameter Kästner. Roy Bridge has made a fine drawing of Kästner, along with La Pérouse, and this is presented in Figure 8.26(e).

This has been something of a 'whistle-stop tour' of the complex environs of the grand crater Langrenus. In truth, one could spend a lifetime making a study of any chosen small part of it!

8.27   MAESTLIN R [4°S, 319°E] (WITH MAESTLIN)

Just 120 km south-south-west of Kepler, Maestlin R is normally camouflaged by Kepler's rays. However, at low angles of illumination Kepler's rays vanish and features of small vertical relief become prominent. Maestlin R is an arc of isolated peaks, the only remains of an ancient crater that was all but obliterated about 3.2–3.8 billion years ago when the lavas that formed the vast expanse of the Oceanus Procellarum flowed across the lunar surface.

The remnants of the crater rim span about 60 km. Figure 8.27(a) shows an excellent drawing of it made by Roy Bridge. The small crater

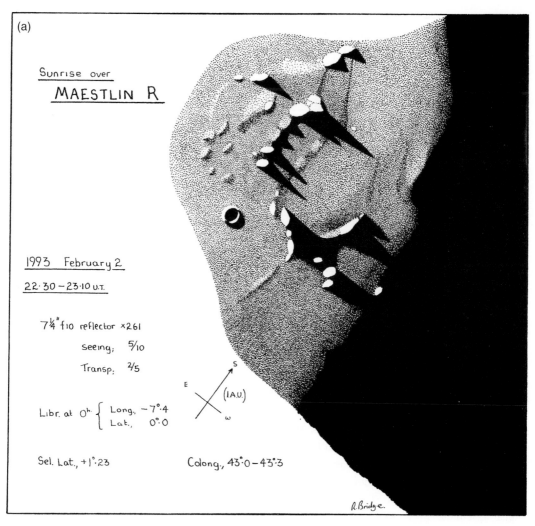

(a)

Sunrise over
MAESTLIN R

1993 February 2

22·30 – 23·10 U.T.

7¼" f10 reflector ×261

seeing; 5/10

Transp; 2/5

Libr. at 0ʰ { Long., −7°·4
Lat., 0°·0

Sel. Lat., +1°·23

Colong., 43°·0 – 43°·3

S
E
(I.A.U.)
W

R. Bridge.

shown on the drawing is Maestlin. It is rather dish-shaped, being 7.1 km in diameter and 1.6 km deep.

The main reason I selected this formation as one to feature is not so much the formation itself but rather that Roy Bridge has made a superb sequence drawing of sunrise over it. This is shown in Figure 8.27(b). Sequence drawings like these are highly instructive. Unless you happen to have a space-probe fitted with high-resolution radar-ranging equipment that you can put into lunar orbit, this technique is the most sensitive available to you for detecting small variations of surface height. Appearances change rapidly at the terminator, as Figure 8.27(b) shows very well. Combined with drawings made on other dates, sunrise and sunset sequences can be used to generate extremely detailed profiles of lunar

Figure 8.27 (a) Maestlin and Maestlin R, drawn by Roy Bridge. Note the north–south orientation of this drawing.

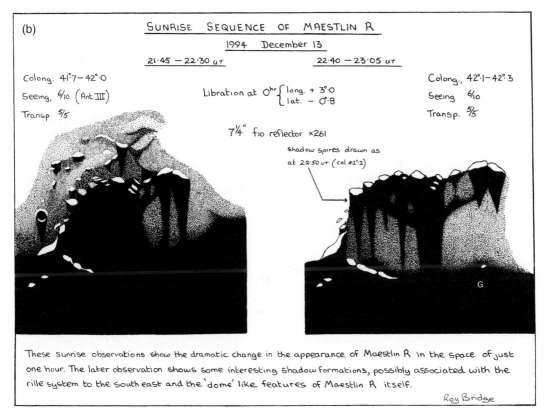

(b)

SUNRISE SEQUENCE OF MAESTLIN R

1994 December 13

21.45 — 22.30 UT   22.40 — 23.05 UT

Colong. 41°.7 – 42°.0   Colong., 42°.1 – 42°.3

Seeing, 6/10 (Ant III)   Seeing 6/10

Transp 5/5   Transp. 5/5

Libration at 0^{hr} { long. + 3°.0
                      { lat. – 0°.8

7¼" f/10 reflector ×261

shadow spires drawn as
at 22.50 UT (col. 42°.2)

G

These sunrise observations show the dramatic change in the appearance of Maestlin R in the space of just one hour. The later observation shows some interesting shadow formations, possibly associated with the rille system to the south east and the 'dome' like features of Maestlin R itself.

Roy Bridge

Figure 8.27 (*cont.*)
(b) Sunrise sequence of Maestlin R, drawn by Roy Bridge. The orientation of this drawing can be ascertained by comparing it to the drawing shown in (a).

surface topography. For instance, the drawings in Figures 8.27(a) and (b) can be used in this way. I will not pretend that your results will be cutting-edge science and that the professional planetologists will be waiting with bated breath for your publications. However, as with all your topographic studies of the Moon, **you** will get to know parts of the Moon in intimate detail.

8.28   MESSIER [2°S, 48°E] (WITH MESSIER A)

Not far from the crater Langrenus, just a little further north and quite close to the western shore of the Mare Fecunditatis, is situated one of the Moon's real oddities: the pair of craters Messier and Messier A. You might just be able to make them out on Figure 8.26(a), back in Section 8.26 dealing with Langrenus. Figure 8.28(a), presented here, shows them in much greater detail. This photograph was taken with the 1.5 m reflector of the Catalina Observatory on 1966 April 6^{d} 08^{h} 00^{m} UT, when the Sun's selenographic colongitude was 97°.3.

The eastern (left in the photograph) crater of the pair is Messier. It is rather oval, being elongated in the east–west direction, with dimensions of about 9 km × 11 km.

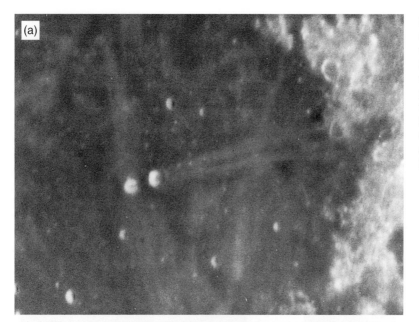

Figure 8.28 (a) Messier and Messier A. The western shore of the Mare Fecunditatis is visible on the right of the photograph. Details in text. (Courtesy Lunar and Planetary Laboratory.) (b) *Orbiter V* view of Messier and Messier A. (Courtesy NASA and E. A. Whitaker.)

Messier A, which used to be called Pickering (I think that it is a pity the name was changed), is rather oddly shaped. The eastern (Messier-facing) side of it is flattened, while the western rim is much more pointed in form. In fact, the crater rim forms the same profile as a hen's egg! It is about 13 km across at it's widest part and is of similar east–west length to Messier. Both craters have highly reflective interiors with dark streaks extending westwards from the central regions up their west walls. There seems to be a thin ridge of raised ground extending from the west rim of Messier to the east rim of Messier A. Figure 8.28(b), an *Orbiter V* photograph, shows the craters in more detail.

Even more weird are the two rays which extend, comet-like, westwards from Messier A. The rays are very prominent under a high Sun. A 60 mm refractor will easily show them at these times. Remarkably the rays stay quite prominent even at very low Sun angles, unlike most other lunar ray systems.

Many of the observers of yesteryear suspected both long- and short-term changes in Messier and its companion. Certainly they can change in relative prominence during the lunation but all these changes are purely optical effects. However, there is one real mystery which surrounds this pair – how were they formed? On this there is no universal consensus. It seems certain that both are the result of a very low-angle (probably about 5°) impact of material striking the lunar surface from the east but was it a case of 'one lump or two'?

Did the impactor hit the Moon to form Messier, bouncing and finally dropping again to form Messier A? Or were there two impactors, perhaps two fragments of a comet, flying side by side? Nobody really knows for sure. The bright interiors and surviving ray pattern must mean the craters were formed in the last few hundred millions of years – why not seek them out yourself and ponder on how they were created.

8.29   MORETUS [71°S, 354°E] (WITH CYSATUS, GRUEMBERGER, SHORT)

Moretus stands like a sentinel at the gateway to the Moon's south polar region. It is a magnificent crater, 114 km in diameter, with beautifully sculpted interior terraces. At the centre of the large, arena-like, floor a central mountain mass rises to a height of about 2.1 km.

Moretus is pictured in Figure 8.29, which is a Catalina Observatory photograph. It was taken on 1967 January 20^d 01^h 52^m UT, when the Sun's selenographic colongitude was 18°.5. As the photograph shows, there are signs of another crater on the southern slopes of the interior. Oddly, all that shows are a series of mountain peaks. Could this be a case of a meteor having struck almost immediately after the great impact

Figure 8.29 Moretus is the large crater in the upper middle of this Catalina Observatory photograph. The formation in the lower-right corner is actually part of the crater Clavius. Details in text. (Courtesy Lunar and Planetary Laboratory.)

which created Moretus itself? My thinking is that the interior of Moretus would still have been largely molten and suffering the aftershocks of the primary impact, so preventing the second crater from fully forming.

The large crater, actually 94 km in diameter, immediately to the northwest of Moretus is called Gruemberger. It is obviously older than Moretus, its outline being much more eroded and its floor more extensively cratered. To the north of Moretus, also encroaching into Gruemberger, is the much fresher looking Cysatus. Cysatus, of diameter 49 km, has a finely terraced interior and a very low central mountain. South of Moretus is the 71 km diameter crater, Short.

The western extremity of the crater Moretus lies on the Moon's central meridian – latitude 0° (and 360°) E. That, together with the convergence of the terminator, enables one to deduce the direction of the lunar south pole. Of course, whether one can see it or not depends on the libration. I wonder how long it will be before a lunarnaut makes the trek and is first to stick a flag in the Moon's most southerly location?

8.30   NECTARIS, MARE [15°S, 35°E] (WITH BEAUMONT, FRACASTORIUS,
          PICCOLOMINI, ROSSE, RUPES ALTAI)

Mare Nectaris is a somewhat undistinguished looking basalt flood-plain in the Moon's south-eastern quadrant, adjoining the Mare Tranquillitatis. With the names meaning 'Sea of Nectar' and 'Sea of Tranquillity' this region of the Moon certainly sounds delightful!

The region is pictured in Figure 8.30(a). As can be seen from the photograph, Mare Nectaris is somewhat irregular in outline, though its nature as a lava-filled basin is not too hard to imagine. Its diameter averages about 350 km. The craters Catharina, Cyrillus and Theophilus are identified

Figure 8.30 (a) The smooth grey plain in the upper-left part of this photograph is the Mare Nectaris. The rugged mountain range (really an escarpment) to the right of the photograph is known as the Rupes Altai. The three major craters to the left of the mountain range are Catharina (upper), Cyrillus (connected to Catharina) and Theophilus (overlapping Cyrillus). Photograph taken by Tony Pacey, using his 10-inch (254 mm) Newtonian reflector on 1991 March 21$^\mathrm{d}$ (time only approximately known – circa 20$^\mathrm{h}$ 00$^\mathrm{m}$ UT). He used eyepiece projection and a 1 second exposure on T-Max 100 film, processed in HC110 developer.

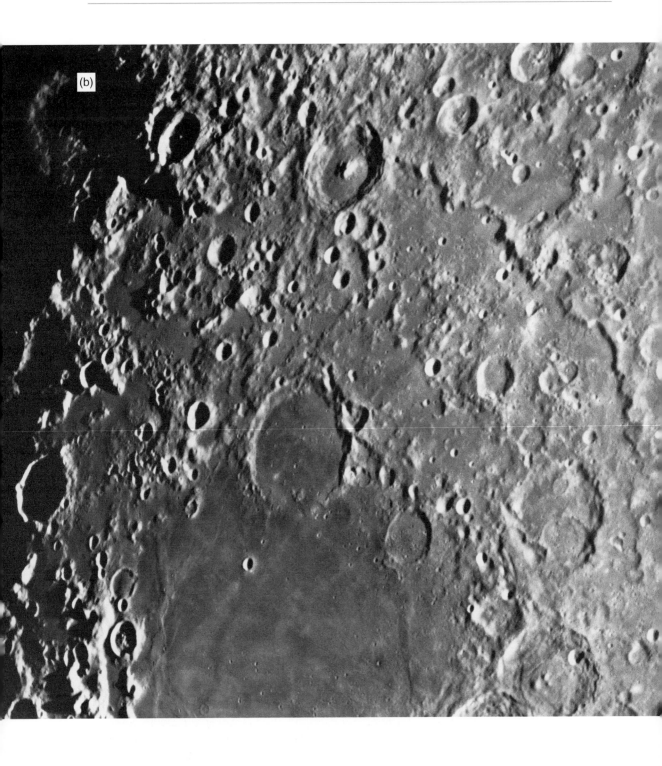

Figure 8.30 (*cont.*)
(b) Wide-angle view encompassing the southern half of the Mare Nectaris and the Rupes Altai. The dominant crater at the top of the photograph is Piccolomini. The flooded, incompletely enclosed, crater at the top of the mare is Fracastorius. The crater Catharina and part of Cyrillus can be seen to the lower right. Photograph taken with the Catalina Observatory 1.5 m reflector on 1965 November 12^d 10^h 31^m UT, when the Sun's selenographic colongitude was 134°.3. (Courtesy Lunar and Planetary Laboratory.)

on the photograph but a detailed discussion of these is reserved to Section 8.44, as is the small crater Mädler just to the east of Theophilus.

The *Apollo 16* Lunar Excursion Module touched down in the hinterlands approximately 300 km west of Theophilus. The rock samples collected were found to be very complex. In many cases the origins of the samples could not be determined with absolute certainty. This is particularly so in the case of Nectaris Basin ejecta. Bearing that in mind, planetary scientists have assigned an age of 3.92 billion years to the Nectaris Basin. Even if the sample identifications are correct, then the analytical techniques used still give an uncertainty of plus or minus 0.05 billion years for the result.

The interval between the formations of the Nectaris and Imbrium basins is now defined to be *Nectarian Period* in the chronology of the Moon. Officially the age range this represents spans 3.92–3.85 billion years. This was a time of heavy bombardment – a real lunar 'Blitzkrieg'. During the Nectarian Period a dozen other basin-forming impacts occurred, together with many of the larger, and now degraded, craters and some light plains formed by basin ejecta and a general heavy reworking of the Moon's regolith.

Also very evident on Figure 8.30(a) is a vast mountain range (more accurately described as an escarpment) encircling the mare to the south and west. This is the Rupes Altai, actually the surviving section of a ring that must have surrounded the Nectaris Basin until soon after its formation. The Nectaris Basin was a multi-ring structure. The best-preserved example of this type of formation is the larger and younger Mare Orientale, situated on the Moon's western hemisphere and mostly on the hidden side.

Figure 8.30(b) shows the southern section of the Mare Nectaris in more detail. Notice also that the lighting is from the opposite direction. The major part of the Rupes Altai is also shown. Figure 8.30(c) is an enlarged section of (b), concentrating on the Rupes Altai. The whole escarpment spans about 500 km and follows a curve of radius approximately 480 km centred on the Mare Nectaris. The reason this formation is really best described as an escarpment, rather than a mountain range, is that the peaks are elevated to very little above the ground to the west. However, the ground falls away sharply on the side facing into the Mare Nectaris, the average drop being about 1.8 km.

It seems that the Rupes Altai is much more a slump-fault than a range of uplifted crustal blocks, as is the case for the Montes Apenninus, bordering the Imbrium Basin.

The formation is striking even through a small telescope when seen at lunar ages of around 5½ days and 19 days, though it rapidly loses its distinctiveness when the terminator moves away from the area. As you might be able to discern in Figure 8.30(b), there are distinct traces of the continuation of the ring beyond the southernmost point and round to the

Figure 8.30 (*cont.*)
(c) Enlarged portion of (b) showing the Rupes Altai. Piccolomini is mostly hidden beyond the top-left corner and Catharina is to the lower right. (Courtesy Lunar and Planetary Laboratory.)

Figure 8.30 *(cont.)*
(d) Enlarged portion of
(b) centred on
Piccolomini. (Courtesy
Lunar and Planetary
Laboratory.)

east of the Mare Nectaris but these are much less obvious than the Rupes Altai proper.

The southern portion of the Rupes Altai terminates in a beautifully sculpted large crater (shown in Figure 8.30(b)). This is the 89 km diameter Piccolomini. Figure 8.30(d) is another enlargement from (b) and is centred on Piccolomini. The walls of the crater rise to nearly 4.5 km above the significantly convex floor. The walls have very fine interior terraces which are noticeably smoothed by erosion and landslips. The northern half of the outer slopes are also rippled with slumps running parallel to the crater rim. The complicated central mountain mass also shows evidence of landslips and much of the floor of the crater is also quite well smoothed. Most remarkable of all, though, is the intrusion into the crater of the external terrain to the north of it. Indeed, the terrae seem almost to have 'poured' into the crater over its southern rim and even 'flowed' some way onto the crater floor. Clearly, Piccolomini has been subject to some very significant seismic shaking since its creation. A truly remarkable object.

The lava flooding that created the mare is thought to have occurred about 3.7–3.8 billion years ago. One of the casualties of these floods was the crater Fracastorius. This 124 km diameter crater is pictured, along with the southernmost section of the Mare Nectaris, in Figure 8.30(e). This is an enlargement of yet another part of Figure 8.30(b). Notice how the northern wall of the crater has been largely eliminated by the lavas. It seems that this section has been more than just buried. Rather, much of the original crater wall has been melted and washed away.

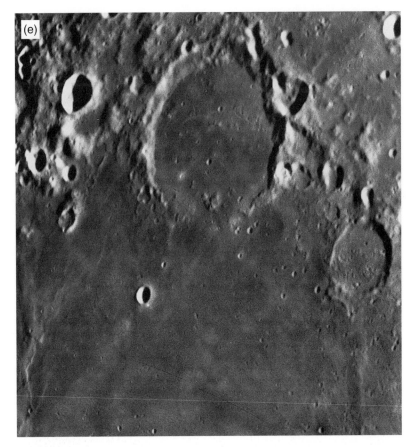

Figure 8.30 (*cont.*)
(e) Enlarged portion of
(b) showing the southern
sector of the Mare
Nectaris. Fracastorius is
in the upper half of the
photograph. The much
smaller, flooded crater
Beaumont is to the right.
The prominent small
crater to the lower left of
Fracastorius is Rosse.
(Courtesy Lunar and
Planetary Laboratory.)

The nearby crater Beaumont is a smaller-scale (53 km diameter) cousin of Fracastorius. In this case there is a small breach in the eastern section of the wall. The tiny craters that pepper the southern section of the Mare Nectaris, and the flooded floor of Fracastorius in particular, make an excellent test for observer, telescope and seeing conditions. How many can you see and record?

The small crater Rosse, 12 km in diameter and 2.4 km deep, situated on the Mare Nectaris about 70 km to the north of Fracastorius, should be easily visible in almost any telescope which can honestly sport the title 'astronomical'. It has a bright interior and is obviously quite 'fresh' by the standards of the local terrain.

All in all, this is a fascinating and highly complex area of the Moon.

8.31   NEPER [9°N, 84°E] (WITH JANSKY)

If you **really** want a challenge, try identifying and observing the crater Neper. It lies so close to the Moon's eastern limb that libration often

Figure 8.31 (a) Neper is the large crater close to the limb of the Moon in this Yerkes Observatory photograph. (Courtesy Yerkes Observatory and E. A. Whitaker.)

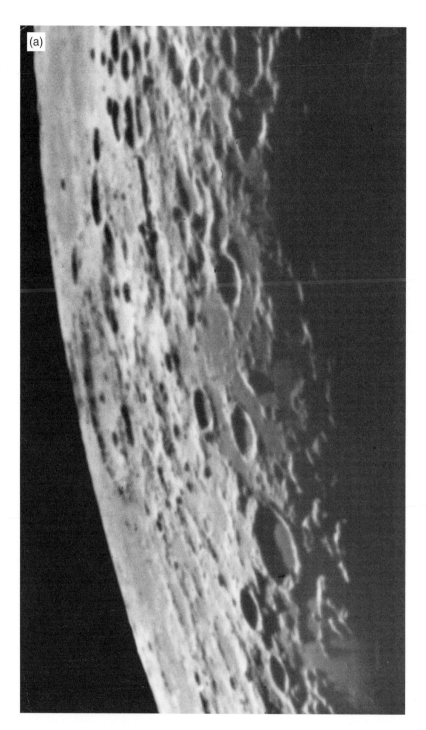

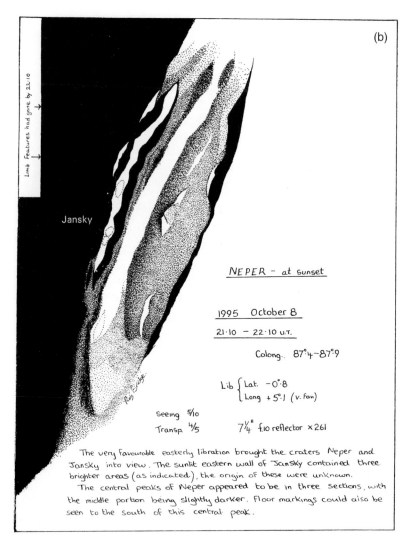

(b)

Limb features had gone by 22·10

Jansky

NEPER – at sunset

1995   October 8

21·10 – 22·10 U.T.

Colong.   87°·4–87°·9

Lib { Lat.   −0°·8
      { Long +5°·1 (v. fav.)

Seeing 5/10
Transp. 4/5   7¼" f.10 reflector ×261

The very favourable easterly libration brought the craters Neper and Jansky into view. The sunlit eastern wall of Jansky contained three brighter areas (as indicated), the origin of these were unknown.
    The central peaks of Neper appeared to be in three sections, with the middle portion being slightly darker. Floor markings could also be seen to the south of this central peak.

Figure 8.31 (*cont.*)
(b) Neper, drawn by Roy Bridge.

conspires to remove it from view just when the lighting is suitable and the sky is clear and steady! Moreover, conditions will only be favourable just after full Moon. The only other time it is seen under a low Sun-angle is when the Moon is a very thin crescent. However, then the Moon is very close to the Sun in the sky. At those times you will only be able to see the Moon against a twilight sky. Even then, its altitude will be very low and the atmosphere is bound to be very unsteady. Figure 8.31(a) shows a view made under such conditions. You may think the photograph does not show very much, even though the libration was obviously rather favourable for observing this feature. In point of fact, it was taken with the largest refracting telescope in the world – the 40-inch (1.02 m) telescope of the Yerkes Observatory (unfortunately I do not have any further details) and getting a view of the crescent Moon as good as this is no mean achievement!

Neper is a great formation, 142 km in diameter, with terraced walls and a central mountain. Roy Bridge has made a superb drawing of this difficult feature and this is presented in Figure 8.31(b). This time the view is of local sunset over the crater. Roy even manages to show part of the large crater Jansky which lies **beyond** Neper and the easternmost part of which actually lies **beyond** 90°E. Jansky is 72 km in diameter.

8.32 PITATUS [30°S, 346°E] (WITH HESIODUS)

Pitatus is a beautiful flooded crater on the southern shore of the Mare Nubium. It is 105 km in diameter, and has highly eroded walls and a peculiarly offset central peak. There are some ridges, hills and rilles on its floor but all are very delicate objects requiring large apertures and steady seeing and, very importantly, just the right lighting to show them up. Figures 8.32(a) to (d) are a series of views at a range of lighting conditions. Figure 8.32(a) is a drawing by Andrew Johnson showing the formation at local sunrise (Sun's mean selenographic colongitude 14°.3 for the period of the drawing). Figure 8.32(b) is another Andrew Johnson drawing, this time made at a mean colongitude of 17°.9. Figure 8.32(c) is a Catalina Observatory (1.5 m telescope) photograph taken at a colongitude of 22°.6 and Figure 8.32(d) is another photograph taken with the same telescope when the Sun's selenographic colongitude was 39°.9. Notice the dramatic changes in the appearance of the formation.

The crater joined to the western flank of Pitatus, shown in Figures 8.32(c) and (d), is the 42 km diameter Hesiodus. The breach in the wall between these two lunar arenas would be a fascinating place for a lunarnaut to explore. I wonder who will be the first lucky person to trek between Pitatus and Hesiodus and when that journey will be made? In the meantime, there is much to interest and challenge the backyard telescopist in these particular lunar formations.

(a)

PITATUS

LN 907

1996 APRIL 26

2035 – 2125 HRS. (UT.)

⊙ ₆ { COLONG. = 14.1° – 14.5°
         LAT. = –0.5°

L = –3.0° }
B = 5.9° } @ 2055 HRS.

Shadows drawn by 2055 HRS.
(C. = 14.3°)

SEEING. (ANT.) III – II
(Moments of good seeing
became tremulous with
arrival of thin clouds @
2115 HRS.)
TRANSP. 3/5

N

210mm F7.5 NEWTONIAN
@ ×195.

NOTES :– Interesting sunrise view of Pitatus. Of note were
the rapidly changing shadows on the floor of Pitatus. Two
regions of the floor were already revealed to the north & south
of the 'central' hill α. The shadows cast by the east rim were
nearly evenly divided by a 'spire' of illuminated floor.
The extensive inner west wall and rim were at this stage a
jumble of brightish 'islands', though the broad nature of
this section of the wall contrasts sharply with the northern
section, which is almost non-existant. The craters Pitatus
G, N, P & Q formed an deviare allignment; more so at a
later stage of illumination.

ANDREW JOHNSON, KNARESBOROUGH, NORTH YORKSHIRE.

Figure 8.32 (a) Pitatus at a colongitude of 14°.3, drawn by Andrew Johnson.

Figure 8.32 (*cont.*)
(b) Pitatus at a
colongitude of 17°.9,
drawn by Andrew
Johnson. Note Andrew's
comments, especially
that about the true shape
of the crater.

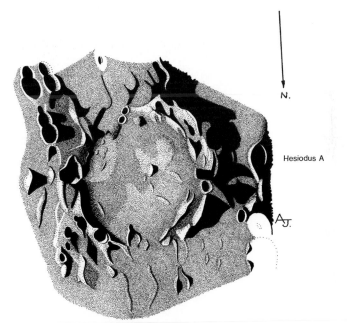

(b) **PITATUS**   L.N. **888**

1994 OCTOBER 13
1810 – 1950 HRS. (U.T.)
COLONG. = 17.4° – 18.3°
    LAT. = −0.7°

GEOC LIBRATION {L = 6.9°
@ 1900 HRS.   {B = −6.8°

210MM F7.5 NEWTONIAN
× 195.
SEEING (ANT.) II – III.
TRANSP. 2/5 (SOME MIST.)

N.

Hesiodus A

A.J.

NOTES// No sign of any of the internal rilles, seeing was not perfect
though. No indication of rille as depicted by Rükle in his atlas;
from Pitatus G to n. of central hills. Wealth of detail in walls –
almost too much to include in one session. Subtle albedo changes
on floor. Suspected dome-like appearance s.w. of central hills.
(NB. floor drawn too round, more elongated e-w.)

KNARESBOROUGH , NORTH YORKSHIRE.

Figure 8.32 (*cont.*)
(c) Pitatus and Hesiodus.
Catalina Observatory
photograph, taken with
the 1.5 m reflector on
1966 May 29^d 04^h 41^m UT,
when the Sun's
selenographic
colongitude was 22°.6.
(Courtesy Lunar and
Planetary Laboratory.)
(d) Pitatus. Catalina
Observatory photograph,
taken with the 1.5 m
reflector on 1966
December 23^d 04^h 54^m
UT, when the Sun's
selenographic
colongitude was 39°.9.
(Courtesy Lunar and
Planetary Laboratory.)

8.33   PLATO [51°N, 351°E] (WITH MONS PICO, MONS PITON, PLATO A)

If any one formation on the Moon's surface is more popular, and consequently more observed, than any other, surely it must be the crater Plato. It is bathed in sunshine from first to last quarter Moon and, as Figure 8.33(a) shows, appears as a dark oval set into the bright strip of rough terrain betwixt the Mare Imbrium (Sea of Rains) and the Mare Frigoris (Sea of Cold). It certainly grabs the attention and Johannes Hevelius named it 'The Great Black Lake'.

Plato marks the northernmost termination of the Montes Alpes, described in Section 8.24, earlier in this chapter. These mountains are the surviving remnants of an inner-ring feature in the Imbrium Basin. Hence Plato is situated on the basin shelf. Since Plato could not have survived the basin-forming impact about 3.85 billion years ago, it must have been formed after that. Given it is clearly flooded with dark basaltic lava, that puts its age at no greater than 3.0 billion years **if** we are correct in our assertion that all the major lava upwelling on the Moon was over by then. Actually, there is some evidence that minor volcanic activity continued on the Moon for another billion years, though the evidence is strong that Plato was formed – and subsequently flooded – in the interval 3.85–3.00 billion years ago. Perhaps of significance is that the composition of Plato's lava is a little different to that of the nearby 'seas', as witness its somewhat darker colouration.

The foreshortening due to its location makes Plato appear oval. Really it is quite circular and regular in outline, spanning 100 km from rim to rim. The walls of this arena reach up to approximately 2 km above the level of the dark floor but the summit peaks are fairly jagged. This is shown to the most beautiful effect when the Sun-angle is very low over the formation. Very striking spire-like shadows then extend across the crater floor (see Figure 8.33(b)). When the crater is very near the terminator the shadows show changes that are apparent after just a few minutes observation.

In fact, the shadows can reach right across the floor from the wall casting them to the wall opposite. This is a consequence of the fact that Plato's floor is one of the flattest large areas on the Moon's surface.

Under close inspection (though only needing a small telescope for this) the walls of Plato shows considerable signs of slumping along its northern and western sections. Most obvious of these is the huge triangular block, known as Plato Zeta, which has broken away from the western wall and slumped inwards, leaving a canyon behind it.

As can be seen in Figure 8.33(a), a number of small craters are situated on Plato's dark floor. Their visibility is heavily lighting- and seeing-dependent. With Plato close to the terminator and in good seeing,

Figure 8.33 (a) The crater Plato (centre right) photographed with the 1.5 m reflector of the Catalina Observatory in Arizona, on 1967 January 20^d 01^h 45^m UT, when the Sun's selenographic colongitude was 18°.4. Part of the Mare Imbrium can be seen above Plato. Just above Plato is the outline of a 'ghost crater'. On the right side of it can be seen part of the Montes Teneriffe, while on the upper part of the ghost crater is situated Mons Pico. Mons Piton is in the upper-left corner of this frame. The crater Plato A is just to the lower right of Plato. (Courtesy Lunar and Planetary Laboratory.)

I have used a power of ×432 with my 18¼-inch (0.46 m) reflector and seen them as perfectly formed bowl-shaped craters with distinctive interior shadows, though these shadows are not as intensely black as elsewhere. The largest, which is just a little off-centre in Plato, is the easiest to see. It is about 3 km in diameter. The crater south-west of this and the pair (appearing as one in poorer seeing conditions) to the north-west of the near-central crater are significantly more difficult. There is a very much smaller crater close to Plato Zeta, which I have glimpsed only very rarely. You might be able to make it out in Figure 8.33(a). It is also well shown in Figure 8.33(c), another of Terry Platt's excellent CCD images. Terry's image is even more impressive when you consider that the lighting is rather higher than the ideal for viewing their crateriform aspect!

At higher Sun-angles these craters usually appear as white disks. In bad seeing these white disks become white 'blobs' which fluctuate rapidly in visibility. On the worst nights they are completely invisible. Of course, it is the atmospheric turbulence which causes them to behave in this way. Even accepting that, a large folklore has been built up about apparently enigmatic variations in visibility of Plato's floor craters. Almost all of the major observers of the past have recorded instances when they considered the floor craters should be visible and yet are

Figure 8.33 (*cont.*)
(b) Plato, drawn by
Roy Bridge.

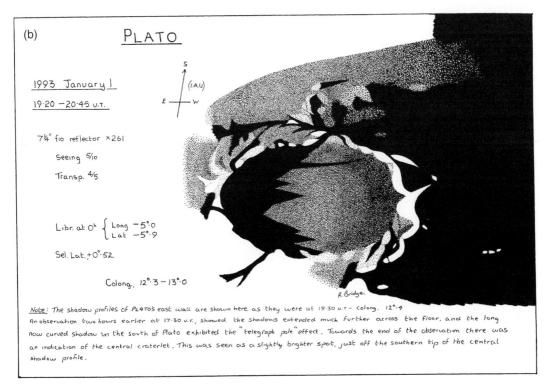

(b) PLATO

1993 January 1

19·20 – 20·45 u.т.

7¼" f/10 reflector ×261

Seeing 5/10

Transp. 4/5

Libr. at 0ʰ { Long −5°·0
{ Lat −5°·9

Sel. Lat. +0°·52

Colong., 12°·3 − 13°·0

R. Bridge

Note: The shadow profiles of PLATO's east wall are shown here as they were at 19·30 u.т – Colong. 12°·4 An observation two hours earlier at 17·30 u.т., showed the shadows extended much further across the floor, and the long now curved shadow in the south of Plato exhibited the "telegraph pole" effect. Towards the end of the observation there was an indication of the central craterlet. This was seen as a slightly brighter spot, just off the southern tip of the central shadow profile.

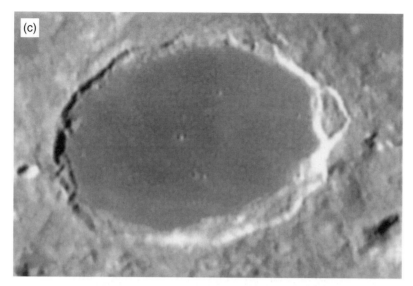

Figure 8.33 (*cont.*)
(c) Plato, imaged by Terry Platt using his 12½-inch (318 mm) tri-schiefspiegler reflector and *Starlight Xpress* CCD camera (other details not available). The author has applied slight image sharpening and brightness re-scaling.

completely absent. Many report a fog-like veil extending over the crater floor and extinguishing the details of all it covers. In all the nearly 40 years that I have been observing the Moon, I cannot say with certainty that I have ever seen the craters much less distinct than I considered they should be for the given seeing conditions. However, I have seen the opposite effect on just a few very rare occasions. For instance, I have been perplexed to see the floor craters as bright white blobs (the near-central one by far the brightest) through my 6¼-inch (152 mm) reflector when all is fuzzy and violently rippling in ANT. V. seeing! Yet most other times they are invisible through telescopes small or large when the seeing is not quite as bad. Puzzling!

Plato is also a 'hot spot' for other types of transient phenomena. Flashes have been occasionally reported; also an apparent blurring of parts of the crater rim while other parts of the crater remain sharp and clear-cut. I have seen this effect myself. Coloured glows are sometimes reported extending along the crater rim. Again I can concur. However, the normal prismatic effect of the Earth's atmosphere can produce exactly these effects. The northern section of the rim of Plato then appears reddish-orange and the southern section appears blue. Is this the cause in every instance? I think so, though I did once see a red glow that seemed very different in colour and extent from the normal spurious colour (which was also present).

Is Plato occasionally the site of genuine TLP or are these variations just illusions, perhaps caused by the normal intertwining of seeing conditions and the complex interaction of sunlight with the formation? I think that is certainly the case in the vast majority of instances. Some people

Figure 8.33 (*cont.*)
(d) Mons Piton, drawn by
Roy Bridge.

(d)

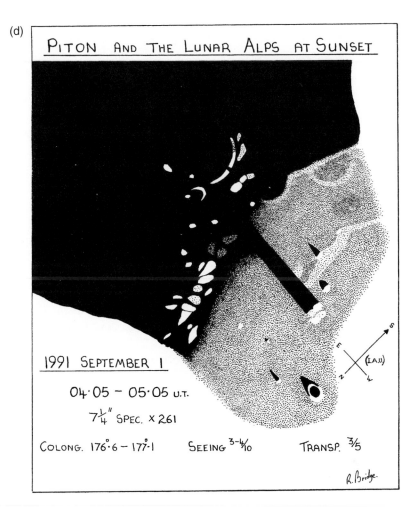

PITON AND THE LUNAR ALPS AT SUNSET

1991 SEPTEMBER 1

04·05 – 05·05 U.T.

7¼" SPEC. × 261

COLONG. 176°·6 – 177°·1   SEEING 3–4/10   TRANSP. 3/5

R. Bridge.

are very sure that is the case in **all** instances. In fact, they dismiss all suggestions of genuine TLP out of hand. I think there is a case for further study – the type of study that involves very careful monitoring of Plato through the telescope. I have more to say about TLP research in the final chapter of this book.

Another effect, that has long ago been demonstrated to be an illusion, is the apparent **darkening** of the floor of Plato as the Sun rises higher over it. Of course, the floor actually brightens with increasing Sun-angles. What is happening is that the rough surrounds of the crater brighten more rapidly than its smooth floor, so increasing the contrast. Various light spots and mottled patterns also appear towards local lunar noon (which is full Moon, as we see it from the Earth). A lighter sector in the south-west, covering about one-eighth of the total floor area, becomes

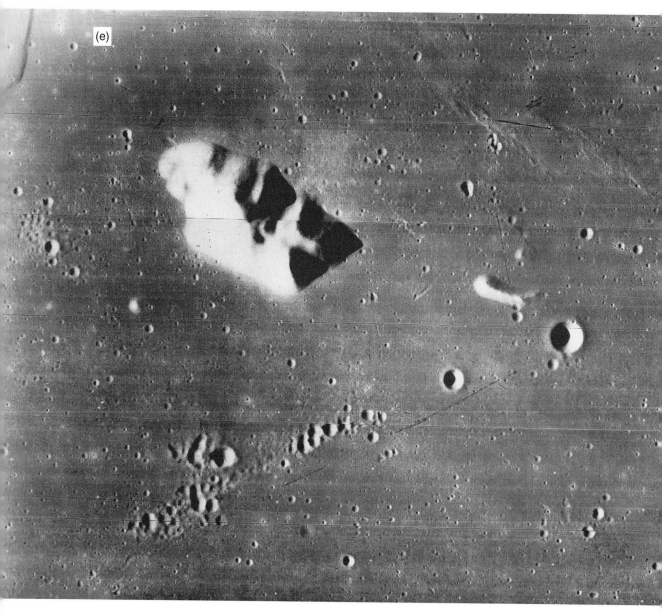

(e)

especially apparent at these times – a frequent cause of 'mists extending across the floor from the crater wall' reports from the uninitiated.

Plato's surrounds are very complex and the nearest sizeable crater is Plato A, 22 km across, situated about 20 km to the north-west of Plato. It is shown in Figure 8.33(a). A number of fissures cut through the hinterlands, most prominent of these being a rille extending eastwards from Plato (the western termination of which occurs a little east of the flanks of the crater).

The strip of rough ground south of Plato is narrowed by the northern part of the outline of a 'ghost crater' extending southwards onto the Mare

Figure 8.33 (cont.)
(e) *Orbiter IV* photograph of Mons Pico. (Courtesy NASA and E. A. Whitaker.)

Figure 8.33 (*cont.*)
(f) Plato (lower right) to
Mons Piton (upper left)
photographed using the
1.5 m Catalina
Observatory reflector on
1966 September 6^d 10^h
44^m UT, when the Sun's
selenographic
colongitude was 167°.3.
(Courtesy Lunar and
Planetary Laboratory.)

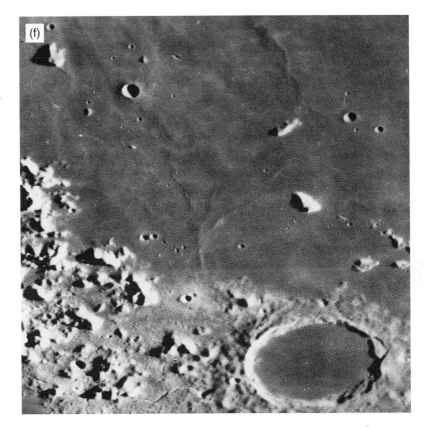

Imbrium. It is mostly visible as a slightly raised ridge in the mare, and is well shown as such in Figure 8.33(a) and in Chapter 5, Figure 5.1. The feature is approximately 115 km in diameter.

On the western flank of the ghost crater are a collection of mountain peaks, the Montes Teneriffe. These reach a height of approximately 2.4 km. The isolated peak on its southern rim is Mons Pico, also about 2.4 km high. It is a very reflective object and seems extraordinarily bright when the upper parts of it catch the sunlight and the immediate surrounds are in the darkness beyond the terminator. This mountain is also the source of various alleged changes and anomalous appearances, most probably all illusory. A remarkably straight line of craters cross the ghost crater a little to the north of Mons Pico.

There are various other isolated peaks poking up through the lavas of the Mare Imbrium. Some of these can be seen in Figure 8.33(a). One of the most notable is Mons Piton, visible in the top-left corner of the photograph. It is 2.25 km high. This mountain is another source of reported changes and odd appearances. Figure 8.33(d) is a drawing of the area made by Roy Bridge (see also the drawing by Andrew Johnson, shown in

Chapter 3, Figure 3.5). Figure 8.33(e) shows an *Orbiter IV* view of Mons Pico. Mons Pico and Piton are shown in Figures 8.33(a) and (f) where they are lit from opposite directions. Notice how their appearances differ between the two views. Is it any wonder that the visual observers of yesteryear considered the mountains subject to change!

8.34 PLINIUS [15°N, 24°E] (WITH DAWES, MENELAUS, PROMONTORIUM ARCHERUSIA, ROSS, MARE SERENITATIS, MARE TRANQUILLITATIS)

In the manner of an ancient gate-keeper, the crater Plinius stands close to the narrow intersection of the Mare Tranquillitatis with the Mare Serenitatis. It is shown, near the end of a lunar day, and still on guard, in Tony Pacey's excellent photograph, which is presented in Figure 8.34(a). Perhaps the smaller crater Dawes, standing a little to the north-east (left and slightly lower in the photograph) is an apprentice guard?

Plinius is actually situated just inside the Mare Tranquillitatis. The mountainous cape immediately west of Plinius is known as the Promontorium Archerusia, the name surviving from Hevelius's map. In his chart Hevelius named what we now call the Mare Tranquillitatis and Mare Serenitatis the Pontus Euxinus, meaning 'Black Sea'. Neither sea might actually be black, but there is certainly a very distinct colour difference between the two maria. Mare Tranquillitatis is clearly very much darker than Serenitatis and it is much bluer. In fact, I find that it takes on a very distinct Prussian blue tint when seen at low power through my telescopes (to repeat yet again, the real colours on the Moon are various shades of brown – but the eye averages the view as white, so producing apparent coloured tints in specific features. The colours themselves might not be true but at least they do indicate the colour **differences**).

Serenitatis, to me, has the faint greenish colour in common with the other lunar maria. I ought to repeat that experience has shown me that my eyes are more sensitive to colours than most (and to my regret I find that this heightened sense is dwindling with age) and you may well go to your telescope and fail to see any colours at all. Proper colorimetric studies do show that the Mare Tranquillitatis **is** much bluer than the other lunar maria. This is because the Tranquillitatis lavas are much richer in titanium than is the norm.

Figure 8.34(b) shows a wide-angle view encompassing both maria that I took myself. It was taken on colour film and even that shows the Mare Tranquillitatis as bluer than the other mare. It is a pity that it has to be reproduced in monochrome here. The Tranquillitatis Basin is probably the oldest of the large basins. Certainly it has a much less well-defined shape than most. It is also masconless. Could that mean the basin was

Figure 8.34 (a) Plinius (the large crater just below centre) and environs, photographed by Tony Pacey. He used his 12-inch (305 mm) reflector for this ½ second exposure on T-Max 100 film on 1992 September $16^d\ 23^h\ 55^m$ UT, when the Sun's selenographic colongitude was 139°.5.

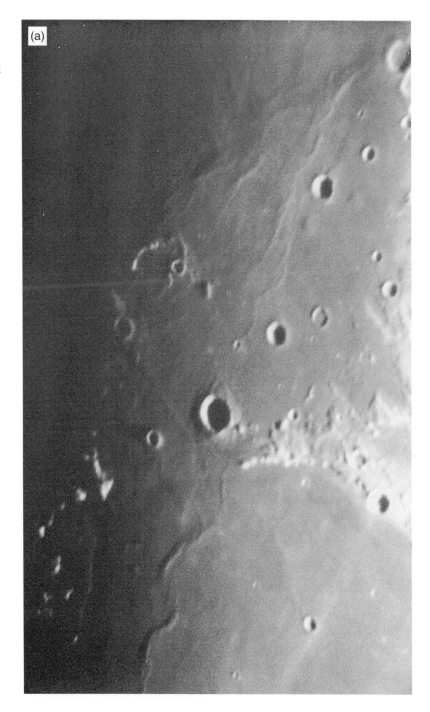

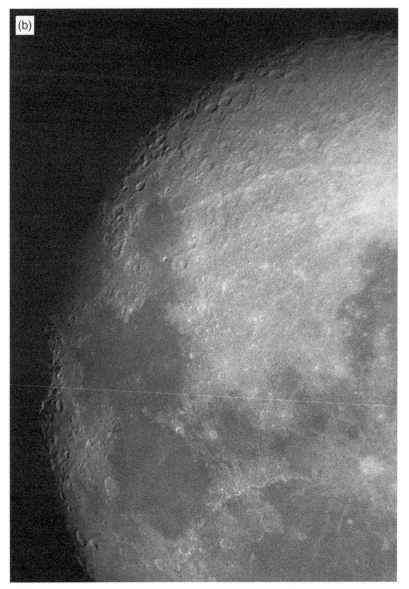

Figure 8.34 (*cont.*) (b) The two main dark areas close to the lunar terminator are Mare Tranquillitatis (upper) and Mare Serenitatis (lower). The small 'sea' attached to the south (top) of the Mare Tranquillitatis is the Mare Nectaris. Photograph taken by the author on 1992 September 14^{d} 22^{h} 46^{m} UT, when the Sun's selenographic colongitude was 114°.5. He hand-held his camera, fitted with a 58 mm lens, to a 44 mm Plössyl eyepiece (EFR = f/7.4) for a 1/1000 second exposure on *3M Colourslide 1000* film.

formed at a time not long after the creation of the Moon, when all but an outer very thin crust was still molten? If so, the Tranquillitatis Basin probably dates back about 4.5 billion years!

The diameter of the Tranquillitatis flood-plain is very roughly 800 km, about 20 per cent larger than that of the Mare Serenitatis. In common with the other lunar maria, the lava flooding commenced about 3.9 billion years ago (less than a hundred million years after the formation of the Serenitatis Basin), and successive lava flows are evident on both maria.

Figure 8.34 (*cont.*)
(c) Plinius and environs,
drawn by Andrew
Johnson.

(c)

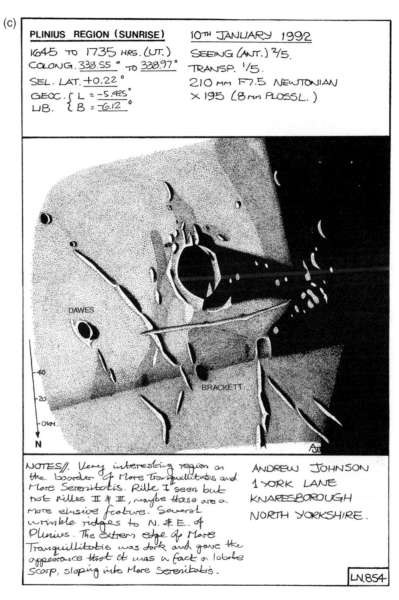

PLINIUS REGION (SUNRISE)   10TH JANUARY 1992

1645 TO 1735 HRS. (UT.)   SEEING (ANT.) 2/5.
COLONG. 338.55° TO 338.97°   TRANSP. 1/5.
SEL. LAT. +0.22°   210 MM F7.5 NEWTONIAN
GEOC. { L = -5.425°   X 195 (8mm PLOSSL.)
LIB. { B = -6.12°

DAWES

BRACKETT

40
20
0KM.

N

NOTES//. Very interesting region on the boarder of Mare Tranquillitatis and Mare Serenitatis. Rille I seen but not rilles II & III, maybe these are a more elusive feature. Several wrinkle ridges to N. & E. of Plinius. The extrem edge of Mare Tranquillitatis was dark and gave the appearance that it was in fact a lobate scarp, sloping into Mare Serenitatis.

ANDREW JOHNSON
1 YORK LANE
KNARESBOROUGH
NORTH YORKSHIRE.

LN.854

The last really significant lava eruptions probably occurred about 3.6 billion year ago. Interestingly, the same titanium-rich lavas that dominate the covering of the Mare Tranquillitatis also encroach onto the southern half of the perimeter of the Mare Serenitatis.

Figure 8.34(a) shows many of the wrinkle ridges on the mare superbly well. Figure 8.34(c) is a drawing of the region executed by Andrew Johnson. It is usual for compressional features on the lunar maria, particularly the wrinkle ridges, to be complemented with tensional faults around the

Figure 8.34 (*cont.*)
(d) Plinius (the large crater on the left), Ross (largest crater above Plinius) and Menelaus (large crater on the extreme right) photographed using the Catalina Observatory 1.5 m reflector on 1966 April 27^d 02^h 54^m UT, when the Sun's selenographic colongitude was 351°.0 (morning illumination). The cape to the right of Plinius is the Promontorium Archerusia. Notice the rilles extending from it and passing below Plinius. (Courtesy Lunar and Planetary Laboratory.) (e) Plinius and environs, photographed using the Catalina Observatory 1.5 m reflector on 1966 September 4^d 10^h 38^m UT, when the Sun's selenographic colongitude was 142°.9 (afternoon illumination). Other details as for (d). (Courtesy Lunar and Planetary Laboratory.)

peripheries. These mainly manifest as graben-type rilles. You might expect the junction between two maria to be especially rich in rilles – and you would be right! Rilles of the graben variety can, indeed, be found at the junction of the two maria and these are very well shown in Figures 8.34(d) and (e), which are Catalina Observatory photographs.

The various illustrations accompanying this section show a day in the life of the crater Plinius. Figure 8.34(c) pictures the crater at sunrise, while Figure 8.34(d) reveals it later on in the local morning, Figure 8.34(e) shows it in the local afternoon, and Figure 8.34(a) shows it at sunset.

The crater is 43 km in diameter and has quite a sharp rim, 2.3 km above the hummocky floor. The interior walls are terraced and rather complex, the outer slopes also being complex and with some radial ridging. The central mountains are very peculiar and can give the impression of being a crater under some illuminations, such as that shown in Figure 8.34(d). Under a very high Sun the central peak becomes rather bright and five bright, rather fuzzy, streaks extend from the central peak up the interior terraces. This appearance reminds me of a spoked wheel.

To the north-east of Plinius, a little to the left and slightly below it on Figure 8.34(a), is the crater Dawes. This, another sharp-rimmed crater, has a diameter of 18 km. Dawes is also shown in Andrew Johnson's drawing (Figure 8.33(c)). The features on its floor are all of low height and so are rather difficult for the backyard telescope-user to appreciate. Approximately south of Plinius, and above it in Figures 8.34(d) and (e), is the 26 km diameter crater Ross. The photographs show the somewhat peculiar profile of this crater. Again referring to Figures 8.34(d) and (e), the large crater to the extreme right on the photographs is the 27 km diameter Menelaus.

Menelaus has spectacularly terraced walls which rise 3 km up to a sharply defined rim. The crater brightens considerably under increasing Sun-angles. Near full Moon it takes on the appearance of a brilliant white ring. It also has an asymmetric ray pattern, the chief component of which is a bright streak which bisects the Mare Serenitatis. Obviously the crater is no more than a few hundred million years old. The asymmetry of the ray pattern and the somewhat off-centre interior mountain complex suggest the impactor arrived at an angle to the surface from the south-east.

Another area of the Mare Tranquillitatis is discussed in Section 8.45. The next section details another part of the Mare Serenitatis.

8.35 POSIDONIUS [32°N, 30°E] (WITH CHACORNAC, DANIELL, LACUS SOMNIORUM, POSIDONIUS A)

The 100 km diameter crater Posidonius is situated on the eastern edge of the Mare Serenitatis (see the last section for more details of this mare) and at the southernmost junction with it and the inset Lacus Somniorum (Lake of

Dreams). Figure 8.35(a) can be used to identify the crater. It appears in the extreme top-left corner of this photograph by Tony Pacey. It is also shown in the wide-angle view presented as Figure 8.34(b) in the last section. There it appears as the large white disk at the left-hand side of the Mare Serenitatis, and you can also see how it is positioned at the southernmost junction of the Sea of Serenity and the Lake of Dreams.

Nigel Longshaw has made a drawing of Posidonius and this is shown in Figure 8.35(b). This drawing is already impressive to the casual glance. Look closely at the hand-written notes accompanying it and you will see that Nigel used a 3½-inch (90 mm) catadioptric telescope to make the observation! This puts the use of large-aperture telescopes into a proper perspective. It is the skill and application of the person behind the eye-piece that really counts.

(a)

Figure 8.35 (a) Posidonius (crater in top-left corner) to the Moon's north pole, photographed by Tony Pacey. Details given in the main text.

Figure 8.35 *(cont.)*
(b) Posidonius (main crater) with Chacornac (adjoining Posidonius to the upper left) and Daniell (bottom crater), drawn by Nigel Longshaw.

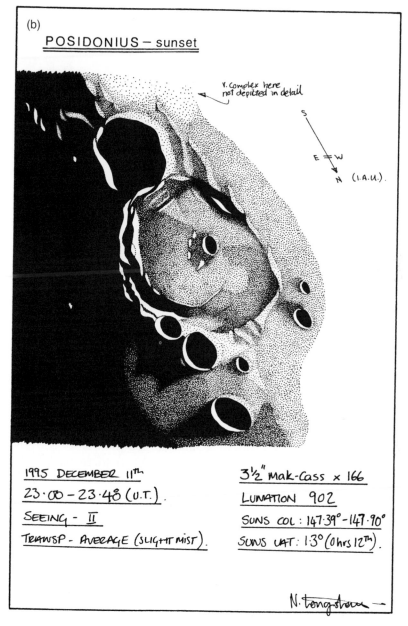

(b)

POSIDONIUS — sunset

V. complex here
not depicted in detail

S

E ⇌ W

N   (I.A.U.)

1995 DECEMBER 11ᵗʰ
23·00 – 23·48 (U.T.)
SEEING - II
TRANSP - AVERAGE (SLIGHT MIST)

3½" Mak-Cass × 166
LUNATION 902
SUNS COL: 147·39° – 147·90°
SUNS LAT: 1·3° (0hrs 12ᵀᴴ)

N. Longshaw —

Figure 8.35(a), already referred to, shows the crater illuminated by the morning Sun. To obtain this photograph Tony Pacey used his 10-inch (254 mm) Newtonian reflector, with eyepiece projection (enlargement factor not given) onto T-Max 100 film, processed in HC110 developer. The ½ second exposure was made at approximately 20^h UT on 1991 March 21^d. The value of the Sun's selenographic colongitude was approximately 330° at the time.

Figure 8.35 (*cont.*)
(c) Posidonius with
Chacornac and Daniell,
photographed using the
Catalina Observatory
1.5 m reflector. Details in
text. (Courtesy Lunar and
Planetary Laboratory.)

Figure 8.35(c) is a photograph of Posidonius taken with the Catalina Observatory 1.5 m reflector on 1966 September 4^d 10^h 03^m UT. This time the crater is shown under late afternoon illumination, the colongitude now being $142°.6$. Figure 8.35(b) shows the formation at sunset (details presented with the drawing).

If you get a feeling of *déjà vu* when looking at the illustrations accompanying this section, then look back to Section 8.22 and you will see why. The crater Gassendi seems, superficially at least, almost to be the twin of Posidonius. Gassendi is just 10 km larger in diameter and each formation has a smaller adjoining crater. In the case of Gassendi, it is Gassendi A. In the case of Posidonius, it is Chacornac. The main similarity, though, is in the interior structures of the craters. Both have hummocky and rille-ridden floors which give the impression of being pushed upwards by forces from below.

Since the craters are of similar size, one can assert that the incoming projectile was endowed with a similar amount of kinetic energy (which depends on both the speed and the mass of it), but can the locations of the two craters also be a factor in determining their final forms? Both craters lie on basin shelves: the Humorum Basin shelf in the case of Gassendi and the Serenitatis Basin shelf in the case of Posidonius. My speculation is that the particular subsurface structure that exists at a basin shelf lends itself to a Posidonius and Gassendi type of crater, given an impactor of similar energy.

Concentrating on the fine details of Posidonius, it has a somewhat west-of-centre crater on its floor. This is the 11 km diameter Posidonius A.

On the east side of this crater is a little ring of mountain peaks. Nigel Longshaw's drawing (Figure 8.35(b)) shows them particularly well. Perhaps these are the surviving remnants of an old, now obliterated, crater? A further mountainous ridge encompasses much of the eastern half of Posidonius. Is this another old and wrecked crater? As Figure 8.35(c) shows very well, much of the floor of Posidonius within this arc is raised upwards. Also there is significant faulting at the interface between this raised ground and the rest of the interior of Posidonius. Clearly the history of this crater is not at all straightforward.

The surrounding walls of Posidonius vary considerably in height, being at their highest (of the order of 2 km) to the east. The largest crater adjoining Posidonius is Chacornac. It is 51 km in diameter. It has a very 'tortured' interior, though it does seem to have been formed after Posidonius. It is shallower than Posidonius, the lowest point near the centre being about 1.5 km below the level of the crater rim. As the illustrations all show, several craters lay on or just north of the northern half of the perimeter of Posidonius. The most interesting of these is Daniell, shown as the lowest crater rendered in Nigel Longshaw's drawing of the area (Figure 8.35(b)). Of course, all the craters in the area appear foreshortened from the Earth because of their position on the Moon's disk. Look carefully at the illustrations and you will see that Daniell is much more oval than the others. Actually it spans about 30 km in the (roughly) north–south direction but is only about 23 km wide measured (roughly) east–west! Daniell is about 2.1 km in depth. I will leave you to ponder on the events and processes which have led to the formation we see today.

8.36 PYTHAGORAS [63°N, 258°E] (WITH BABBAGE)

To find out if you can see Pythagoras, first locate the Sinus Iridum. Then look radially from it towards the limb of the Moon. If the area is in sunlight you will locate the crater Pythagoras very close to the lunar limb. The crater is best seen a day or so before full Moon. Before then it will not be sunlit. After that the higher Sun will cause the details to wash out, especially so since the sunlight will be coming from approximately the same direction as we are looking towards it and thus providing no shadow relief for us to see.

This 128 km diameter crater would be a magnificent spectacle if it were further round to our side of the Moon. As it is, it is impressive enough during that brief window of opportunity each lunation. Andrew Johnson has made a drawing of this formation under these (sunrise) conditions (see Figure 8.36(a)). If you are wondering about the sunset view, unfortunately the Moon is then a waning crescent and will be so close to the Sun in the dawn sky that the seeing is bound to be bad.

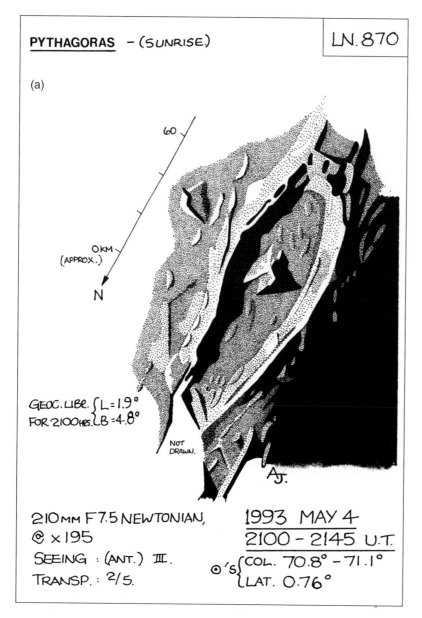

PYTHAGORAS – (SUNRISE.)   LN. 870

(a)

60

OKM
(APPROX.)

N

GEOC. LIBR. { L = 1.9°
FOR 2100 HRS. { B = 4.8°

NOT
DRAWN.

A.J.

210 MM F7.5 NEWTONIAN,   1993 MAY 4
@ ×195   2100 – 2145 U.T.
SEEING : (ANT.) Ⅲ.   ⊙'s { COL. 70.8° – 71.1°
TRANSP. : 2/5.   { LAT. 0.76°

Figure 8.36 (a)
Pythagoras, drawn by
Andrew Johnson.

Another view of this crater is provided in Figure 8.36(b). This is a photograph taken with the 1.5 m reflector of the Catalina Observatory in Arizona on 1966 October 28^{d} 06^{h} 34^{m} UT. The Sun's selenographic colongitude was 79°.3, the Sun being just a few degrees higher than for Andrew Johnson's drawing.

Figure 8.36 (*cont.*)
(b) Pythagoras,
photographed using the
1.5 m reflector of the
Catalina Observatory.
Details in text. (Courtesy
Lunar and Planetary
Laboratory.)

Although foreshortened, it is quite easy to make out the crater's somewhat hexagonal outline. The spectacular terraced walls soar to 5 km above the arena-like floor of the formation. Examine the crater closely and you will find plenty of evidence of substantial landslips. The central mountain cluster is multi-peaked and reaches up to a height of about 1.5 km.

Upper right of Pythagoras in Figure 8.36(b), and extending into the top-right corner of the photograph, is the peculiar formation named Babbage. It seems to be the fusion of two main craters. As well as two large (32 km and 14 km diameter) craters within it, Babbage also contains much interesting detail. I will leave a detailed study of it to you. This is a fascinating area of the Moon, though one that is certainly not the easiest for the telescopist.

8.37   RAMSDEN [33°S, 328°E] (WITH RIMAE RAMSDEN)

The crater Ramsden, 24 km in diameter and about 2 km deep, is fairly unremarkable. It is situated on the Palus Epidemiarum, which is itself connected to both the Mare Nubium and the Mare Humorum (see Section 8.22). The crater appears in the centre of Figure 8.37(a), which is a Catalina Observatory photograph. It was taken with the 1.5 m

Figure 8.37 (a) Ramsden and Rimae Ramsden. Catalina Observatory photograph. Details in text. (Courtesy Lunar and Planetary Laboratory.)

Figure 8.37 *(cont.)*
(b) Ramsden and Rimae
Ramsden, drawn by
Roy Bridge.

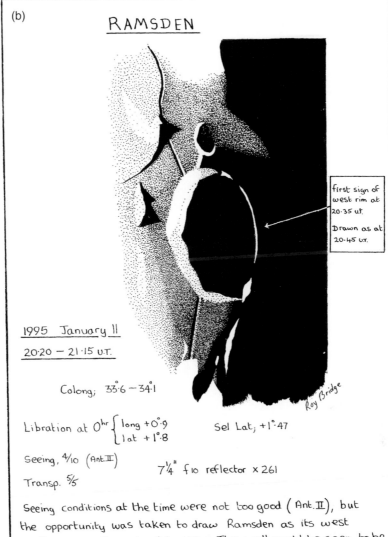

(b)

RAMSDEN

first sign of
west rim at
20·35 u.t.

Drawn as at
20·45 u.t.

1995 January 11
20·20 — 21·15 u.t.

Colong; 33°·6 — 34°·1

Libration at 0^{hr} { long +0°·9
lat +1°·8

Sel Lat; +1°·47

Seeing, 4/10 (Ant.Ⅱ)

Transp. 5/5

$7\frac{1}{4}"$ f 10 reflector × 261

Roy Bridge

Seeing conditions at the time were not too good (Ant.Ⅱ), but
the opportunity was taken to draw Ramsden as its west
wall was just coming into view. The wall could be seen to be
broken in both the north and south. The southern gap was
possibly due to the intersection of Rille Ⅴ. The poor seeing
and close proximity to the terminator prevented any
further study of the other rilles. (see also, Nigel Longshaw's
observation, the same evening, T.N.M. Vol. 7, No 2.)

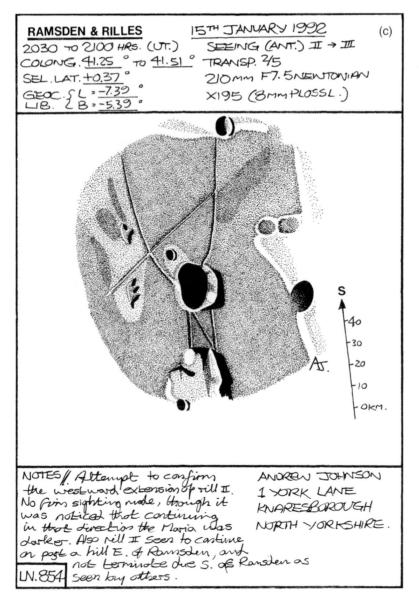

Figure 8.37 (*cont.*)
(c) Ramsden and Rimae
Ramsden, drawn by
Andrew Johnson.

reflector on 1966 December 23^d 04^h 54^m UT, when the Sun's seleno-graphic colongitude was 39°.9.

The major feature of interest is the system of rilles, the Rimae Ramsden, associated with the crater. This particular system spans about 130 km but the area as a whole is rille-ridden. Figure 8.37(b)

Figure 8.37 (*cont.*)
(d) Ramsden and Rimae
Ramsden, another
drawing by Andrew
Johnson.

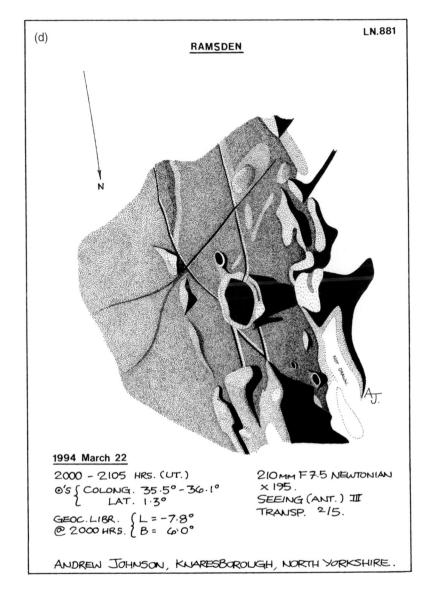

(d)

LN.881

RAMSDEN

N

Not Drawn

A.J.

1994 March 22

2000 – 2105 HRS. (UT.)

⊙'s { COLONG. 35.5° – 36.1°
{ LAT. 1.3°

GEOC. LIBR. { L = –7.8°
@ 2000 HRS. { B = 6.0°

210 MM F 7.5 NEWTONIAN
× 195.
SEEING (ANT.) III
TRANSP. 2/5.

ANDREW JOHNSON, KNARESBOROUGH, NORTH YORKSHIRE.

shows a sunrise drawing of the formation by Roy Bridge, while
Figure 8.37(c) and (d) are two further studies made by Andrew Johnson
under higher angles of illumination. Figure 8.37(e) is a photograph
obtained by the *Orbiter IV* probe. Can you work out the sequence of events
that has led to the present vista? I will leave you to study the area and
offer the illustrations here as a starting point.

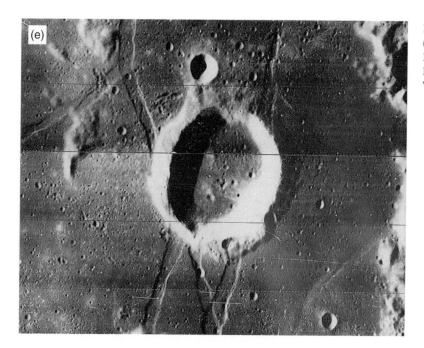

Figure 8.37 *(cont.)*
(e) *Orbiter IV* view of Ramsden. (Courtesy NASA and E. A. Whitaker.)

8.38 REGIOMONTANUS [28°S, 359°E] (WITH PURBACH, THEBIT, WALTER)

Regiomontanus inhabits a very complex area of lunar highlands. Being close to the meridian, the area is perhaps best studied around the times of last and first quarter Moon.

The region is pictured in Figure 8.38(a), which is a photograph taken using the Catalina Observatory 1.5 m reflector. It was taken on 1966 May 29^d 04^h 41^m UT, when the Sun's selenographic colongitude was 22°.6. Regiomontanus is the large and rather oval formation pictured a little above the centre of the photograph. It measures 126 km east–west and 110 km north–south. Notice the eroded walls and the floor peppered with small craters. Does this give you any idea of the age of the crater? The irregular walls rise in places to about 1.7 km above the floor. The attention-grabbing feature of Regiomontanus has to be the off-centre 'central' mountain with its summit crater. Known as Regiomontanus A, this little crater is 5.6 km wide and 1.2 km deep. What do you think about its origin?

Below Regiomontanus in Figure 8.38(a), and overlapping it, is the 118 km diameter 'walled-plain' Purbach. Its very rough walls extend upwards to nearly 3 km above the inner arena. Many interesting details reside inside this formation. What do think about its age?

Above Regiomontanus on Figure 8.38(a) (and not quite completely shown in this view) is the even larger 'walled-plain' Walter. It is also

Figure 8.38 (a) Regiomontanus and environs. Catalina Observatory photograph. The large crater at the top is Walter. The large crater below that is Regiomontanus. Note the off-centre mountain within it, and the mountain's summit crater. Just below Regiomontanus, and encroaching into it, is the large crater Purbach. Near the bottom right of the frame are the overlapping craters Thebit (the largest), Thebit A and Thebit L (the smallest). Further details in text. (Courtesy Lunar and Planetary Laboratory.)

Figure 8.38 (*cont.*)
(b) Regiomontanus and
environs shortly after
sunrise. Details in text.
(Catalina Observatory
photograph – courtesy
Lunar and Planetary
Laboratory.)

Figure 8.38 (*cont.*)
(c) Thebit, drawn by
Nigel Longshaw.

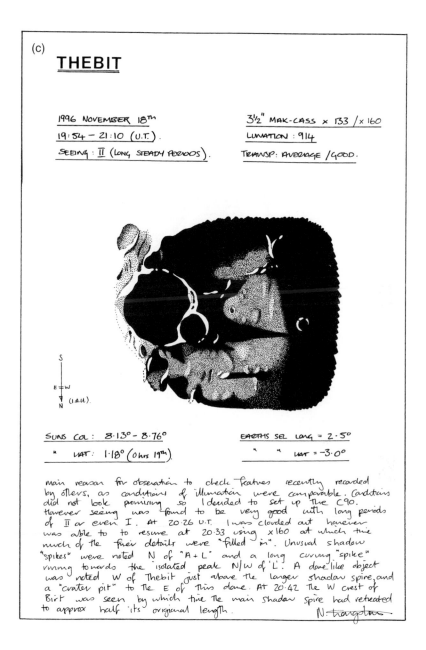

(c)

THEBIT

1996 NOVEMBER 18TH
19:54 – 21:10 (U.T.).
SEEING : II (LONG STEADY PERIODS).

3½" MAK-CASS × 133 /× 160
LUNATION : 914
TRANSP: AVERAGE /GOOD.

SUNS COL : 8·13° – 8·76°
" LAT : 1·18° (0 hrs 19TH)

EARTHS SEL LONG = 2·5°
" " LAT = –3·0°

main reason for observation to check features recently recorded
by others, as conditions of illumination were comparable. Conditions
did not look promising so I decided to set up the C90.
However seeing was found to be very good with long periods
of II or even I. At 20.26 U.T. I was clouded out however
was able to to resume at 20.33 using ×160 at which time
much of the finer details were "filled in". Unusual shadow
"spikes" were noted N of "A+L" and a long curving "spike"
running towards the isolated peak N/W of 'L'. A dome like object
was noted W of Thebit just above the longer shadow spire, and
a "crater pit" to the E of this dome. At 20.42 the W crest of
Birt was seen by which time the main shadow spire had retreated
to approx half its original length.
N. Longshaw.

somewhat 'squashed' in the north–south direction; 132 km, as opposed
to its east–west span of 140 km. Its eroded walls rise up to just over 4 km
above the rough and hummocky floor. The walls are very broad and are
divided by valleys along the southern section. Of particular note is the
cluster of craters on its north-east (lower left in the photograph) quadrant. Any ideas on the evolution of this and the other craters in the area?

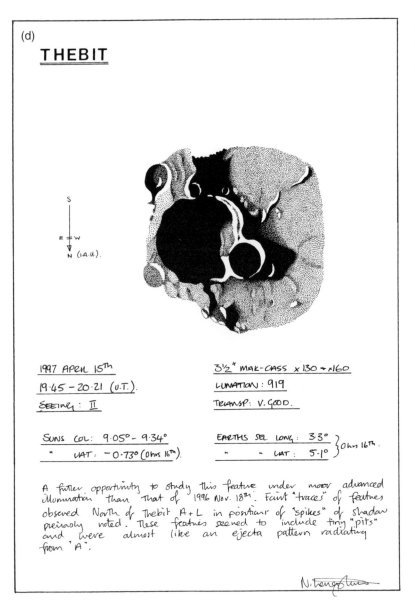

(d)

<u>THEBIT</u>

S

E ≠ W

N (I.A.U.)

1997 APRIL 15TH

19·45 - 20·21 (U.T.).

SEEING : II

3½" MAK-CASS × 130 + ~160

LUNATION : 919

TRANSP: V. GOOD.

SUNS COL: 9·05° - 9·34°

" LAT: -0·73° (0hrs 16TH).

EARTHS SEL. LONG : 3·3°

" " LAT : 5·1° } 0hrs 16TH.

A further opportunity to study this feature under more advanced illumination than that of 1996 Nov. 18th. Faint "traces" of features observed North of Thebit A+L in positions of "spikes" of shadow previously noted. These features seemed to include tiny "pits" and were almost like an ejecta pattern radiating from "A".

N. Longshaw

Figure 8.38 (*cont.*)
(d) A further study of Thebit, by Nigel Longshaw.

Figure 8.38(b) is another Catalina Observatory, 1.5 m telescope, photograph. It shows Walter in its completeness. It also shows the area very shortly after local sunrise, the selenographic colongitude here being 7°.1. The photograph was taken on 1967 January 19^d 02^h 45^m UT.

In the lower-right corner of Figure 8.38(a) (and (b)) is pictured a rather beautiful little arrangement of overlapping craters. The largest of these, at 55 km diameter, is Thebit. It is 3.3 km deep and has a rather rough floor.

This crater is intruded upon by the 20 km diameter Thebit A. It has a smoother, almost bowl-like, profile and a small flat floor, at a level 2.7 km below that of the crater rim. Thebit A is, itself, invaded by a yet smaller crater. This one is now officially known as Thebit L (you may find other designations elsewhere) and is 12 km across. It is very shallow and has a tiny central crater within it.

Proponents of the endogenic theories of crater production made much of this formation in order to support their views. As well as the sequence of craters of diminishing size supporting their idea of a diminishing scale of volcanism occurring along a fault to produce the formation, they cited the undeniable evidence of the perfection of the crater outlines up to the point of the junction between them. They argued that any explosive event would have produced a shaking down of the walls of the earlier craters. However, when we look at lunar craters at much higher resolutions than is usually obtainable by conventional observations through our atmosphere we **do** see a degree of disturbance; enough not to need an endogenic creation mechanism for the intruding crater. The fact that intruding craters are almost always smaller than the craters they break into is explained by the fact that the smaller number of large impactors was more rapidly used up, leaving the greater number of smaller rocky (and icy?) fragments to subsequently pepper the Moon. So, the mystery disappears. In fact, there was never really a mystery at all.

Nonetheless, Thebit has always attracted the attention of observers. Figure 8.38(c) and (d) are two excellent studies of the formation by Nigel Longshaw. The whole area is rich in detail and full of interest. I commend its study to you.

The southern part of the crater rim shown at the bottom of Figure 8.38(a) belongs to Arzachel. I included it so that you might see how this region connects with that detailed in Section 8.4. Immediately west of Thebit is the formation known as 'The Straight Wall', more properly the Rupes Recta. This is discussed in Section 8.43.

8.39 RUSSELL [27°N, 284°E] (WITH BRIGGS, BRIGGS A, BRIGGS B, EDDINGTON, KRAFFT, SELEUCUS, STRUVE)

The crater Russell is situated near the western edge of the Oceanus Procellarum and is very close to the north-eastern limb of the Moon. Roy Bridge has made an excellent drawing of sunrise over this formation and this is presented in Figure 8.39(a). This formation is the remains of an ancient 'walled-plain' type crater, about 99 km in diameter. A wider view of the area is shown in Figure 8.39(b), in which it can be seen that Russell is connected, via its missing south wall, to another great ring structure. This one we now call Struve. In older maps it is called Otto Struve and

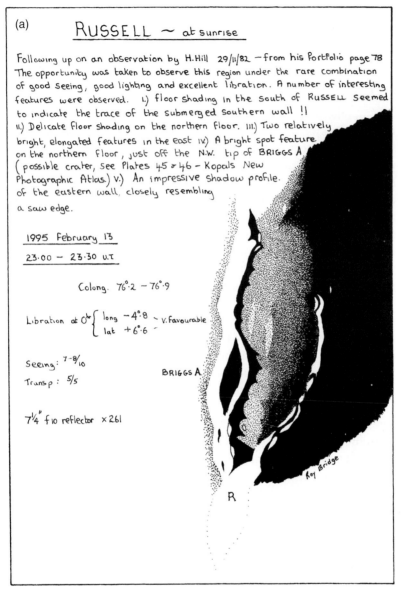

Figure 8.39 (a) Russell, drawn by Roy Bridge.

Russell is sometimes denoted as Otto Struve A. On other old maps, Russell is taken to be part of Otto Struve. As well as A now being called Russell, the adjoining crater now commemorates all three astronomers Struve (Friedrich G. Wilhelm von Struve, his son Otto Wilhelm von Struve, and grandson Otto Struve). It is perhaps appropriate that the crater with three peoples' names is so large. It spans 183 km.

Figure 8.39(b) is a photograph taken with the Catalina Observatory 1.5 m reflector on 1966 October 28^d 06^h 28^m UT, when the Sun's selenographic

Figure 8.39 (*cont.*)
(b) Russell (bottom), connected to Struve (large formation on the terminator) and Eddington (adjoining it and sharing its wall with Struve) on the left of Struve. To the left of Eddington is the crater Seleucus. Above Eddington is the crater Krafft. Of the two craters near the bottom left, the upper one is Briggs and the lower (smaller) one is Briggs B. Briggs A is actually on the eastern (left in this photograph) rim of Russell. Other details in text. (Catalina Observatory photograph – courtesy Lunar and Planetary Laboratory.)

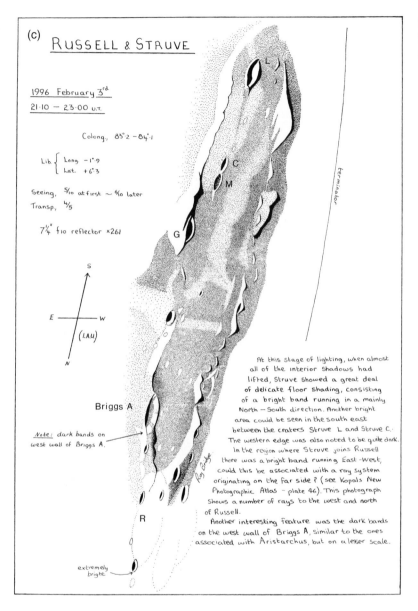

(c) RUSSELL & STRUVE

1996 February 3rd
21·10 — 23·00 u.t.

Colong, 83°·2 — 84°·1

Lib. { Long. −1°·9
      Lat. +6°·3

Seeing, 5/10 at first ∼ 6/10 later
Transp, 4/5

7¼" f10 reflector ×261

(I.A.U)

L

C

M

G

terminator

Briggs A

Note: dark bands on west wall of Briggs A.

Roy Bridge

R

extremely bright

At this stage of lighting, when almost all of the interior shadows had lifted, Struve showed a great deal of delicate floor shading, consisting of a bright band running in a mainly North — South direction. Another bright area could be seen in the south east between the craters Struve L and Struve C. The western edge was also noted to be quite dark. In the region where Struve joins Russell there was a bright band running East — West, could this be associated with a ray system originating on the far side ? (see Kopals New Photographic Atlas − plate 46). This photograph shows a number of rays to the west and north of Russell.
Another interesting feature was the dark bands on the west wall of Briggs A, similar to the ones associated with Aristarchus, but on a lesser scale.

Figure 8.39 (cont.)
(c) Russell and Struve, drawn by Roy Bridge.

colongitude was 79°.3. Notice how the floors of the craters obviously share the curvature of the Moon's surface (revealed by the shading). This is hardly surprising, since they are all flooded with mare lavas.

Figure 8.39(c) is another of Roy Bridge's splendid drawings. This one shows the wider area, encompassing Russell (at the bottom), and Struve. Notice that Roy has included the letter designations of many of the smaller craters. He has also picked out some radial dark bands running up the

Figure 8.39 (*cont.*)
(d) Russell, Struve (old name 'Otto Struve') and Eddington, drawn by Andrew Johnson.

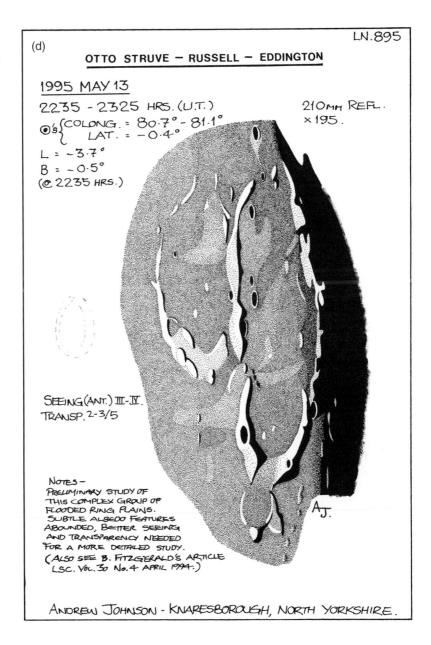

interior of the crater Briggs A, situated on the eastern rim of Russell. You might look for these yourself. Briggs A has a diameter of about 23 km.

Locate Russell, then Briggs A near the bottom of Figure 8.39(b) and you will see two craters to the left (east) of Briggs A. The upper crater is Briggs. It is 39 km in diameter and has a most interesting interior. I will leave its examination to you. The lower one is Briggs B. It is 25 km in diameter.

Attached to the eastern flank of Struve, and sharing part of its rim, here significantly enhanced, is another flooded and ruined old ring. On modern maps this one is called Eddington. Confusingly, on some old maps this crater is known as Otto Struve A. It is 134 km in diameter. Figure 8.39(d) shows a splendidly detailed, and yet wide-angle, view of Russell, Struve and Eddington by Andrew Johnson.

A short distance east of Eddington is the 43 km diameter prominent crater Seleucus. It is well shown in Figure 8.39(b). Notice the unusual profile of this crater. It is very deep for its size, the depth being approximately 3 km.

The prominent crater near the top of Figure 8.39(b) is the 51 km diameter Krafft. Notice the pretty cluster of small craters around it, like bees around a honey pot.

The notes I have given here are very brief. In common with the later sections of this chapter, my intention is just to highlight some interesting areas of the Moon for you to investigate for yourself. I am sure that you will agree that in this region there is plenty to investigate.

8.40   SCHICKARD [44°S, 305°E] (WITH LEHMANN)

Schickard is a vast, partially flooded crater of the 'walled-plain' variety. It is 227 km in diameter and is situated near the Moon's south-western limb. A portrait of this great edifice is presented in Figure 8.40(a). It was

Figure 8.40 (a) Schickard. See text for details. (Catalina Observatory photograph – courtesy Lunar and Planetary Laboratory.)

Figure 8.40 (cont.)
(b) Sunrise over
Schickard, drawn by
Andrew Johnson. Note
the drawing is orientated
with west very
approximately
uppermost.

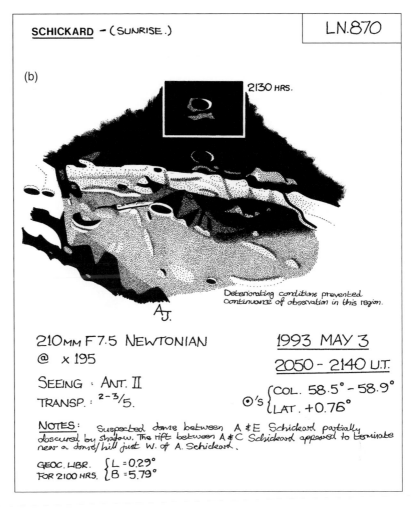

SCHICKARD – (SUNRISE.)

LN.870

(b)

2130 HRS.

Deteriorating conditions prevented
continuance of observation in this region.

A.J.

210MM F7.5 NEWTONIAN
@ x 195

SEEING : ANT. II
TRANSP. : 2-3/5.

1993 MAY 3
2050 - 2140 U.T.

⊙'s { COL. 58.5° - 58.9°
      { LAT. +0.76°

NOTES : Suspected dome between A & E Schickard partially
obscured by shadow. The rift between A & C Schickard appeared to terminate
near a dome/hill just W. of A. Schickard.

GEOC. LIBR.   { L = 0.29°
FOR 2100 HRS. { B = 5.79°

taken with the Catalina Observatory 1.5 m reflector on 1966 September
10^d 12^h 02^m UT, when the Sun's selenographic colongitude was 216°.8.

BAA Lunar Section members have long been interested in Schickard
and the variety of objects that litter its floor. Probably the selenographer
who has, more than most, made this formation his own is Keith Abineri.
He began his telescopic studies in April 1946 and continued to work at
this feature until the end of the twentieth century. In the last years of his
work he applied himself to the examination of *Orbiter* and *Clementine*
space-probe imagery. He has published many papers in the BAA *Journal*
and BAA Lunar Section publications, such as *The New Moon*. You might like
to search these out for yourself. As well as being interesting in their own

(c)

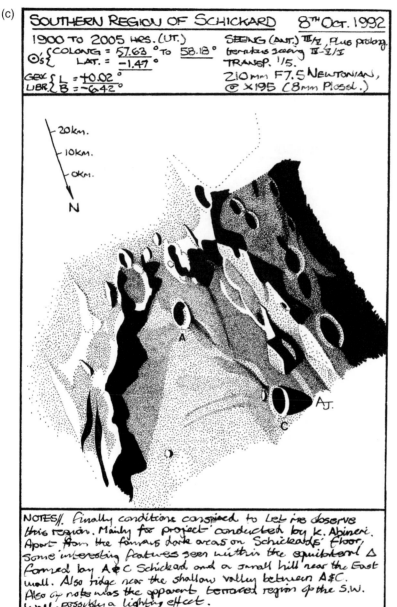

Figure 8.40 (*cont.*) (c) The southern half of Schickard, drawn by Andrew Johnson.

Figure 8.40 (*cont.*) (d) The
northern half of
Schickard, drawn by
Andrew Johnson.

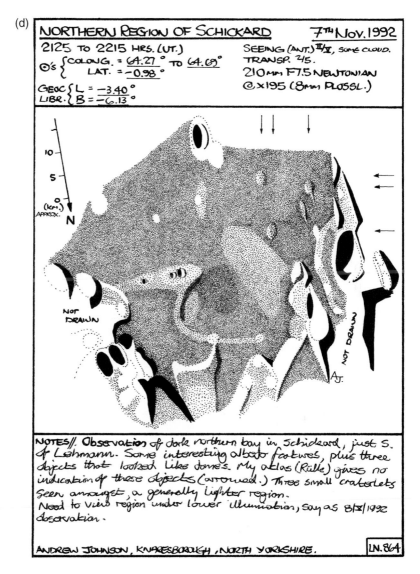

right, you might find some of these instructive with regard to the methodology he adopted.

Figure 8.40(b), (c) and (d) are recent telescopic studies of sections of this formation undertaken by Andrew Johnson.

A roughly triangular swath of lighter-hue material crosses the floor of Schickard from the south-west (where it is widest) to the north-east, the rest of the crater floor being very dark. This appearance is best seen under a high Sun, though indications of it are visible in Figure 8.40(a).

The rim of Schickard is interrupted by various craters and the largest abutting crater, at 53 km diameter, is the highly eroded Lehmann. It is shown in the bottom-right corner of Figure 8.40(a).

8.41 SCHILLER [52°S, 320°E]

Not far from the giant crater Schickard (see previous section), is a real lunar oddity. At first glance Schiller appears like a considerably elongated crater. It is 179 km long and spans 71 km at its widest point. Figure 8.41(a) to (d) illustrate the crater under a series of lightings from early morning (a) to sunset (d). They are photographs taken with the Catalina Observatory 1.5 m reflector and the details of date, time and selenographic colongitudes are given in the accompanying captions. Figure 8.41(e) presents a much more detailed view, this having been obtained by the *Clementine* space-probe.

The usual explanation given in print for this formation is that it is the fusion of two craters. However, I disagree. I think that Schiller was formed from at least three, possibly four, fused craters or, perhaps, by at least three, possibly more, large projectiles arriving virtually simultaneously. My reasoning stems from the outline of the formation. To me, the southern rim has a smaller radius than that of the adjoining main section. The narrower northernmost section also constricts sharply

Figure 8.41 (a) Schiller, photographed using the 1.5 m reflector of the Catalina Observatory in Arizona, on 1966 April 2^d 08^h 03^m UT, when the Sun's selenographic colongitude was 48°.7. (Courtesy Lunar and Planetary Laboratory.)

Figure 8.41 (*cont.*)
(b) Schiller, photographed using the 1.5 m reflector of the Catalina Observatory in Arizona, on 1967 February 22^d 03^h 43^m UT, when the Sun's selenographic colongitude was 61°.0. (Courtesy Lunar and Planetary Laboratory.)
(c) Schiller, photographed using the 1.5 m reflector of the Catalina Observatory in Arizona, on 1966 January 6^d 05^h 45^m UT, when the Sun's selenographic colongitude was 80°.9. (Courtesy Lunar and Planetary Laboratory.)

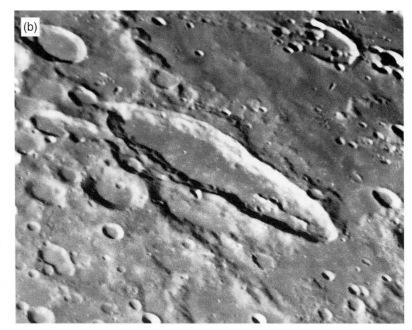

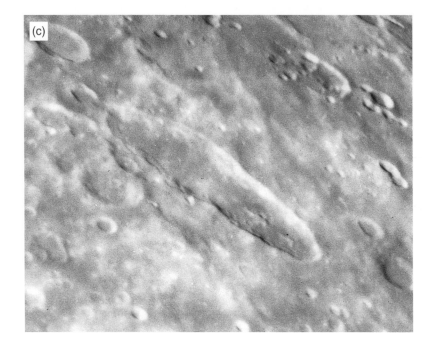

Figure 8.41 (*cont.*)
(d) Schiller, photographed using the 1.5 m reflector of the Catalina Observatory in Arizona, on 1966 September 10^{d} 12^{h} 02^{m} UT, when the Sun's selenographic colongitude was 216°.8. (Courtesy Lunar and Planetary Laboratory.)

where it ends in an arc of smaller radius. Moreover, the northern section sports two 'central peaks' both of which are highly elongated into ridges running along the long axis of the southern part of Schiller. The northern half of Schiller doesn't quite follow the same axis. It slightly 'kinks' a little to the east.

Putting the, albeit superficial, evidence together, I most favour the idea that a tight cluster of projectiles, whether they be cometary or asteroidal fragments, impacted the Moon at a low angle. The direction would obviously be along the long axis of the formation – but from which direction? The fact that just the northern end of Schiller has a 'central peak' type of formation might be significant. Was any part of the current formation already in existence on the Moon before the impacts that created the rest of it, or was it all formed in one go?

Of course, I must make it very clear that the foregoing is nothing official. It is merely my speculations on the subject. One thing is certain, though: there is much more to Schiller than the simple 'fusion of two craters' idea usually peddled. What do you think about it?

Figure 8.41 (*cont.*)
(e) *Clementine* image of
Schiller. (Courtesy
NASA.)

8.42   SIRSALIS, RIMAE [14°S, 320°W] (WITH SIRSALIS, SIRSALIS A)

Situated near the Moon's western limb, the Rimae Sirsalis is just about the longest of the lunar rilles. Its length is 330 km. It is even more noteworthy in that it is one of the straightest over such a long length. Most of it is just one rille, the Rima Sirsalis, but it does have a few branches and extensions. Hence the more exact name of Rimae Sirsalis.

Figure 8.42(a) shows the main part of it in a photograph taken using the 1.5 m reflector of the Catalina Observatory, in Arizona. It was taken on 1966 February 4^d 07^h 02^m UT, when the Sun's selenographic colongitude was 74°.2. The rille can be seen running down the centre of the photograph. It is a graben; a slump feature. What caused it? The outer fringes of the Orientale Basin are not too far away. However, the rille runs neither radial to it, nor tangential. So, was the Orientale impact the cause?

Many amateurs have observed this feature as a sport, trying to trace the limits of its extension north and south. Four serious studies of the rille are presented here in Figure 8.42(b)–(e). From the first two, they cover sections going progressively northwards along it.

The significant pair of overlapping craters shown close to the centre of Figure 8.42(a) are Sirsalis and Sirsalis A. The complete one is Sirsalis. It is 44 km in diameter and is about 3 km deep. The central mountain is small, really just a hill, but is nonetheless prominent. Sirsalis A is actually slightly larger, at 49 km diameter, than Sirsalis though it is significantly overlapped by Sirsalis.

(a)

Figure 8.42 (a) Sirsalis and Sirsalis A (the overlapping craters below the centre) and Rimae Sirsalis (running down the middle). Details in text. (Catalina Observatory photograph – courtesy Lunar and Planetary Laboratory.)

Figure 8.42 *(cont.)*
(b) Sirsalis and Rimae
Sirsalis, drawn by
Andrew Johnson.

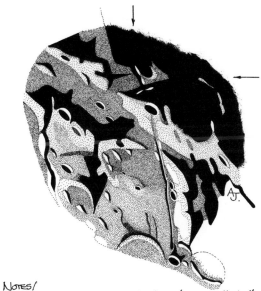

(b)

SIRSALIS & RILLE

1995 APRIL 12

2000 - 2110 HRS. (U.T.)

☉'s { COLONG. 61·1° – 61·7°
         LAT. 0·4°

GEOC. LIBR. { L = -6·3°
@ 2000 HRS. { B = 5·2°

210mm F7·5 NEWT.
@ × 195

SEEING. (ANT.) II
TRANSP. 2/5

AJ.

NOTES/
    Further observation in search of confirmation that the Sirsalis rille
does reach Sirsalis J, climbing the surrounding slope of the crater.
This observation was made under a lower angle of illumination, and
as the rille was indeed seen, and that it lacked any real shadows,
perhaps these conditions indicate that the rille may be quite
shallow over this section of its path? Arrows indicate possible
extension of rille beyond the terminator.
(Shadows shown as at 2025 HRS.)

ANDREW JOHNSON, KNARESBOROUGH, NORTH YORKSHIRE.

Figure 8.42 (*cont.*)
(c) Sirsalis and Rimae
Sirsalis, another drawing
by Andrew Johnson.

Figure 8.42 *(cont.)*
(d) Rimae Sirsalis and De
Vico A, drawn by Roy
Bridge.

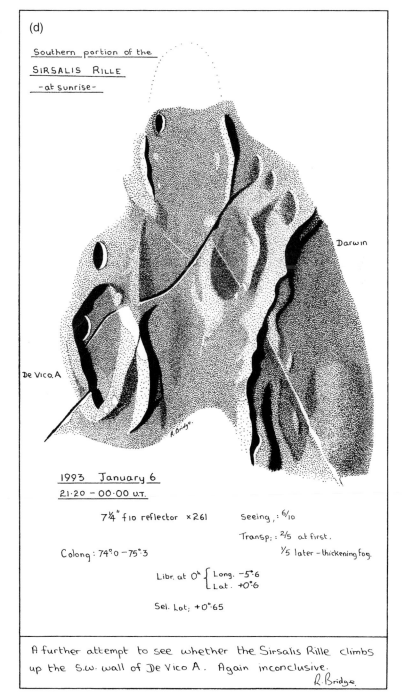

(d)

Southern portion of the
SIRSALIS RILLE
-at sunrise-

De Vico A

Darwin

R.Bridge.

1993 January 6
21·20 – 00·00 u.t.

7¼" f10 reflector ×261   Seeing, : 6/10

Transp; : 2/5 at first.

Colong : 74°·0 – 75°·3   1/5 later – thickening fog.

Libr. at 0ʰ { Long. –5°·6
             { Lat . +0°·6

Sel. Lat; +0°·65

A further attempt to see whether the Sirsalis Rille climbs
up the S.w. wall of De Vico A. Again inconclusive.
                                        R. Bridge.

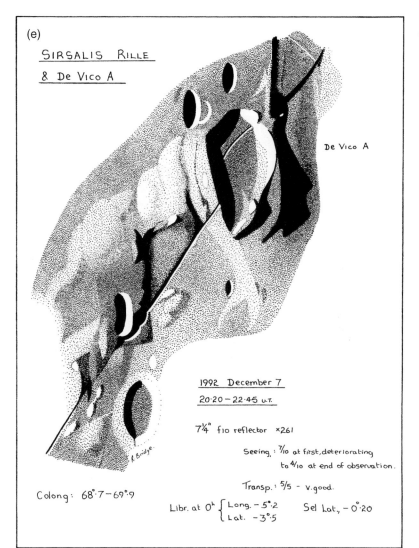

(e)

SIRSALIS RILLE
& De Vico A

De Vico A

1992 December 7
20·20 — 22·45 u.t.

7¼" f10 reflector ×261

Seeing : ⁷/₁₀ at first, deteriorating
to ⁴/₁₀ at end of observation.

Transp. : ⁵/₅ - v.good.

Colong: 68°·7 — 69°·9

Libr. at 0ʰ { Long. - 5°·2   Sel Lat., - 0°·20
            { Lat. - 3°·5

R. Bridge.

Figure 8.42 (*cont.*) (e) De Vico A and the northern section of the Rimae Sirsalis, drawn by Roy Bridge.

This observation follows up a query brought to my attention by Harold Hill. Does the rille climb up the S.W. wall of De Vico A? Unfortunately the observation was inconclusive. I found it extremely difficult at the time to say whether it did or not. There certainly seemed to be some indication of it doing so, but the area in question is small and the seeing wasn't perfect. It may have been just illusionary with the eyes attempting to join up the rille on either side of the wall!

R. Bridge.

8.43 'STRAIGHT WALL' (RUPES RECTA) [22°S, 352°E]
(WITH BIRT, BIRT A)

Properly known as Rupes Recta, this formation is so widely and popularly known by its old name of the 'Straight Wall' that I have entered it here under that name.

One of Tony Pacey's excellent photographs shows it well (Figure 8.43(a)). Tony used his 10-inch (254 mm) Newtonian reflector and a 4 mm Orthoscopic eyepiece to project the image onto FP4 film. The ½ second

Figure 8.43 (a) The 'Straight Wall' formation, more properly called Rupes Recta, photographed by Tony Pacey. Appearing here as a thin black line, the overlapping craters Thebit, Thebit A and Thebit L are just to its left and the small but distinctive Birt, with Birt A on its rim, is just to its right. Other details given in the main text.

Figure 8.43 (*cont.*) (b) Rupes Recta, photographed using the 1.5 m reflector of the Catalina Observatory in Arizona. Details in text. (Courtesy Lunar and Planetary Laboratory.)

exposure was given at 1988 April 24^d 22^h 00^m UT. At the time the Sun's selenographic colongitude was 356°.8.

Rupes Recta is situated near the eastern border of the Mare Nubium, just west of the distinctive overlapping craters Thebit, Thebit A and Thebit L (see Section 8.38). Tony Pacey's photograph shows the general environs, including the Thebit trio and Regiomontanus (upper-left corner) and Arzachel, Alphonsus and part of Ptolemaeus (in the lower left). These areas are detailed in Sections 8.38 and 8.4, respectively. The crater Pitatus is shown in the upper-right corner of Figure 8.43(a) and this feature is described in Section 8.32. Just west (right in Figure 8.43(a)) of Rupes Recta is a small but distinctive crater, Birt, with a small crater on its rim, Birt A. Birt is a bowl-shaped formation 17 km across and 3.5 km deep. Birt A has a diameter of 6.8 km and a depth of 1 km.

The 'Straight Wall' is not particularly straight and it most certainly is not a wall. It comes into daylight just after first quarter Moon and appears as a thin black line when illuminated from the east (local morning). As with most lunar relief features, it washes out near full Moon but then appears as a thin light line towards the late afternoon, then being illuminated from the west. This shows that there is a difference in height of the ground to either side of it.

Despite appearances, the feature is not a sheer cliff-face of extraordinary height. Rather it is a fairly gentle slope linking the higher ground to the east with the lower plain to the west. The average slope is of the order of 7° and its height is not much more than 240 m. However the formation is remarkable in view of its length, which is about 110 km.

What caused it? Is it a case of the ground to the east being uplifted, or the ground to the west slumping downwards? Opinions are still divided but the most popular view is that the ground to the east was buckled upwards under compressional forces across this part of the Mare Nubium.

Take a careful look at Figure 8.43(a) and you should be able to discern the outline of an almost entirely obliterated crater. Look at the border of the terrae to the east of the Rupes Recta. Thebit actually lies across the old rim. The lavas of the Mare Nubium have melted away most of its western rim but traces of it can still be made out on the mare. You will see that Rupes Recta spans much of the diameter of this old 'ghost' crater. Could that be significant?

I will leave you to ponder on this intriguing formation but, to get you started, you might like to consider the other buried ring at the southern end of it and the rille that runs approximately parallel to it, a little to the west and passing just beyond Birt. A close-up view is provided in Figure 8.43(b), which is a Catalina Observatory photograph, taken using the 1.5 m reflector on 1966 May 29^d 04^h 41^m UT, when the Sun's selenographic colongitude was 22°.6.

8.44   THEOPHILUS [11°S, 26°E] (WITH CATHARINA, CYRILLUS, MÄDLER, MONS PENCK)

One of the most striking arrangements of craters on the Moon must be that of Catharina, Cyrillus and Theophilus. They first come into sunlight about 5–6 days after new Moon and then present the spectacle shown in Figure 8.44(a). Figure 8.44(b) shows the area under a higher Sun. Even then the grouping still looks impressive. The craters are shown lit from the opposite direction in Figure 8.44(c), which is a view you will see about 19–20 days after new Moon. The craters fill with shadow and are finally extinguished at a lunar age of about 19½ days (see Figure 8.44(d)).

In part, the dramatic appearance of the craters is heightened by the scarp Rupes Altai framing the group to the west. The scarp is a raised ring of mountains uplifted at the time of the impact that created the nearby, and now lava-flooded, Nectaris Basin. Cyrillus, Catharina and Theophilus are thus sandwiched between the Mare Nectaris, to their east, and the Rupes Altai, to their west. The rest of the Mare Nectaris region is discussed in Section 8.30.

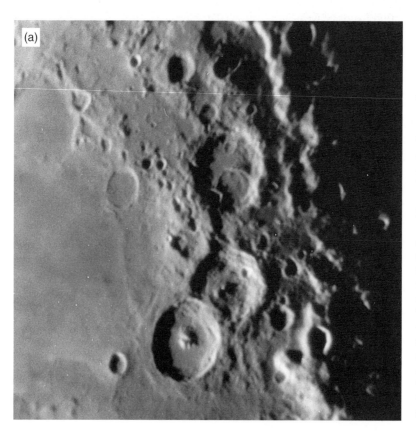

(a)

Figure 8.44 (a) Theophilus (bottom large crater), Cyrillus (above and adjoining Theophilus) and Catharina (above Cyrillus), photographed by Tony Pacey using his 10-inch (254 mm) Newtonian reflector on 1991 March 21^d. The approximate time of the exposure was 20^h UT and the approximate value of the Sun's selenographic colongitude was 330°. The ½ second exposure was made on T-Max 100 film, processed in HC110 developer. The small crater on the left of Theophilus is Mädler, which is actually sited on the Mare Nectaris.

Figure 8.44 (*cont.*)
(b) Theophilus and
environs photographed
using the 1.5 m reflector
of the Catalina
Observatory on 1966
September 3^d 09^h 19^m
UT, when the Sun's
selenographic
colongitude was 130°.0.
(Courtesy Lunar and
Planetary Laboratory.)

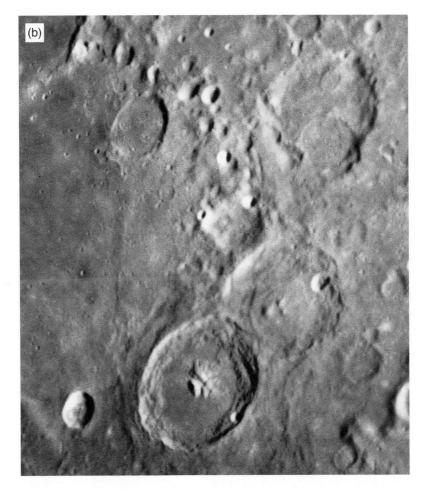

The southernmost of the trio of craters is Catharina. It is 97 km in diameter. Its rough walls are very irregular in outline and heavily crater-spattered. They reach up to 3.1 km above the crater. Notice the interior details. There is a story here, but I will leave its investigation to you.

Catharina is clearly the most eroded, and thus the oldest, of the trio but the middle crater, Cyrillus, is not much younger as far as lunar chronology goes. Notice the apparent channel connecting Cyrillus to Catharina. In reality the connection is the result of further impacts, though a degree of ground-slumping is also evident.

Cyrillus is 93 km in diameter and is of similar depth to Catharina, though its gentler interior slopes do give the impression of it being deeper than it really is. Notice the prominent double-peaked mountain near the centre of Cyrillus. It is instructive to compare the floors of the

craters Cyrillus and Catharina. What are their similarities and what are their differences – and how do you explain them?

Obviously the youngest of the three craters is the 100 km diameter Theophilus. It is a spectacular object in its own right, even divorced from the presence of Catharina and Cyrillus (which it invades). The outer flanks of this crater are very complex and the crater rim rises up to about 1.2 km above the level of its outer surrounds. The interior is rather deep, extending to 3.2 km below the level of the outer surrounds. As the accompanying illustrations show, the interior slopes of Theophilus are very complex. Clearly the initial terracing has been much degraded by localised landslips. Notice the smooth central arena that surrounds the magnificent central mountain cluster. The highest peaks of this complex soar up to about 2 km above the level of the crater floor. I wonder when it will be that a lunar mountaineer will climb the main mountain and from the summit view the incredible spectacle of the surrounding unearthly landscape?

Various observers have noted odd appearances in Theophilus and many consider it to be a 'TLP hot spot'. I have never seen anything untoward in it, myself.

At the position where the Rupes Altai passes closest to Theophilus is a prominent mountain peak reaching further eastwards than the rest of the range at this point. This feature is named Mons Penck and it is the subject of the drawing by Andrew Johnson shown in Figure 8.44(d). Figure 8.44(e) presents one of Terry Platt's very good CCD images of Theophilus.

The Mare Nectaris lavas encroach right up to the eastern flanks of Theophilus and the significant crater to the east of Theophilus, called Mädler, actually sits in the junction of the Mare Nectaris with the Mare Tranquillitatis. It has a somewhat distorted outline, its diameter averaging about 28 km. It is very deep for its size, measuring about 2.7 km vertically from floor to rim.

Just a little north of this area of grand craters and spectacular mountains is a little grouping of very small and apparently insignificant craters that are, nonetheless, fascinating in their own right and may even hold a secret or two. These are the subject of the next section.

(d)

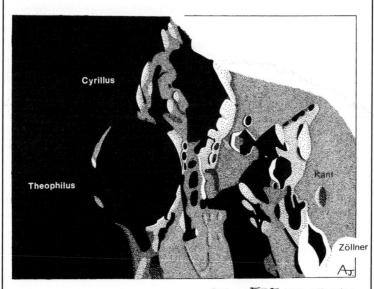

Mons Penck (sunset)

LN.888

1994-OCTOBER 24

2215 – 2310 HRS. (U.T.)
COLONG. = 153·2° – 153·7°
    LAT. = −0.37°
GEOC. LIBR. { L = −4.75°
@ 2215 HRS. { B = 4·87°

SEEING (ANT.) II
TRANSP. 3/5.

Cyrillus

Theophilus

Kant

Zöllner

A.J.

210MM F7.5 NEWTONIAN
@ × 195 (8MM PLOSSL.)

NOTES:
    Clouded out by 2325 HRS. Glimpsed between clouds around 2330 HRS; shadow from Mons Penck had linked-up with craters along west flanks of Theophilus. (Cyrillus M region) consequently looked enormous!
    Although seeing good; consisánt Antoniadi II, transparency only useable at best, hence contrast down. Hoped to attempt a short sequence, to follow shadow development.

KNARESBOROUGH, NORTH YORKSHIRE.

Figure 8.44 (cont.)
(d) Theophilus, Cyrillus and Mons Penck, drawn by Andrew Johnson.

Figure 8.44 (*cont.*)
(e) Theophilus. CCD
image by Terry Platt,
using his 12½-inch
(318 mm)
tri-schiefspiegler
reflector and *Starlight
Xpress* CCD camera. Other
details not available. The
author has applied slight
image sharpening and
brightness re-scaling.

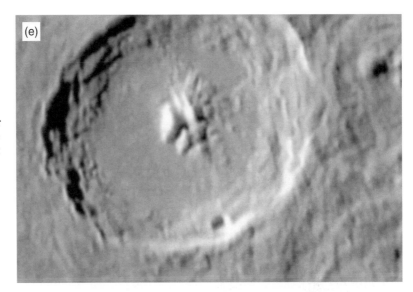

8.45 TORRICELLI [5°S, 28°E] (WITH CENCORINUS, MOLTKE, RIMAE HYPATIA, TORRICELLI A, B, C, F, H, J AND K)

Less than two hundred kilometres north of Theophilus, on the Mare Tranquillitatis, is an interesting little group of features. These are shown in Figure 8.45. This figure and Figure 8.44(b) in the previous section are reproductions of parts of the same Catalina Observatory photograph and the details are given in the previous section. (See the accompanying caption to Figure 8.45 for the identification of the features listed.)

Despite its small size (about 20 km diameter), Torricelli does tend to catch the eye because of its 'keyhole' shape. In fact it seems to be the fusion of two craters, one smaller than the other.

The crater Cencorinus is only 3.8 km wide and yet it is very attention-grabbing when seen under a high Sun. Its interior is very highly reflective and it is surrounded by a patch of bright ejecta. It is normally written that Aristarchus is the brightest crater on the Moon. Undoubtedly that is the case when one takes into account its much larger size. **Area for area**, though, I think Cencorinus is brighter when it is seen under the highest angles of illumination. Cencorinus is probably one of the freshest craters on the Moon that is big enough to resolve with a backyard telescope.

Another prominent, though much less brilliant, little crater is Moltke. It is a bowl-shaped edifice with a diameter of 6.5 km and a depth of about 1.3 km. Space-probe images show it to have a razor-sharp rim and a bright, smooth, interior. It also possesses a fairly bright ejecta nimbus. The eastern end of the Rimae Hypatia passes between Moltke and the hinterland just

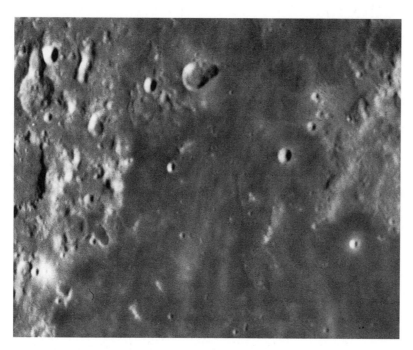

Figure 8.45 The 'keyhole' crater Torricelli (top centre), Cencorinus (bright crater near bottom-left corner), and Moltke (bright crater, two-thirds of the way down close to the right-hand side), photographed by the Catalina Observatory 1.5 m reflector. Same details as for Figure 8.44(b). The crater half way along a line between Torricelli and Moltke is Torricelli C. Notice the little arc of craters just above Torricelli C. From top to bottom the three main ones are Torricelli K, J and H. Of the two craters just to the left of Torricelli, the larger one is Torricelli A and the smaller one Torricelli F. Below these craters is the crater Torricelli B, notable as the origin of many recent reports of Transient Lunar Phenomena. (Courtesy Lunar and Planetary Laboratory.)

to the south of it. You might be able to see faint indications of it on Figure 8.45, though the lighting is not suitable to show it at its best.

The identified craters Torricelli A, B, C, F, H, J and K have diameters of 11, 7, 11, 7, 7, 5 and 6 km, respectively. Of these, Torricelli B is particularly interesting. Space-probe images show it to be rather conical in profile and the interior is rather asymmetrically surfaced with deposits of varying composition. In particular the *Clementine* images show a brilliant streak of material (which I think is rich in feldspar – not enough room here for me to explain why – though I do not at the present time know if this is the official view) 'splashed' up the north-east interior from the centre to the rim. Much of the south-west of the interior flank seems to be also, though less richly, covered in the same material. This explains the brilliant 'blob' in the corresponding position that I have observed in this crater on many occasions through my own telescopes.

Torricelli B seems to be one of the more definite TLP 'hot spots'. It, and the other craters around it, show variations in brightness and prominence during the progress of a lunation. All perfectly normal and understandable. The crateriform aspect of Torricelli B is most obvious when the Sun shines at a low angle over it and much of its interior is filled with shadow. Under a higher Sun it becomes a greyish disk, brightest close to the time of full Moon. Every so often, though, Torricelli B looks either much brighter or much duller than one expects that it should at the given point in the lunation. I also find that its colour occasionally changes from its normal white and takes on a very strong blue caste. When the colour is strong the crater can even take on a purplish halo! I have also watched the crater erratically varying in brightness with the variations happening on time-scales of the order of minutes. Meanwhile the surrounding features, such as the craters Moltke and Cencorinus stay sensibly constant in appearance. Naturally any variations of the order of a second or two, and less, could well be caused by atmospheric turbulence – but variations of brightness on time-scales of minutes?

The BAA Lunar Section members have kept an eye on this crater since the first anomalies were noticed in January 1983. In particular one member, Mrs Marie Cook, has diligently observed Torricelli B visually and with the use of coloured filters from then right up to the time of writing. She has made many hundreds of observations of Torricelli B, using Moltke and Cencorinus as comparisons. Her results do seem to confirm my impression of occasional erratic changes of colour and brightness and it is a pleasure to pay tribute to her work here.

Are these changes illusory, or is there a real physical process happening on the Moon? Certainly spurious colour and bad seeing conditions can, and do, effect the appearances of features. I am sure that this is the correct explanation for the vast majority of reports of supposed TLP. For instance, the brilliant Cencorinus is often bedevilled by spurious colour (prismatically created by the Earth's atmosphere).

Sometimes Cencorinus shows an anomalous brightness change, as occasionally does Moltke, though neither show significant colourations other than that caused by spurious colour. An intriguing area of the Moon, ripe for observation and research.

8.46 TYCHO [43°S, 349°E]

As Figure 8.46(a) shows, the crater Tycho is an impressive formation when seen under a low angle of solar illumination. Figure 8.46(b) is a drawing of this crater made by Andrew Johnson and Figure 8.46(c) is one of Terry Platt's CCD images. The crater is 85 km across and is very deep, as lunar craters go. The vertical height of the crater rim above its deepest point is 4.8 km. The interior terraces leading down from the rim to the central arena are spectacular by any standards, as Figure 8.46(c) probably shows best even though it was made under the conditions of a higher Sun-angle. As one might expect, space-probe images show the crater in greater detail. Figure 8.46(d) is a stunning view obtained from *Orbiter V*. The central mountain massif soars upwards to a point 1.6 km above the crater floor.

Figure 8.46 (a) Tycho photographed on 1967 January 20^d 01^h 52^m UT, using the 1.5 m reflector of the Catalina Observatory in Arizona. At the time of the photograph the Sun's selenographic colongitude was $18°.5$. (Courtesy Lunar and Planetary Laboratory.)

As I said, the crater is spectacular enough when seen under a low angle of illumination. If anything, it becomes even more so under a high Sun. True, the interior relief disappears but the crater then becomes a white disk with a brilliantly white ring marking the crater rim and an equally brilliant 'blob' marking the central mountain. The crater itself is then surrounded by a darkish halo. Beyond this commences a magnificent ray system, which radiates across much of the Moon's visible face (see Figure 8.46(e)).

The ray systems associated with craters had long puzzled selenographers and a wide range of theories were 'cooked up' to explain them. We now know that they were generated by the impact explosion that created each parent crater. The rays are composed of a very fine sprinkling of pulverised ejecta (mainly glassy beads) spattered ballistically across the Moon. In the space of a few hundred million years the rays fade away because of the effect of solar-wind bombardment, micrometeorite impacts and, particularly, the 'gardening' (churning) of the topsoil that results from micro meteorites and the diurnal thermal stresses.

Figure 8.46(f) was, despite its appearance, taken from Earth. In fact, a Catalina telescope photographic image was projected onto a white sphere and this was photographed, to produce a 'rectified' view of Tycho and its rays. Note the zone of avoidance in the ray system, indicating the direction of the incoming projectile that created Tycho.

The obvious rarity of large craters with ray systems lends support to the idea that no large meteorites have struck the Moon for a very long

Figure 8.46 *(cont.)*
(b) Tycho, drawn by
Andrew Johnson.

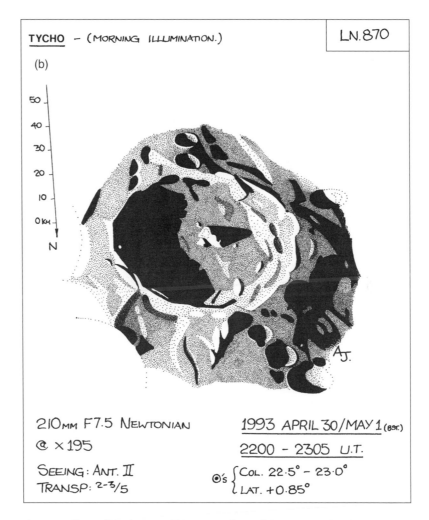

time. In fact, with the possible exception of the crater Giordano Bruno, Tycho is reckoned to be the youngest large crater on the Moon. One of the rays from Tycho passes across the *Apollo 17* landing site. It was from the samples brought back that the nature of the rays was finally settled. Also the impact that created Tycho was determined as having happened about 100 million years ago.

There have been a number of reports of Transient Lunar Phenomena associated with this crater but I have never seen anything I cannot explain as being due to the conditions and/or lighting effects. One relevant observation of mine might be that of 1995 March 10^{d} 22^{h} UT (mid-time – I observed for several hours). Using my 18¼-inch (0.46 m) reflector, at ×144, I could see that parts of the shadow inside the crater were not quite as deep black as the rest, see Figure 8.46(g). The effect was

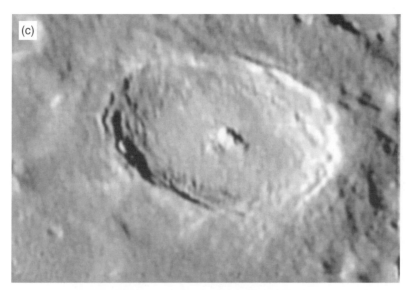

(c)

Figure 8.46 (*cont.*)
(c) Tycho. CCD image by
Terry Platt, using his
12½-inch (318 mm) tri-
schiefspiegler reflector
and *Starlight Xpress* CCD
camera. No other details
available. The author has
applied slight image
sharpening and
brightness re-scaling.

delicate but, nonetheless, too strong to be anything but real. I telephoned
other BAA members and got them to examine the black shadow in Tycho
(but without revealing the details) and was not surprised to get back
confirmation.

How black are the shadows in craters? They certainly **look** very black
most of the time. However, if you were sitting in one of these shadow
regions you would certainly have no trouble in seeing details around you.
For one thing, there would be a weak glow thrown down by the stars in
the black sky. Much more than that, there would be a very bright Earth
shining down on you (if we can see your locale from the Earth, it follows
that you can see the Earth from where you are on the Moon). Also, and
this is the factor of relevance to my Tycho observation, you will be able to
see beyond the shadow-covered ground to the moonscape which is brilliantly illuminated by sunlight. I think the pools of slightly less deep-
black shadow in my Tycho observation were mostly created from light
reflected off the central mountains which were sticking up into the
sunlight, this being enhanced by the light reflected from the encircling
crater walls. Of course, the central peak also interrupted some of the light
reflected back from the far walls, hence the darker divide between the
two 'grey' regions. In how many craters can you see details within the
'black' shadows?

Figure 8.46 (*cont.*)
(d) *Orbiter V* image of
Tycho. (Courtesy NASA
and E. A. Whitaker.)

Figure 8.46 (*cont.*)
(e) Tycho and its ray
system dominate this
photograph of the Moon,
taken by Tony Pacey.
He used his 12-inch
(305 mm) reflector to
image the Moon onto PanF
film at the telescope's
f/5.4 Newtonian focus
(no additional optics
used). The 1/500 second
exposure was made
on 1992 November
$11^d\ 21^h\ 45^m$ UT, when
the Sun's selenographic
colongitude was $100°.9$.

Figure 8.46 (*cont.*)
(f) Rectified image of
Tycho and its ray system.
(Courtesy Lunar and
Planetary Laboratory.)

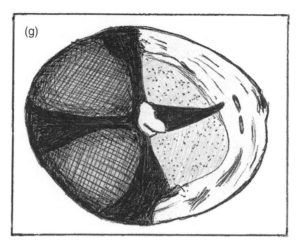

Figure 8.46 (*cont.*)
(g) Tycho. Sketch by the author (details in text) showing the areas of 'less deep-black' shadow. The effect is here grossly exaggerated for the sake of clarity. In reality the patches were very ill-defined and hard to see, scarcely lighter than the deep-black shadow. At the time of the observation the value of the Sun's selenographic colongitude was 7°.8.

8.47   WARGENTIN [50°S, 300°E] (WITH NASMYTH, PHOCYLIDES)

This is the largest (84 km diameter) of a very rare breed of lunar craters indeed – ones filled to the brim with basaltic lava. A CCD image of it by Terry Platt is shown in Figure 8.47(a) and Figure 8.47(b) shows a drawing made by Andrew Johnson that highlights the tree-like pattern of wrinkle ridges on its surface. Wargentin is joined along its lunar southern edge by the vast (114 km diameter) flooded crater Phocylides, while adjoining both is the overlapped (by both) remnants of the once 77 km diameter Nasmyth. The whole grouping is best seen a little before full Moon. Figure 8.47(c) and (d) show Wargentin together with Nasmyth and Phocylides under this sort of lighting (details given in the accompanying captions). I commend you to seek out Wargentin and have a look at it yourself. It is strange to see this lunar 'cup that runneth over'.

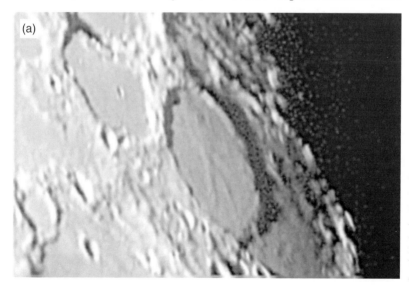

Figure 8.47 (a) Wargentin, imaged by Terry Platt using his 12½-inch (318 mm) tri-schiefspiegler reflector and *Starlight Xpress* CCD camera. The author has applied slight image sharpening and brightness re-scaling. No other details available.

Figure 8.47 *(cont.)*
(b) Wargentin, drawn by
Andrew Johnson.

(b)

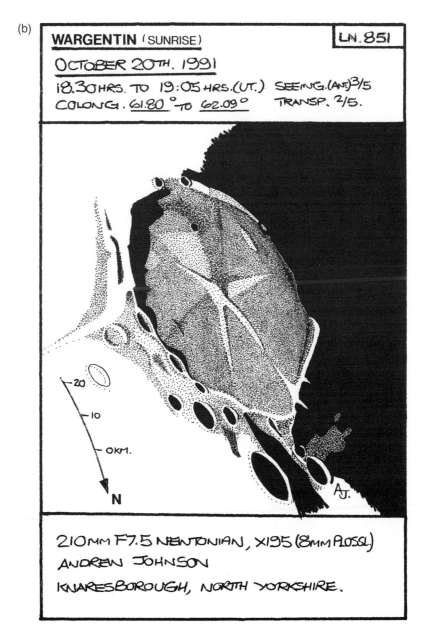

WARGENTIN (SUNRISE) LN.851

OCTOBER 20TH. 1991

18.30 HRS. TO 19:05 HRS. (UT.) SEEING. (ANT) 3/5
COLONG. 61.80° TO 62.09° TRANSP. 2/5.

20
10
0 KM.
N

A.J.

210 MM F7.5 NEWTONIAN, X195 (8 MM PLOSSL)
ANDREW JOHNSON
KNARESBOROUGH, NORTH YORKSHIRE.

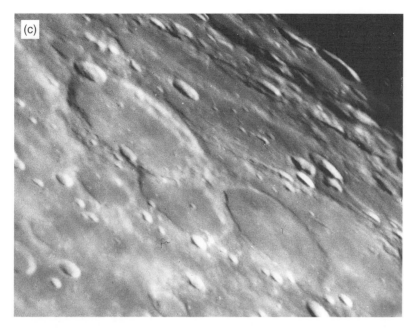

Figure 8.47 (*cont.*)
(c) Wargentin (lower right), Phocylides (largest crater – upper left), and Nasmyth (large crater attached to, and overlapped by, both Wargentin and Phocylides), photographed using the 1.5 m reflector of the Catalina Observatory. The photograph was taken on 1966 January 6^d 05^h 45^m UT, when the Sun's selenographic colongitude was 80°.9. (Courtesy Lunar and Planetary Laboratory.)
(d) Wargentin, Phocylides and Nasmyth, photographed using the 1.5 m reflector of the Catalina Observatory on 1967 February 22^d 03^h 43^m UT, when the Sun's selenographic colongitude was 61°.0. (Courtesy Lunar and Planetary Laboratory.)

8.48   WICHMANN [8°S, 322°E]

In the previous 47 sections we have explored many formations replete
with spectacle and grandeur. I have deliberately finished this chapter with
a region of the Moon that seems obscure and even uninteresting to the
casual glance. Situated on the south-eastern sector of the Oceanus
Procellarum, the little (10.6 km diameter) crater Wichmann certainly
does not grab your attention as you peer through the telescope eyepiece.

Figure 8.48 (a) Wichmann
region, drawn by Andrew
Johnson.

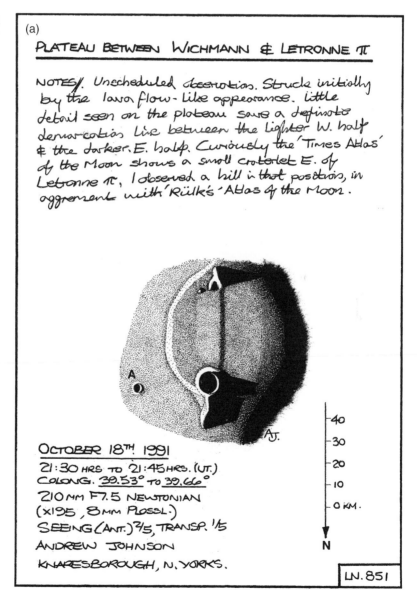

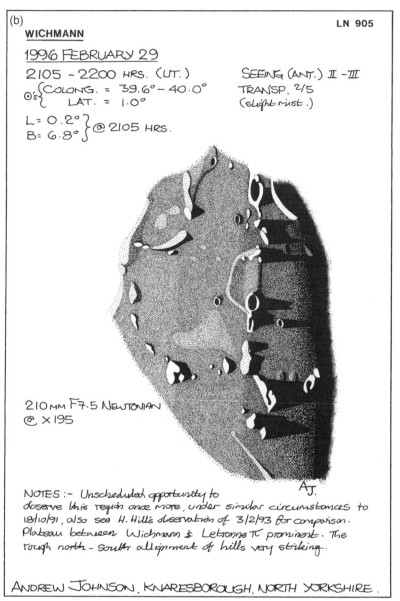

Figure 8.48 (cont.)
(b) Another study of the Wichmann region of the Moon, by Andrew Johnson.

However, what do you see when you look – when you **really** look? You see all sorts of interesting detail – wrinkle ridges, chains of mountains, the raised plateau on which the crater stands. Andrew Johnson has looked – **really** looked. Figure 8.48(a) and (b) show two of his drawings. What are the origins of these features? How do they relate to the Moon as a whole?

The point I am driving at is that the 'old favourite' Moon features may have a great attraction but there is a great deal to be learned from the more obscure areas. The whole of the surface of the Moon tells a story – the story of how it all came to be as it is. I hope that you will feel inclined to investigate the lunar surface for yourself – if so, you will find much of value in some of its less 'well-trodden' regions.

8.49   WEBCAM GALLERY

We finish our tour of the Moon with a small gallery of webcam images. They were all taken by my friend Martin Mobberley and are of formations not, apart from Atlas, discussed earlier. They do not feature on the key map (Figure 7.1 on page 154) but their selenographic co-ordinates are given in the figure captions.

Figure 8.49 (a) Atlas (47°N, 44°E, diameter 87 km). The image was taken by Martin Mobberley on 2005 September 21^d 02^h 34^m UT using a 250 mm f/6.3 Newt @ f/50 + *Lumenera* LU 075M camera. The Sun's selenographic colongitude was 123°.5 at the time of exposure. (See also Section 8.15.)

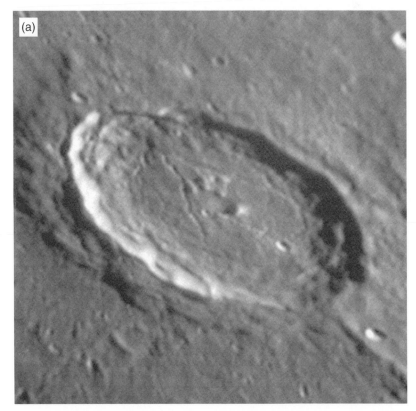

(a)

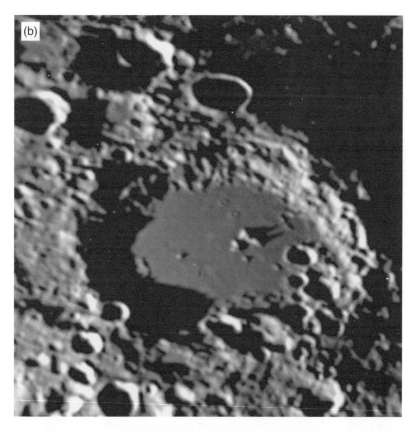

Figure 8.49 (*cont.*)
(b) Longomontanus (50°S, 22°W, diameter 145 km). The image was taken by Martin Mobberley on 2005 April 18^d 22^h 08^m UT using a 245 mm Newt. *AtiK* HIS webcam + I band filter (700–900 nm) f/22, 160 frames × 0.1 s. The Sun's selengraphic colongitude was 28°.2 at the time of exposure.

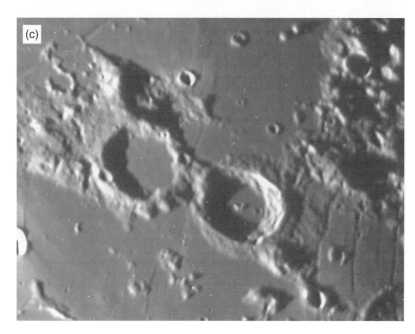

Figure 8.49 (*cont.*)
(c) Mercator (29°S, 24°W, diameter 47 km), Campanus and Rimae Hippalus. The image was taken by Martin Mobberley on 2004 March 1^d 19^h 22^m UT using a ToUcam Pro at 10f ps & 1/25th, with a 178 frame stack. 0.3 m LX200 @ f/22. The Sun's selenographic colongitude was 32°.3 at the time of exposure.

Figure 8.49 (*cont.*)
(d) Mutus (64°S, 30°E,
diameter 78 km). The
image was taken by
Martin Mobberley on
2005 September
$22^{d}\ 01^{h}\ 47^{m}$ UT using
a 250 mm f/6.3 Newt
@ f/50 + *Lumenera* LU
075M camera. The Sun's
selenographic
colongitude was 135°.2 at
the time of exposure.

We may have covered a lot of lunar ground together but there is plenty more of the Moon left to explore. I do hope that you will feel inclined to continue to explore the Moon's unearthly vistas for yourself.

Figure 8.49 (*cont.*) (e) Rugged terrain near Vieta (29°S, 56°W, diameter 78 km). The image was taken by Martin Mobberley on 2005 April 21^d 21^h 21^m UT using a 245 mm Newt *AtiK* HIS webcam + I band filter f/22, 230 frames × 0.2 s. The Sun's selenographic colongitude was 64°.4 at the time of exposure.

CHAPTER 9

TLP or not TLP?

One of the few areas of study for the amateur lunar specialist that remains of genuine scientific use is the highly controversial one of *Transient Lunar Phenomena* (*TLP* – Americans use the term *Lunar Transient Phenomena*, *LTP*). The reason for the controversy is that since the 1950s many people have made quite ridiculous claims of frequent weird happenings on the Moon's surface. The credibility of the subject has suffered greatly from the fanaticism of these people. Now, many people dismiss the whole subject out of hand. Some go as far as to deride all who would study the phenomena.

Most of the serious, long-term, students of the Moon – both amateur and professional – are wisely cautious about supposed transient lunar phenomena but accept that there is some evidence to support the view that there is, at the very least, something worthy of investigation. In this chapter I present some of the evidence and explain how you might take part in this study yourself.

9.1 THE MYSTERY UNFOLDS

Observers, most significantly regular observers, of the Moon report occasional odd appearances, including short-lived glows (sometimes coloured) and mist-like obscurations, involving small areas of the Moon's surface. This is not a new phenomenon. Indeed, reports go back centuries but by the twentieth century, when it was established that the Moon has essentially no atmosphere, most astronomers were of the opinion that all the observed oddities could be explained as mere tricks of the eye.

However, a professional astronomer – Dr Dinsmore Alter – obtained evidence of something more than a simple illusion happening on the Moon in 1955. He used the 60-inch (1.52 m) reflector at Mount Wilson to photograph areas of the Moon in near-infrared and near-ultraviolet–blue light. He did this by using appropriate filter and photographic emulsion

357

combinations. While all remained clear in the infrared photographs, he found that several of the UV–blue images showed a loss of detail on part of the floor of the lunar crater Alphonsus (while the surrounds remained clear). Some researchers wondered if Dr Alter had recorded the presence of a temporary mist emanating from that part of the crater. Infrared light would penetrate a tenuous mist, while the ultraviolet would not do so easily. A few amateur and professional astronomers became interested. One of them was Nikolai Kozyrev, in Russia.

Kozyrev used the 50-inch (1.27 m) Cassegrain reflector of the Crimean Astrophysical Observatory to take regular spectrograms of the Moon's surface. He monitored the Moon through the guiding eyepiece of the spectrograph while he did so. His efforts were rewarded on 3 November 1958 when he obtained spectrographic evidence of a real physical transient event at the Moon's surface.

On that night he was conducting his normal programme when, at just after 01^h UT, he noticed that the central peak of the crater Alphonsus was enveloped in a reddish haze. (See Section 8.4 for more about this crater, including photographs of it). He set the entrance slit of the spectrograph across the image of the central peak of the crater and began taking spectra while he continued to monitor the view through the guiding eyepiece (the slit jaws being reflective allowed guiding and the precise selection of the part of the image sampled by the spectrograph).

During the next couple of hours Kozyrev saw the peak of Alphonsus become very bright and white. Between $03^h 00^m$ and $03^h 40^m$ UT the appearance of the crater returned to normal and Kozyrev ceased taking spectra. When the spectrographic plates were processed several of them showed an anomaly that afflicted the part of the spectrum – the stripe running through the centre of it – formed by the light from the central peak. Meanwhile the parts of the spectrum formed from the other parts of the crater sampled by the slit showed nothing but the normal features of sunlight reflected from the Moon's surface. The spectra taken when the central peak appeared bright showed strong emission bands. Actually these were identified as being the 'Swan bands' produced when molecular carbon vapour, C_2, is excited into emission. Other spectral features were present (indicating other chemical components in the gas), blending with the spectrum of carbon. The last spectra taken showed that all had returned to normal, in accordance with the visual impression.

Kozyrev's observation aroused world-wide interest. You will find an account of it in the February 1959 issue of *Sky & Telescope* magazine. It is titled 'Observation of a volcanic process on the Moon', reflecting his interpretation of what he had observed. Few others accepted this explanation even back then. Now we are sure that all lunar volcanic activity ceased millions of years ago.

Kozyrev gives a more detailed account of his procedures, reductions and conclusions in a paper 'Spectroscopic proofs for existence of volcanic processes on the Moon' in *The Moon – Symposium No. 14 of the International Astronomical Union*, edited by Kopal and Mikhailov for the Academic Press, 1962. Not an easy reference to obtain; perhaps an academic library/inter-library loan service can obtain a copy for you. It is worth reading, even though Kozyrev still gives his interpretation of volcanism in it. Most scientists thought at that time (and still do today) that what Kozyrev had recorded was the relatively quiescent effusion of a gas from the lunar surface.

In the same book, Kozyrev's paper is followed by another: 'Microphotometric analysis of the emission flare in the region of the central peak of the crater Alphonsus on 3 November 1958', by A. Kalinyak and A. Kamionko of the Pulkovo Observatory in Russia. The authors conducted a detailed analysis of Kozyrev's spectrum and concluded that the event was indeed a gas release and the gas was excited to fluoresce under the action of solar radiation. They determined that the temperature of the gas was less than 480 °C (maybe much less) and was of rather similar composition to the gas found in the heads of comets. They confirmed the presence of the Swan bands of carbon and deduced that the pressure of the gas was certainly less than one-hundred-millionth of a millimetre of mercury (this is about one-hundred-thousand-millionth of the sea-level pressure of the Earth's atmosphere), and perhaps much less than that.

There is a fascinating exposé of the American efforts to verify and understand Kozyrev's spectrum given in the October 1996 issue of *Sky & Telescope*. The article is entitled 'The lunar volcanism controversy'. The once sceptical American astronomers, particularly the famous Gerard P. Kuiper, changed their views when they eventually had the chance to study the original spectrographic plates for themselves.

Some amateur and professional astronomers kept a watch on Alphonsus for any signs of after-effects subsequent to the 1958 episode. There were a few reports of red patches seen on the floor of the crater, though no photographic evidence was ever secured and nothing remained of them after some months had passed.

Kozyrev continued his programme of lunar observations and he 'turned up' a couple more spectrographic anomalies – one in Alphonsus on 23 October 1959 and one in the crater Aristarchus (see Section 8.7 for more about this crater) on 1 April 1969 – though the results were not as definite as for the 1958 event. In the first, all that was identifiable was a brightening at the red end of the spectrum. The spectrum of the Aristarchus event sported the same general reddening but with bands of molecular nitrogen and of the molecular species CN just about identifiable.

Sightings of red glows were reported by astronomers using the 24-inch refractor of the Lowell Observatory and others using the 69-inch (1.75 m) reflector at the Perkins Observatory in the USA. Many amateur groups, particular the Lunar Sections of the British Astronomical Association (BAA) and the Association of Lunar and Planetary Observers (ALPO), began looking for TLP. Observers soon began reporting anomalies. Undoubtedly most of these are explainable as being due to things other than real events at the surface of the Moon (actually I would go as far as to say that **the vast majority of anomalies are not genuine TLP** – more about this in Section 9.5). I must say, for instance, that I strongly suspect that the incidents of 'red glows' I have just cited were really just *spurious colour* produced in the Earth's atmosphere. More about that later. There are, though, a small number of cases which are not so easy to dismiss.

Barbara Middlehurst and her colleagues in the USA and Patrick Moore in the UK were independently compiling catalogues of the reported instances of TLP but they eventually published a combined catalogue in 1967. Patrick Moore updated it in 1971, the number of entries then standing at 713.

A pattern seemed to emerge from the hundreds of TLP reports. TLP were not distributed randomly. There was an apparent preference for 'events' to occur around the borders of the lunar maria and within certain craters. The highland areas were avoided. The crater Aristarchus came out as the 'hottest' TLP spot on the Moon with about a third of all reported events involving this crater.

Aristarchus has even been the subject of a space-borne observation of an anomaly. The three astronauts aboard *Apollo 11* reported a glow localised in the wall of the crater on 19 July 1969. At $18^h 45^m$ UT the crew could first see an illuminated area to the north of the spacecraft. As they drew nearer they confirmed their first impression that it was inside Aristarchus. Part of the transcription of their communication to Mission control (it is difficult to make out, so the transcription is incomplete) is:

> It's getting to be about zero phase. One wall of the crater (Aristarchus) seems to be more illuminated than the others. It is definitely brighter than anything else I can see. It's an inner wall of the crater ... there doesn't appear to be any colour involved in it ... it (is) the inner part of the west-north-west, the part that would appear more nearly normal ... looking at it from the Earth. ...

Earth-based astronomers also confirmed the presence of the brightening, as well as reporting other anomalies in Aristarchus on other dates around the same time. Actually it was because of the earlier ground-based observations that the astronauts were asked to keep a look out for anything unusual.

During the *Apollo 16* mission Ken Mattingly, in the orbiting Command Module, witnessed several flashes from the lunar surface (though I do not have any further details about this) and *Apollo 17* astronaut Harrison Schmidt witnessed a flash emanating from the region of the lunar crater Grimaldi. This is another TLP 'hot spot'. See Section 8.20 in Chapter 8 for more about this crater.

Another TLP 'hot spot' (at least it was until the 1970s, even if rather less so since then) is the crater Gassendi. See Section 8.22 for photographs and details about it. Transient bright points of light were seen by Walter Haas (10 July 1941) and H. P. Wilkins (17 May 1957) but the most significant 'event' involving Gassendi occurred on 30 April 1966. First detected by Peter Sartory, for about four hours a reddish wedge-shaped streak was seen to span the central peak to the south-west rim of the crater. Several independent observers witnessed this phenomenon. One of these was Patrick Moore and he describes it as "the most unmistakable red event I have ever seen on the Moon".

Back in those days there was an assumption that all TLP should appear as red glows. One BAA Lunar Section member, Peter Sartory, designed a device, called a 'Moonblink' which consisted of a unit containing red and blue filters that was plugged into the telescope before the eyepiece. Turning a knob brought either filter into the optical path. By manually alternating between the red and blue filters while watching through the eyepiece, any red patch on the Moon would appear light when seen through the red filter but would show up as darker when seen through the blue filter. Many positive 'blinks' were obtained by observers using this arrangement.

Meanwhile, in the USA, a network of professional astronomers was set up under the auspices of NASA using more sophisticated versions of this device (utilising an electronic detector rather than a human eye – see the July 1967 issue of *Icarus* for an account of this work).

One of the team, Winifred S. Cameron, remained involved in research into TLP for decades after. She was a prolific author of papers and articles on the subject. One reference you might especially like to seek out is an article, entitled 'Lunar Transient Phenomena', in the March 1991 issue of *Sky & Telescope*. One of the cases she cites in the article involves the crater Pitatus (see Section 8.32 for illustrations and for details about this crater) and she shows two photographs taken by Gary Slayton of Fort Lauderdale, Florida, of a bright 'blob' in the crater which moves between the time the two photographs were taken (the times are not given, though the date was 5 September 1981). This is not the only instance of moving lights on the Moon being reported by Earth-based observers.

Returning to the subject of the 'blink' devices, there are problems with them. In particular, *spurious colour*, described later, is an effect which has its cause in the Earth's atmosphere and almost all 'blinks' were undoubtedly due to this, rather than anything real on the Moon. Eventually observers realised that the majority of TLP were actually not even red, only a minority of visual anomalies showing any significant colour at all. Blink devices have now rather gone out of fashion.

Since I have only room enough to cite a few examples of TLP, I think this would be a good point to interrupt this potted history to summarise the main types of the visual anomalies observed.

9.2   CATEGORIES OF TLP

The following are the types of visual anomaly most often reported. In each case the area of the Moon affected is usually only a few kilometres square at most.

Short-term albedo changes

These are unusual increases or decreases in the apparent brightness of the Moon's surface. The change might last for a few hours, though most times for rather less than a hour. Sometimes the change occurs and then is fairly steady – say the brightness of a small patch of the Moon increases, then remains at roughly the same elevated level for a while only then to decrease steadily back to the normal value once more. At other times the brightness pulsates. When this happens the brightness usually varies rather erratically. These brightness fluctuations can happen on time-scales as short as a second or two.

Obscurations of surface details

A small patch of the Moon might appear blurred or indistinct while the surrounds remain sharp and clear-cut. The effect often lasts for about an hour. Significantly, the blurring is sometimes seen to start out very localised and distinct but then spreads out, becoming less obvious as it does so – suggestive of a cloud of vapour thinning out.

Coloured effects

Sometimes coloured effects are seen as an accompaniment to brightness changes, or to surface obscurations. Sometimes they are seen on their own. Most observed anomalies do not show significant colours. However, there have been rare instances when the colours have been very vivid. In my own experience, regions showing short-term brightness changes tend to show a bluish tint if any colour is visible at all.

Flashes of light

These are the rarest of all reported TLP. They appear as bright (sometimes very bright) flashes of light or brief twinkles of light against the Moon's surface. Occasionally flashes are seen along with other types of TLP in progress. Since the first edition of this book there has been incontrovertible proof of flashes occurring on the Moon. They have been recorded on video and, moreover, a very rational explanation has been found for at least some of them. More on that later.

9.3 THE MYSTERY CONTINUES

The growing interest of amateur and professional astronomers in Transient Lunar Phenomena through the 1950s and 1960s continued into the 1970s and 1980s. Unfortunately, as I said in the introduction, the subject also attracted more than its fair share of cranks. These people tended to go to their telescopes and see all manner of coloured effects, plumes of smoke erupting from craters, etc. Many rushed into print with half-baked, and ill-informed, ideas about the mechanisms producing these supposed manifestations.

Not only have these people brought the derision of many astronomers on the whole subject, their 'reports' pollute the data, making any serious study difficult. It is for that reason that I give low weighting to the results of statistical studies based on the complete TLP databases. Various links, such as with the solar activity cycle, with the position of the Moon in its orbit (and so the passage of the Moon through the Earth's magnetotail), with moonquake activity, etc., all have been variously 'proved' and 'disproved' by researchers.

Fortunately we do have **some** good evidence and data. It is the work of the professional astronomers which is most credible. For instance, astronomers of the Tokyo Astronomical Observatory have conducted many long-term studies of the Moon's surface brightness and the way the Moon's surface polarises the incident sunlight it reflects (some vibration angles of the light-waves are reflected more than others).

In 1970 they observed an event involving Aristarchus which they described in *The Moon* – issue 2 (1971). Their paper is entitled 'An anomalous brightening of the lunar surface observed on March 26, 1970', and is authored by Naosuke Sekiguchi. While on their normal programme of photometric (brightness) and polarimetric observations with a 36-inch (0.91 m) reflector, Sekiguchi and co-workers found, on that date, the region around Aristarchus became 0.3 magnitude brighter than normal. At the same time its colour index decreased by 0.1. In other words, as well as becoming over 30 per cent brighter, the region became significantly bluer. The change in optical polarisation that was recorded only in the Aristarchus region also adds weight to the assertion that a genuine TLP was recorded. In his paper, Sekiguchi refers to some of the many other papers which detail other professional observations of this phenomenon. He also suggests that the TLP he recorded may be related to a major solar flare that occurred 29 hours before his observation.

Much of Professor Antoine Dolfus' work over his professional life has been the careful collection and analysis of photo-polarimetric (brightness and polarisation) measures of the Moon and planets. During observations using a video polarimeter fitted to the Meudon Observatory 43-inch (1.07 metre) Cassegrain reflector he encountered a transient anomaly near the centre of the crater Langrenus. On 30 December 1992 the normal view of Langrenus was disturbed by two small patches that were about 10 per cent brighter than the surrounds and the light from which was significantly polarised. The next two nights were cloudy but Professor Dolfus was able to observe again on 2 January 1993. The patches were again present, though diminished and altered in shape. Professor Dolfus ascribes the patches to fine dust levitated from the lunar surface, possibly by outgassing from the lunar crust. You will find an illustrated account of this in an article 'A TLP in Langrenus crater' by Dr Richard McKim in the December 1999 issue of the *Journal of the British Astronomical Association*. In his article, Dr McKim cites 12 further references you might care to follow up.

I will finish this historical overview, sketchy as it is, with the briefest possible account of my own involvement in TLP research and a summary of the 'state of play' at present.

In 1979, after returning to live in my parents' home (in Seaford, East Sussex, UK) at the end of my undergraduate years and taking up a teaching post in Hove, I joined the Lunar Section of the British Astronomical Association and straight away resumed making lunar observations with my 6¼-inch (158 mm) and 18¼-inch (0.46 m) reflecting telescopes. There existed a 'telephone alert network' whereby an observer reported any suspect appearances to a central co-ordinator. The co-ordinator then contacted all the available active observers who had joined the network. In the years that followed I observed the Moon whenever I could. I moved to my own house in Bexhill-on-sea, in East Sussex, in 1983 (having started teaching at a local sixth form college in 1982). In 1995 I added an 8½-inch (216 mm) reflector to my equipment.

I moved to a semi-rural village in Norfolk in 1999 and joined the Breckland Astronomical Society (BAS), in addition to still being an active member of the BAA. Thanks to the energetic and very welcoming members of that provincial society, I am able to make use of their 20-inch

(0.5 m) computer-controlled telescope in a very well-built observatory, in addition to my own equipment at my home. Currently, though, I live an inconvenient three-quarter of an hour's drive away from the BAS observatory. My observational work (lunar and in other fields) continues to this day, though since the 1990s a long-term physical illness has seriously reduced the amount I can manage to do.

I have responded to many 'alerts' issued by the BAA in the years since 1979. In some cases I could not confirm any visual anomaly. In some I could, though almost always concluding that the effect was caused by a mechanism other than anything happening on the Moon (see Section 9.5 for more about this).

In just a few instances, there seemed to be something genuinely anomalous at the surface of the Moon. I instigated a few 'alerts' myself, though in some of these cases I suspected (but could not be sure of) causes other than a genuine TLP. I should perhaps make it clear that the person providing the alert would never give more than the general location of the supposed 'event'. Any corroboration, or otherwise, was sought after the event by checking the written reports and any accompanying sketches.

My earlier move to Bexhill proved expedient when the opportunity arose to be a Guest Observer at the Royal Greenwich Observatory, just a twenty-minute drive away. Thanks to introductions by Patrick Moore, and because I am qualified as an astronomer, I was afforded the very great privilege of using the RGO's equipment.

From January 1985 to March 1990 I had the run of the 'Equatorial Group' of telescopes and additional facilities in the main buildings. My main purpose there was to repeat Kozyrev's efforts in the hope of obtaining a good-quality spectrum of a TLP in action.

The main instrument I used was the 30-inch (0.76 m) coudé reflector with its elaborate high-dispersion spectrograph (Figure 9.1). Another useful instrument there was the 36-inch (0.91 m) Cassegrain reflector in the adjacent dome (Figure 9.2). I used to set up both telescopes for lunar observing sessions. The Cassegrain reflector was better for monitoring purposes and it took only moments to move through the adjoining corridors from one telescope to the other.

Of course, I took many standard spectra for comparison purposes with the hope of recording the spectrum of a TLP in action. The equipment would now be considered old-fashioned, in that the spectra were recorded on 7-inch × 1-inch (178 mm × 25 mm) photographic plates – which I had to cut from standard 10-inch × 2-inch stock of *Kodak* IIaO plates, subsequently processing them after the exposures were made in the telescope's spectrograph. Today's newly graduated astronomers would consider this a quaint art practised in a bygone age!

Figure 9.1 The 30-inch (0.76 m) coudé reflector at the Herstmonceux site of the former Royal Greenwich Observatory. The light from the telescope emerges from the door in the back of the telescope tube to be intercepted by the mirror at the top of the tall tripodal gantry. The light is then passed down into the head of the spectrograph. The various parts of the spectrograph are arranged on three floors of the building!

Figure 9.3(a) shows a print I made from one of the plates. Some details are given in the accompanying caption but suffice it to say here that you can see it is of very high spectral resolution. I scanned the plates using the RGO's temperamental plate density scanning instrument (PDS) which

Figure 9.2 The author can be seen below the 36-inch (0.91 m) Cassegrain reflector at the Herstmonceux site of the former Royal Greenwich Observatory.

produced 3 metre long tracings (plots of intensity versus wavelength) from each plate. A very small part of one of these is shown in Figure 9.3(b). The tracings allowed for detailed measurements of any spectrum I suspected of showing an anomaly. For a fuller description of the equipment and procedures see the July–September 1987 issue of

(a)

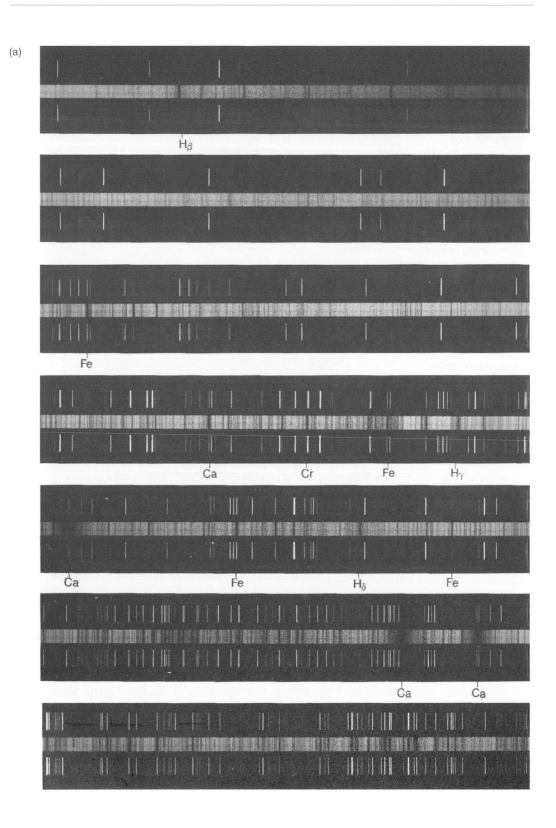

Figure 9.3 (a) High-resolution spectrum of sunlight reflected from the surface of the Moon, taken by the author, using the 30-inch (0.76 m) coudé reflector and high-dispersion spectrograph of the Royal Greenwich Observatory. The wavelength range covered is approximately 355 nm (left-hand side of the bottom strip) to 504 nm (right-hand side of the top strip), the wavelength increasing from left to right along each strip. There is a small amount of overlap between each strip. Above and below the main spectrum is a copper–argon emission spectrum exposed on the plate at the same time as the Moon spectrum for calibration purposes.

Astronomy Now magazine, or the March 1985 and December 1985 issues of the BAA Lunar Section publication, the *Lunar Section Circular*.

Now the big question – did I spectrographically record a TLP? I am afraid I did not.

On 27 September 1985 I was busy in the dome of the 30-inch (0.76 m) telescope when a 'TLP alert' came through by telephone. The seeing was extremely bad and I had already taken a standard spectrum of Aristarchus. In response to the alert I set the telescope so that Torricelli B, the subject of the alert was on the slit. Actually this was difficult to do since it was nearly invisible in the violently unsteady seeing. I made the required exposure (9 minutes in that case). When I processed the plate, and subsequently scanned it, I found that it contained nothing but the usual spectrum. I must say that there was little hope of success on that night, even if Torricelli B was undergoing a genuine TLP, as only for a very small amount of the exposure was the light from this small crater actually dropping through the entrance slit of the spectrograph. Most of the time the turbulent distortions made Torricelli B miss the slit and so the spectrum was mainly built up from the light of the surrounding moonscape.

The next night a thick fog hung over Bexhill and I decided to stay at home. However, the telephone rang – there was another 'TLP alert', again concerning Torricelli B. After a hazardous drive to the observatory I set up the equipment and gave the plate a much longer exposure than normal (to try and offset the effects of the fog). Even so, the processed and scanned plate proved useless; no anomalies were detectable, save the 'noise' generated by the thinness of the image.

Frustratingly, one of the few periods the RGO telescopes were not available to me because of other usage coincided with some significant TLP activity. All I could do was to use my own telescopes. One of these instances concerned Torricelli B. On 31 May 1985 I was using my 18¼-inch (0.46 m) reflector when, at $20^h\ 23^m$ UT, I noticed that Torricelli B was bright and mauve coloured and sported a mauve 'halo'. I telephoned the co-ordinator. By $20^h\ 29^m$ UT I was back at the eyepiece but the colour had gone! The crater then seemed to be pure white in hue and varying erratically in brightness, with a mean period judged to be about 2 seconds. The ambient seeing conditions caused a general lapsing of focus with a mean period estimated at about 5 to 10 seconds, while the turbulent rippling of the image had a period of about half a second. At $20^h\ 34^m$ UT I noted that the crater had taken on an intermittent pinkish tinge. From then until close of observations at $22^h\ 18^m$ UT I noticed definite variations in the apparent brightness of the crater that seemed not to be the result of the atmospheric conditions. The two other observers in the BAA Lunar Section available under clear skies that night independently recorded

(b)

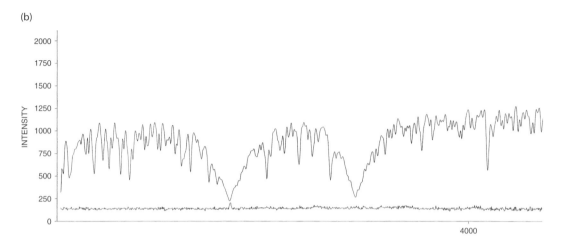

similar brightness and colour variations, along with 'star-like flashes' occurring inside the crater. They agree to the nearest minute about the times of the flashes. I saw no flashes.

The most significant TLP I have ever witnessed occurred the day before, 1995 May 30^d. Patrick Moore, in company with Paul Doherty, noticed an anomalous brightening of the west wall of Aristarchus, plus some blurring on the north-west wall. Myself and another five members of the BAA Lunar Section were alerted and observed this crater (remember – no details of the anomaly were ever passed on during an alert; merely the general location of the suspected anomaly).

I had bad seeing conditions but by 20^h 53^m UT I noticed that in brief flashes of good imaging a pink tinge was present along the northern sector of the crater interior. At 22^h 08^m UT I noticed that there was an odd appearance to the shadow on the north-west wall (see Figure 9.4) – a notch formed in it! By 22^h 54^m UT an intense ruby-red colouration developed on the inside of this 'notch'. The seeing deteriorated to the point that I had to pack up by 23^h 08^m UT. All who took part agreed on most of the points about the anomaly, and the timings – and two others agreed with me about the 'notch' in the interior shadow – and all the reports were entirely independently made!

My anguish about not having the use of the RGO telescopes on those two nights was intense. There were other frustrating occasions when 'TLP alerts' were telephoned to me but the skies were not clear over Herstmonceux. So, I had failed to secure a spectrum of a TLP in action. Nonetheless, I enjoyed myself immensely using the professional telescopes for those five years and I did conduct other successful observational projects with them. It was a great experience and it only came to an end when the RGO moved to Cambridge. Indeed, I was the very last

Figure 9.3 (*cont.*) (b) A small part of a tracing from the spectrum. The two broad dips in the spectrum of moonlight (upper trace) correspond with the broad calcium (Ca) absorption features indicated at the right-hand end of the second strip from the bottom in (a).

Figure 9.4 A visual anomaly recorded by the author in Aristarchus (see text for details).

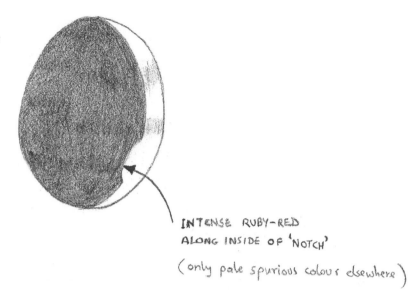

INTENSE RUBY-RED
ALONG INSIDE OF 'NOTCH'
(only pale spurious colour elsewhere)

person to use the Herstmonceux telescopes while they were under the aegis of the RGO. Now the RGO has closed down altogether – a piece of Britain's scientific heritage cast into the garbage can by our politicians.

For me, it was back to the backyard telescopes. I did build a spectrograph onto my largest telescope (Figure 9.5 – and see my book *Advanced Amateur Astronomy* for a fuller description) but the illness I have already mentioned was beginning to tighten its grip and I found myself able to do less and less observing. Nonetheless, amid changes to the BAA Lunar Section I even took over as co-ordinator of the TLP observing group, which by then was flagging. I managed to re-activate the group but, after a number of years my health and circumstances forced me to give up this work also.

The present situation is that there are a few scattered groups active around the world in the field of TLP observation and study, though I cannot say that there is as much enthusiasm as there once was. The BAA Lunar Section continues with the TLP work, now having joined forces with ALPO, though with fewer active members than it had in its heyday.

So much for the potted history. I hope I have said enough to whet your appetite. Of course, the burning question arising from all this is – what causes Transient Lunar Phenomena?

9.4 WHAT MIGHT BE THE CAUSE(S) OF TLP?

If we only ever had one genuine TLP, the example which Kozyrev observed and recorded the spectrum of in 1958, I would have concluded that Kozyrev had witnessed the after-effects of a small piece of cometary

(a)

Figure 9.5 (a) The spectrograph the author built onto his 18¼-inch (0.46 m) reflector can be seen arranged along the back of the 8-feet (2.45m) long tube of the telescope. (b) Spectrum of sunlight reflected from the Moon, obtained by the author, using his spectrograph.

(b)

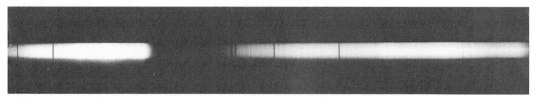

material striking the Moon at, or very close to, the central peak of Alphonsus. I would have reckoned that the explosively uplifted surface materials produced the initial reddish cloud. When this cleared it might have left the gases from the vaporised impactor to fluoresce under the solar radiation (the sunlight plus the solar wind bombardment).

However, we do not just have the one TLP. To date there have been about a thousand anomalies recorded, though I reckon that instances of genuine TLP definitely number less than a hundred and almost certainly number less than fifty. The greatest difficulty with the comet impact theory is explaining why it is that certain locations on the Moon's surface seem especially prone to TLP. Surely cometary impacts, if they were the correct explanation, should produce a random distribution of TLP-type events?

Could it be that the Moon has, at least in certain locations, large quantities of sub-surface comet-type ices and these are released through fissures? This idea seems ridiculous in the extreme. True, there is the questionable apparent detection by the *Lunar Prospector* probe of ice near the polar regions but there is certainly no evidence for comet-type ices below the surface anywhere else on the Moon.

Much more reasonable is the case for radioactive gases and the remnants of gases left over from the Moon's more volcanically active ancient past emanating through fissures from deep below the surface. Interestingly, the seismometers left on the Moon have provided data which indicate that TLP sites tend to lie above moonquake epicentres (though see my earlier note of caution about supposed statistical correlations).

I also think that the solar wind and radiation from the Sun might play a part in producing TLP. Simple calculations show that the average solar-wind flux conveys insufficient energy to produce any visible manifestations at the surface of the Moon. However, the solar wind is very gusty and the most intense blasts **do** convey sufficient energy. I have already referred to Sekiguchi's observation of 26 March 1970, and the fact that he draws attention to a major solar flare that occurred 29 hours earlier as the possible source of the fluorescence he recorded.

Certainly this fluorescence might just be in the surface rocks. In fact, we have long known that there is more to moonlight than simply reflected sunlight. A really significant study of this was undertaken by the 'Manchester Group' in the 1960s. Z. Kopal's article in the May 1965 issue of *Scientific American* is well worth the trouble of obtaining. He, along with T. W. Rackham, photographically recorded a number of luminous events at Pic du Midi. Some of the photographs are presented in the article. The lunar craters Copernicus, Kepler, Plato and Aristarchus were particularly affected. At times, temporary enhancements in brilliance up to about

80 per cent of that due to the incident sunlight were detected. Kopal presents a detailed analysis in his article and concludes that corpuscular radiation from the Sun is responsible.

It is worth noting that not everybody accepts that the Kozyrev 1958 spectrum is the result of a gas release. For instance E. J. Öpik, in *Advances in Astronomy and Astrophysics*, Volume 8 (1971) (edited by Z. Kopal), points out that the region of bright emission did not encroach into the part of the central peak of Alphonsus in shadow. The demarcation visible on the spectrum is sharp, which is not what one would expect from a cloud of gas. Öpik concludes that fluorescence from the solid lunar surface occurred, rather than any gas emission.

I have great faith in the lunar-rock fluorescence theory as the explanation for many (of the few genuine) TLP. However, I still wonder if gas effusion from the lunar surface also forms part of the story. For instance, we know for sure, as a result of the *Apollo* missions, that radon gas escapes from below the lunar surface. The Apollo Alpha Particle Spectrometers (AAPS) aboard the *Apollo 15* and *Apollo 16* Command Modules – which orbited the Moon while two of the crew worked on the lunar surface – detected alpha particles with the signature energy spectrum which links them to the radioactive decay of radon gas.

Even better, three American scientists, Paul Gorenstein, Leon Golub and Paul Bjorkholm, have conducted a detailed analysis of the AAPS results and found that the sites of maximum emission coincided with the established 'hot spots' of TLP (this correlation, at least, seems strong enough to be definite!). Their paper 'Radon emanation from the Moon, spatial and temporal variability' is presented in *The Moon*, issue 9 (1974). Grimaldi, Alphonsus and the edges of the lunar maria figure particularly highly. The authors found that the largest recorded radon emissions occurred from the area of Aristarchus – the 'hottest' spot of the lot!

It strikes me as plausible that any effusing radon gas might be excited into fluorescence in the same way as for the surface rocks. Perhaps other gaseous species are brought up along with the radon gas.

The solar-wind particles would cause the gas to fluoresce by colliding with the gas atoms/molecules and causing electrons in the atoms to be temporarily excited to higher energy levels. When they de-excite, the electrons hop down the energy-level rungs in the atoms, so producing a characteristic spectrum. This is the way a conventional fluorescent lamp works (such as a sodium street-light). Interestingly, like sodium vapour, radon gas produces a nearly monochromatic spectrum. The main emission in the visible spectrum occurs at a wavelength of 434.96 nm (multiply this figure by 10 if you prefer wavelengths expressed in Ångstrom units). This is in the blue–violet part of the spectrum.

If the Sun produced an extra strong gust of solar-wind particles which arrived at the Moon just as a particular location effused radon gas, then the interaction between them could easily lead to the sort of blue fluorescent effect that has been observed in features such as Torricelli B. If this seems too coincidental, remember I think genuine TLP occur only rarely, and most of the genuine ones are explainable as surface fluorescence.

I also conjecture that if any gas release was particularly 'violent' (though still extremely feeble by terrestrial standards) then maybe some of the finest (perhaps colloidal-sized) particles could be swept up from the lunar surface to produce a temporary hazy obscuration over a small area of the Moon. Of course, the local topography would determine whether or not a cloud of dust could be raised by a gas escape. One might expect this gas cloud to be either white or to appear reddish in hue, depending on the sizes of the levitated particles. The light scattered by the cloud would also be partially polarised, the extent of the polarisation depending on the particle sizes.

Finally, ionised gas atoms/molecules in motion (and particularly those interacting with fine solid particles in suspension or motion) could lead to charge separation and a resultant build-up of electrical potential difference. The eventual discharge through the tenuous gas (a form of sheet lightning) might well account for the rarely observed bright flashes and sparkles.

Actually, there is one incontrovertible cause of lunar flashes – meteor impacts. These were first established as fact during the night of 1999 November 17/18 during the annual Leonid meteor shower. J. Kelly Beatty gives an account of the events of that night in the June 2000 issue of *Sky & Telescope* magazine. There were at least six yellowish flashes witnessed and confirmed by at least two or more observers in the USA and recorded on video tape. They occurred on the unlit western side of the 10 day old Moon. The brightest flashes were estimated at magnitude 3 and certainly lasted less than 0.03 second as they showed up on only one video frame each. The brightest flashes must have resulted from rocky meteors about 8 inches (20 cm) across.

So much for meteor strikes, what about the flashes that occur **after** the beginning of other transient phenomena and seem to be associated with them, or multiple flashes from the same location? In these cases one must look for another cause, perhaps the one I have previously muted.

Other people have differing ideas about the causes of TLP, such as piezoelectric discharges from the Moon's surface due to stresses and strains in the surface rocks caused by the diurnal temperature changes, and triboelectric discharges, again caused by the piezoelectric effect but with moonquakes providing the stresses and strains. Winifred Cameron provides an extensive review of TLP in her paper 'Lunar Transient Phenomena (LTP): manifestations, site distribution, correlations and possible causes' in *Physics of the Earth and Planetary Interiors*, volume 14 (1977).

Those who would equate TLPs with fairies at the bottom of the garden are right to criticise all the bad science and uncritical observational practise of the fanatical and misguided people involved over the years. However, to also ignore the good work that has been done and the credible evidence that has been gathered is just as much bad science. I must say that I also think that many people on both sides of the argument are way too aggressive and entrenched about what is merely a range of interesting but extremely feeble effects very occasionally to be seen happening on the Moon.

9.5   POSSIBLE CAUSES OF BOGUS TLP

I am sure that the vast majority of observed anomalies are nothing to do with any real physical process at or near to the surface of the Moon. For explanations of these we need to look much nearer home. There are three sources of the spurious reports of TLP. These are: the Earth's atmosphere; the telescope; the observer. There are a number of different mechanisms operating in each of these sources. Any one, or a combination of these, can result in what I call 'bogus TLP'.

The atmosphere

One of the major pitfalls for the observer is the presence of spurious colour. The Moon's image is composed of light–dark boundaries. You will often see these fringed with colour. We are all familiar with the effect a triangular glass prism has on a thin beam of white light passing through it. The light is deviated by the prism in a direction towards its base. The light is also sorted into its component wavelengths (colours), the amount of deviation being slightly different for each of the component colours. We call this effect *dispersion*. The violet rays are bent slightly more than the red ones.

The Earth's atmosphere acts rather like a prism turned upside down (base uppermost). As well as astronomical objects appearing slightly elevated – making sunrise happen slightly early and sunset happen slightly late – the light-rays from them are slightly dispersed as a result. Obviously, for each light–dark boundary in the image a full 'rainbow' is produced. However, the orientation of the boundary and the image structure of its immediate surrounds play a part and usually only part of the full 'rainbow' is obvious. For instance, most lunar craters appear with a bluish fringe along their southern rims, the opposite rim being fringed with red. Plato often shows this effect, though many craters do not show both the half-rainbows with equal prominence. In the case of

Plato the red colouration along the northern rim is usually more obvious than the blue colour fringing the southern rim. Some craters, for example Aristarchus, normally display spurious colour of the opposite orientation; blue to the north and red to the south.

This spurious colour effect is usually greatest when viewing the Moon (or other celestial body) when it is low over the horizon. However, it also varies with the ambient atmospheric conditions: temperature, humidity, air pressure and the presence of aerosols and particulates.

The implications for TLP hunting are obvious. Time and time again that 'red glow' enveloping a lunar feature (crater wall, central peak, etc.) will turn out to have been generated by our atmosphere. I would urge you to get to know what coloured effects are usual for a given feature. I have already mentioned Plato and Aristarchus. Another interesting example is the crater Lassell. On many nights, especially near full Moon, Lassell seems enveloped by a bluish haze, while the mountain mass just a short distance to the north-west seems to be covered by an orange glow.

Image turbulence, or more properly *scintillation*, is also a real nuisance. Sometimes the image is soft but fairly steady. At other times it is undulating violently. Most often the two effects occur together. Again, the result is different for different lunar features. Has part of the crater wall really 'gone soft' or is it just that the fine terracing present at that location has run together to give a blurred or 'foggy' appearance? There is no substitute for experience when it comes to deciding whether a real anomaly might be present.

The telescope

No telescope is perfect. Even putting aside any mechanical faults, the optical system will have its limitations, both in design and in manufacturing tolerances. I cover some points in Chapter 3 and in Appendices 1 and 2 but lack of space here causes me to refer you to my book *Advanced Amateur Astronomy* for a much fuller and more detailed discussion of telescope optics, their faults and some ways of evaluating and correcting them.

Suffice it to say here that lateral chromatic aberration (colour-fringing) is usually the most misleading error experienced by the TLP observer. Longitudinal chromatic aberration manifests as a general softening of the image when seen near the centre of the field of view. Away from the centre it assumes the lateral form. Observations made with refractors could be suspect because of this but usually it is the telescope's eyepiece which produces the worst effect. So, you might well experience the greatest trouble if using a reflector of low focal ratio with a cheap eyepiece.

Critically examine your eyepieces in use. Do you find that the light–dark boundaries in the image, for instance along the edge of a shadow-filled crater, become fringed with colour as you move the telescope to place the crater near the edge of the field of view? This is classic lateral chromatic aberration. Near the centre of the field of view the image will probably appear colour-fringe free. If you have some high-quality coloured filters try the effect of using them. Does the image seem to sharpen when using the filter? If so, you can be fairly sure that the image is being degraded by the presence of longitudinal chromatic aberration.

The remedy is either to stick to making monochrome observations using coloured filters, or to invest in a high quality Barlow-type lens or eyepieces with better correction.

The observer

We all make mistakes. Our eyes are imperfect and the brains behind them are even more so. When you consider how the complex lunar vista changes with lighting it is not hard to understand how the observer can be caught unawares by apparently strange appearances. Most times those 'strange appearances' turn out to be quite normal for particular lighting angles and seeing conditions. Yet again, experience is the key to deciding whether the way a feature appears to you is truly anomalous or not. Also, a good photographic atlas is always a help whatever your level of experience.

9.6   TLP OBSERVING PROGRAMME

Once in a while an apparent anomaly will appear which is prominent enough to be seen through a small telescope, so it is difficult to put a definite lower limit to the size of telescope required. Obviously, though, anything larger than 6-inch (152 mm) aperture is desirable, provided it is of good quality. Reflectors are normally to be preferred over refractors, provided they have eyepieces that are well corrected for chromatic errors.

If you are a novice observer, it is essential that you first observe through a number of lunations to gain some knowledge of what the Moon really looks like under various lighting angles. Even then, I would recommend concentrating on just one or two specific features and only expand your repertoire when you have gained enough experience to be sure what you are looking at really is normal or not. Perhaps the notes in Chapter 8 of this book may help you get started.

When I go to the telescope to carry out lunar monitoring, I adopt a definite strategy. Though this might vary depending on the conditions, I normally split observing sessions into two main activities. First I use a low power, such as $\times 144$ on my 0.46 m telescope, and I 'raster-scan' the whole of the lunar surface, both the sunlit and the darkened hemispheres. This might take about 15 minutes. By 'raster-scanning' I mean east–west sweeps across the lunar surface, for each sweep setting the telescope a little higher in declination. The bands swept out then overlap a little, ensuring full coverage. Actually I switch off the telescope drive and let diurnal motion provide the east–west motion for me. I carefully scrutinise all the lunar features as they pass through the field of view, looking for any abnormalities.

I then carefully scrutinise any features which I think look suspicious. Perhaps I might momentarily leave the telescope to check charts or photographs of the area in question (under as similar lighting as possible). Obviously, I might keep the area under scrutiny for a while – one can still be flexible within one's 'plan of action'.

Assuming all is normal, as it is the vast majority of times, I then spend some time examining other specific features. My list includes: Aristarchus, Torricelli B, Plato, Proclus, Alphonsus, Messier and Messier A, Tycho. All these are TLP 'hot spots'. Not all will be in sunlight at any one time (except near full Moon) but some features, especially Aristarchus, can often be located in Earthshine.

I then proceed to re-scan the Moon with higher magnifications, spending some time to re-scrutinise my selected features, and so on as the session continues. Most nights I do not increase the magnification beyond $\times 207$, anyway, because of the turbulent seeing conditions. If the image you are looking at is at all 'soft', then there is certainly no point in going to greater powers.

Obviously there is a real danger of observational selection by concentrating on specific features. At least making scans of the rest of the lunar surface will tend to dilute this effect.

I recommend joining a society with a TLP observing group. Then if suspicions are aroused you can telephone a central co-ordinator with your suspicion – but only give the general location, otherwise you might prejudice the subsequent analysis. The co-ordinator can then raise an 'alert' among the other participating members.

You can do valuable work just by visually scanning for TLP. If you can also use other techniques you may be able to make an outstanding contribution. Photography (see Chapters 4 and 5) is an obvious extension of purely visual work. You could include the use of coloured filters with all the imaging methods. Photometry is possible (especially with images saved on a computer). With colour filters in use this becomes colorimetry – the relative brightness in specific wavebands. Use a polarising filter and you can do polarimetry. If you can build or obtain a spectrograph (*S-BIG* commercially produce one for the amateur market) you could follow in the steps of Kozyrev. However, please be aware that you might have to wait for a very long time, probably even years, before you see anything likely to be genuine TLP. There is enough good evidence to convince me that Transient Lunar Phenomena really happen – but they also **rarely** happen!

I have enjoyed observing the Moon for over thirty seven years. Despite my current incapacity, I can still do some telescopic observation and I am hoping to do much more again as my fitness improves in the future. Maybe I can go on for another two to three decades at the telescope eyepiece? I feel sure that by then there will once again be live television broadcasts of humans walking on the Moon's surface.

In this book I have tried to 'shoehorn' as much useful information into the available space as possible. I hope that I have also said enough to persuade you to obtain whatever telescope you can and turn it to the Moon. You will find a world of fascination and mystery among the dramatic mountain ranges and the eerie plains and craters. I hope you will experience for yourself the Moon's "magnificent desolation".

APPENDIX 1: TELESCOPE COLLIMATION

A telescope's optical performance can be severely impaired if its optics are even slightly out of alignment. Some observers insist that every time you go out to your telescope you should re-collimate it. That is not what I do. I leave my permanently stationed telescopes for many months on end without even checking them. In my defence, I find that it is extremely rare for me to have to ever make even the slightest of adjustments. It all depends on what materials the telescope is made from and how robust it is. For instance, a telescope homemade from wood is likely to need frequent adjustments owing to the unstable nature of the material.

Of course, it is important to check any portable equipment each time it is set up anew. So, if you are forced to use portable equipment then checking the collimation of your telescope becomes a necessarily frequent chore. The following notes, concerned with the main types of telescope, may be of help. In all cases I must leave you to become familiar with the types of adjustments provided (perhaps push–pull screws, or nuts and springs, or nuts and lock nuts) and their locations on your own telescope.

COLLIMATING A NEWTONIAN REFLECTOR OF FOCAL RATIO f/6 OR LARGER

The first step is to make or buy a 'dummy eyepiece'. This is really no more than a plug which fits into the telescope drawtube and which has a small hole drilled exactly centre in the top of it. The best size for the hole is about 2 mm. This plug is inserted into the drawtube, replacing the eyepiece, and the function of the small hole is to ensure that your eye is steered onto the axis of the drawtube. When the collimation is successfully completed this axis will also coincide with the optical axis of the telescope.

You could make the dummy eyepiece from an old 35 mm film canister but do make sure that the hole you drill in its base is exactly centred. Alternatively if you have an old high-power eyepiece that is no longer used you could remove its lenses and use that.

We will begin by assuming that the axis of the drawtube is exactly perpendicular to the side wall of the telescope tube. It certainly should be if the telescope has been commercially manufactured. How to check for this and correct any error is covered later.

Start by pointing the telescope at the daytime sky or at a light col-oured wall, or an illuminated screen, wall, or curtained window if you are working at night. The secondary mirror mounting should have some provision for adjustments that will allow it to rotate and to move laterally up and down the axis of the telescope tube. Use these adjustments until you see, when looking through the dummy eyepiece, that the outer edge of the secondary mirror appears concentric with the inner edge of the bottom of the drawtube. The view you see should look rather like that illustrated in Figure A1(a).

If you rack out the drawtube this will make the secondary mirror appear to nearly fill your view of the bottom of the drawtube and so will help you to be more precise in your judgement.

Once you have successfully got the secondary mirror looking concen-tric with the drawtube you can make any fine adjustments to the tilt and the rotation of the secondary mirror cell until you see the reflection of the primary mirror appearing concentric within the inner edge of the secondary mirror (see Figure A1(b)). As before you can rack the drawtube in or out in order to get the reflection of the primary mirror nearly filling the secondary mirror. Even small inaccuracies will then easily show up.

All that is left is to adjust the tilt of the primary mirror cell until the reflection of the secondary mirror is nicely centred within it (as in Figure A1(c)). I always find the secondary support vanes of help here. Any slight error in the tilt of the primary mirror makes the reflections of the vanes very obviously unequal, as seen through the dummy eyepiece.

You could visually check through everything again, refining your adjustments if necessary. Your telescope is now adequately collimated.

COLLIMATING A NEWTONIAN REFLECTOR OF FOCAL RATIO LESS THAN f/6

The lower the focal ratio of the primary mirror, the more critical is the collimation of the telescope. So, if we are to get the best results, we should use a refined technique to achieve the best possible collimation.

Added to the foregoing, there is another slight complication we encounter with low-focal-ratio Newtonian reflectors. We can collimate our telescope with the secondary mirror concentric, as before. However, the area of the field of view where the image is unvignetted (fully illuminated by all the rays arriving from the primary mirror) will be a little offset from the centre of the field of view.

To counter this, the secondary mirror ought to be offset a little away from the eyepiece and an equal distance towards the primary mirror. However, with focal ratios of more than f/3.5 for a 300 mm aperture, more than f/4 for a 400 mm aperture, and more than f/5 for a 500 mm

Figure A1. Collimating a Newtonian reflector.
(a) The view through the 'dummy eyepiece' after adjusting the position of the secondary mirror, making the visible edge of the secondary mirror concentric with the edge of the drawtube. (b) The view after fine-adjusting the tilt and rotation of the secondary mirror, to make the reflection of the outer edge of the primary mirror concentric within it.
(c) The view after adjusting the tilt of the primary mirror, to make the reflection of the secondary mirror-mount concentric.

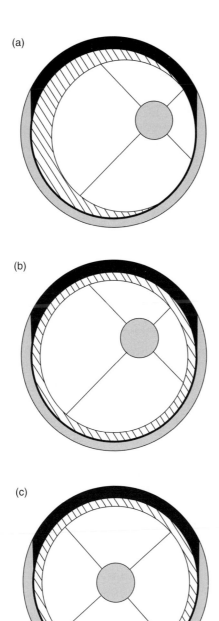

aperture, the adjustments will be less than 7 mm in each direction. The fall-off in image brightness due to the vignetting will be quite small, especially if the secondary mirror is large enough to fully illuminate a patch at least a couple of centimetres across. In that case, if you have any provision for making the offset adjustment (or if you are building your own telescope, or modifying it yourself to minimise outfield vignetting) I would say do not bother offsetting.

However, if you are using a commercial large reflector of low focal ratio then the manufacturer may well have set the secondary mirror off-centre within the telescope tube, leaving you with no choice. You must ensure that the secondary mirror is also offset the correct distance down the axis of the telescope tube in order that the axial rays from the centre of the primary mirror are turned through exactly 90° and passed exactly along the axis of the drawtube.

Of course, you will realise that the rays can still be turned through 90° with the secondary offset away from the eyepiece but not equally offset down the telescope tube. This simply depends on having the secondary mirror set at exactly 45° to the optical axis of the primary mirror. **However, the normal collimation procedure will produce an error in that case: the mirrors will <u>appear</u> collimated to you but only when the secondary is actually tilted by less than 45°. The primary mirror will also be incorrectly tilted and the end result will be that the optical axis will pass at an angle through the drawtube.** Hence the need for the dual offset.

So, carefully measure the position of the secondary mirror to see if it is centred in the telescope tube, or if it is offset and by how much. If it is offset, then you must ensure that the mirror is offset by an equal distance down the telescope tube.

How do you do this? There are various ways. In my view the best method is by direct measurement. Measure the minor-axis diameter of the secondary mirror (so you can determine where its centre is), the diameter of the bottom end of the drawtube and the distance this is from the sky end of the telescope tube (rack the focuser inwards so you can make the measurement). From this you can determine the distance of the exact centre of the axis of the drawtube from the top of the telescope tube. Putting a straight edge across the top of the telescope tube you can measure down to the top edge of the secondary mirror. When the mirror is correctly set at 45° its centre will be this distance plus the minor-axis radius from the top of the telescope tube.

If you want the secondary mirror centred with the axis of the drawtube simply make the distances from the top of the telescope tube to the centre of the secondary mirror, and to the centre of the drawtube, exactly the same. If you want, say, a 4 mm offset then move the secondary mirror a further 4 mm down towards the primary mirror. **Remember, the secondary mirror's offset towards the primary mirror must be equal to the measured offset in the direction away from the eyepiece**.

Having sorted out the correct positioning of the secondary mirror we can now make the, hopefully very small, adjustments to bring the telescope into proper collimation. We have a choice. We can, in daylight, run through the same procedure as outlined in the last section and then take our telescope out under the stars to check for any remaining inaccuracy based on what we see through the eyepiece; then we make the very fine adjustments that may be needed to finish the job. The second option is to first run through the same procedure as before and then, still in daylight or using an illuminated screen or wall, apply another piece of kit to bring the collimation to a higher degree of accuracy.

The first option may sound straightforward but making mechanical adjustments to the telescope at night is tricky. Also you want to be observing, rather than fiddling with the telescope at this time (unless, of course, it is a portable telescope set up at the beginning of your observing session in which case you have no choice). In the next section I cover star-testing a Cassegrain telescope and refining the alignments this way. You can follow the same procedure for fine-tuning your Newtonian telescope.

The second option is usually much preferable but does require you to have placed a mark on the primary mirror (something you may feel nervous about doing – but you may be lucky and have a mirror the manufacturer has centre-marked for you). Here we go with the second option . . .

Carefully measure the exact centre of your telescope's primary mirror and mark a small spot at this point. It has to be correctly placed to ± 1 mm (albeit you are relying on the manufacturer getting the optical centre of the mirror at this same point). A spirit marker will suffice to make the mark. **Do please keep your fingers, palms, sleeves and cuffs away from the mirror surface. As well as the obvious hazards of scratches and dust, the oils, moisture and salt from your skin will badly affect the reflective coating**.

While you are taking on this invasive procedure you might as well check to see that the primary mirror is properly centred within the telescope tube to better than ± 2 mm. If it is not and there are no provisions for adjustment then all you can do is to bear this fact in mind when deciding about any offsets for the secondary mirror. Improve the centring of the primary mirror if there is any way of doing it. Perhaps there are provided radial adjusting screws or bolts? You might consider using shims but do not squeeze the glass of the mirror or you will see the star images distorted as a result.

Next go through the collimation procedure as described in the last section. You can be fairly quick about it this time and no dummy eyepiece will be needed for this rough set-up. Then insert a carefully made 'sighting tube'. This is a tube about 18 cm long which fits snugly into the drawtube. One end is closed with an accurately made dummy eyepiece. The other end (this end is inserted into the drawtube) is open but is crossed with wires. The wires are mutually at right angles and cross at the exact centre of the tube. Sighting tubes can be purchased commercially. If you make your own, you must ensure that the sight hole and the crosswires are accurately positioned.

Using the sighting tube, make any fine adjustments to the secondary mirror's tilt and rotation so that the refection of the spot on the primary mirror appears exactly under the intersection of the crosswires.

For the next adjustment you may need to use a lamp to throw some light into the telescope tube so the crosswires are illuminated enough to be seen via reflection in the telescope mirrors. Fine adjust the tilt of the primary mirror until the reflection of the crosswires you see exactly coincides with the ones at the end of the sighting tube (and so the intersection of the reflected crosswires also coincides with the centre-spot on the primary mirror). Finally check through the procedure again. If you were to star-test your telescope now any viewed errors remaining would likely be the fault of the optical manufacturer.

COLLIMATING A CASSEGRAIN REFLECTOR

As before, it is a good idea to check that both the primary and the secondary mirrors are properly centred within the telescope tube. It is true that both could be off-centre, even by different amounts, as long as both mirrors can be tilted enough to bring their optical axes into coincidence. However, in that case the focuser drawtube would also have to be titled by an amount and direction to suit. If the two mirrors are properly concentric to start with the drawtube can then be perpendicular to the base-plate and all will be well (assuming it is also mounted properly centred on the base-plate). Cassegrain telescopes are often provided with adjusters for this purpose. If so, do your best to get all the components concentric with the telescope tube to ± 1 mm.

Next point the telescope at a light coloured wall and insert a sighting tube into the drawtube. Looking through it adjust the tilt of the secondary mirror until the reflection of the primary mirror in it appears exactly centred. Then adjust the primary mirror until the reflection of the secondary mirror in it is also exactly centred.

If the secondary support vanes are four in number and are equally spaced (as will normally be the case) then you can use the crosswires on the sighting tube to achieve a little extra precision. Simply rotate the sighting tube until the crosswires are brought into near coincidence with the secondary support vanes. By tinkering up the adjustments you should be able to get the crosswires exactly to coincide with the reflection of the support vanes.

Now your telescope will be very close to its optimal collimation. Nonetheless, a Cassegrain reflector is rather more finicky than most other types of telescope and it is as well to be prepared for some **very** slight further adjustments when you actually try it out under a clear sky.

Here is how to go about the final refinements to the collimation. Plug in an eyepiece that will deliver a magnification of several hundred after pointing the telescope at a test star. The star ought to have as high an altitude as possible so that it is not too badly affected by atmospheric turbulence. If the telescope drive is rather erratic (as many are) you should choose the star Polaris for this test.

Once the test star is centred defocus it slightly while watching for any asymmetry. If the expanding disk of light becomes oval and is less bright in one direction, the telescope is still very slightly misaligned. You should be aware that this test assumes the optics are of excellent quality. If the mirrors are even slightly astigmatic then the images will display a distortion which will be rather hard to tell from misalignment.

Now comes the really tricky bit – made easier if you have a willing assistant. With the telescope still pointed at the test star adjust the tilt of the secondary mirror until the star moves a little in the direction that the out-of-focus image is at its faintest and most distended. This will be an adjustment so slight that tightening the one screw without slackening the other two may even suffice, depending on the robustness of the mounting of the secondary. Next adjust the tilt of the primary mirror by just enough to bring the star image back into the centre of the field of view. For this to work the telescope must not have been joggled out of its alignment with the test star while the adjustment to the secondary was made – an incredibly difficult thing to achieve in practice.

If all has gone well the slightly out-of-focus star disk will now appear rather more circular and evenly illuminated. Continue this tricky procedure until you achieve the most circular and evenly illuminated star disk you can manage. Check that it remains so for all positions of the focuser.

COLLIMATING A REFRACTOR

If the manufacturer has not provided any adjustments for the 'squaring on' of the object glass then it is still worth checking the alignment but you will either have to put up with any misalignments or return the instrument to the manufacturer. If any adjustments are provided then the procedure is rather similar to that for the fine tuning of the Cassegrain reflector.

Select a good 'test star' as before. Centre it in the field of a high-power eyepiece. Slightly defocus it and watch for any expanding asymmetry in the expanding disk of light. Try to ignore the different colours you will see as the disk of light expands. This is quite normal behaviour for a refractor's object glass. All you are interested in detecting is any non-circularity and uneven distribution of light. Alter the tilt of the object glass until the re-centred star image looks as symmetrical as possible both a little inside and a little outside the best focus position.

COLLIMATING MAKSUTOV AND SCHMIDT–CASSEGRAIN TELESCOPES

Fortunately the modern ones usually come with instructions on how to collimate them. They do vary in design. Sometimes only one of the optical components are adjustable. The classical Maksutov has an aluminised spot on the centre of the inside surface of the correcting plate. The Rumak version of the Maksutov has a separately mounted secondary mirror between the corrector plate and the primary mirror, the mount of which is often bonded to the centre of the corrector plate. This latter arrangement is also the norm for the Schmidt–Cassegrain telescope.

My advice is to always follow the manufacturer's instructions. If you have acquired an instrument without instructions then do the following: Insert the dummy eyepiece and alter the tilt of whichever component happens to be adjustable until all the reflections appear concentric. If both the primary mirror and the secondary mirrors are adjustable then treat the unit in the same was as for a Cassegrain telescope. To finish, fine-tune as I have previously described for the Cassegrain reflector.

COLLIMATING A SCHMIDT–NEWTONIAN TELESCOPE

Meade's line of Schmidt–Newtonian telescopes come with full instructions on monitoring and adjusting the collimation. In brief, the procedure is the same as for a Newtonian telescope except that the secondary mirror is not readily accessible to you for adjustment. The *Meade* instruments come complete with a factory-made collimation spot at the centre of the primary mirror. The only optical adjustments you should normally make are to adjust the tilt of the primary mirror. It is a good idea to star test and, if necessary, fine-adjust the collimation in the field.

COLLIMATING A MAKSUTOV–NEWTONIAN TELESCOPE

Normally these telescopes have factory-fixed corrector plates with the Newtonian-style secondary mirror in its holder fitted to the centre of it.

So, only the primary mirror is adjustable. Treat the collimation as you would for the latter stages of a classical Newtonian telescope.

'SQUARING ON' THE EYEPIECE FOCUSING MOUNT

If the drawtube is not aligned to the optical axis of the telescope, the focal planes of the eyepiece and the telescope will be tilted with respect to each other. The action of focusing the telescope can then only achieve coincidence of these planes along one line. Star images will be in sharp focus anywhere along this line but will become increasingly out-of-focus with increasing distance from it.

With Cassegrain, Schmidt–Cassegrain and Maksutov telescopes draw-tube alignment is easily checked using a sighting tube. Look through it and check that the crosswires appear centred against the secondary mirror in each case. For checking a refractor make a cap for the object glass with a small hole drilled exactly central. Do the intersection of the crosswires coincide with this hole?

Mechanically checking a Newtonian telescope's focuser is more difficult. It involves removing the secondary mirror and cell and marking a small spot on the far wall of the telescope tube **exactly** opposite the centre of the drawtube. The position of this spot can only be determined by very careful measurement. With the sighting tube inserted, do the crosswires line up exactly with this spot?

Focuser–drawtube misalignment is perhaps best checked by star testing. Firstly, centre the test star and focus it as carefully as possible. Try watching as you rack the focuser from fully in to fully out. As the disk of light changes its size does it also appear to shift across the field of view? If so, try to determine (by waggling the end of the drawtube if necessary) whether this is merely 'slop' in the sliding fit or if there is a progressive lateral shift of the image which can only be due to misalignment.

I should say here that commercial Maksutov and Schmidt–Cassegrain telescopes achieve focusing by moving the primary mirror slightly along the optical axis. You will experience a cyclical to-and-from motion of the image as you change focus due to the inevitable slight mechanical imperfections of the mechanism. It is also difficult to do anything about any drawtube misalignment if you do find any.

Investigate further by moving the telescope in order to place the star at different positions in the field of view. Do you notice any change in the appearance of the star as it is moved across the field of view? Do bear in mind that it will be normal to see some image degradation radially away from the centre of the field of view. This will be especially evident close to its edge. However, is this degradation symmetrical about the centre of the field of view?

Cassegrain telescopes (now rarely sold to amateurs but second-hand ones do come on the market) often come with provision for adjusting the focuser alignment. Most other telescopes do not, so you will either have to put up with the defect or resort to packing the mounting of the focuser with shims in order to achieve proper alignment.

COLLIMATING A STAR-DIAGONAL

At one time scorned by many, star-diagonals have now become almost mandatory with today's marketed refractors, Maksutovs, and Schmidt–Cassegrain telescopes. The best examples of these units have some provision for adjusting the tilt of the mirror/prism.

The test for proper alignment is simple but should only be made after the telescope is properly collimated (including drawtube alignment). Centre a star in a high-power eyepiece (better still if it has fitted cross-wires) and carefully focus. Check again that it really is properly centred. Replace the eyepiece with the star-diagonal and put in the same eyepiece. Refocus. Does the star still appear centred? If not adjust the tilt of the mirror/prism until swapping between diagonal plus eyepiece and eyepiece alone produces no apparent image shift.

OTHER AIDS TO COLLIMATION

You do not really need anything more than I have described in the foregoing notes. However, there are some devices to help you further in collimating your telescope that you can purchase if you really want to.

For instance there is the *Cheshire Eyepiece* and the *Autocollimation Eyepiece*. Their principles of operation are too involved for me to describe here but you can rest assured that their manufacturers always provide full instructions with them. If you want to know more about them, can I point you to an article in the March 1988 issue of *Sky & Telescope* magazine.

The *Laser Collimator* is a very popular device these days. It is inserted into the drawtube and produces a thin beam of laser light which is passed through the optical system of the telescope. The idea is that the secondary mirror is adjusted until the spot of laser light falls exactly onto the centre of the primary mirror (on the centre-spot if there is one). The primary mirror tilt is then adjusted until the laser light passes back into the device, where some of it is passed through a partially reflective mirror and appears on a 'target' which you view from the side. Supposedly this means that the laser beam has passed back exactly along its outgoing path and so ensures that the telescope is properly collimated.

Unfortunately there are hazards. One is the slight but very real physical hazard of catching a dose of laser light in your eye. The second is that a drawtube misalignment, or a secondary-mirror misalignment, can produce a situation whereby you adjust until the spot of light hits the target but the beam has **not** actually passed back along its outward path. You will then have actually set your telescope with a definite misalignment between the axes of the secondary and primary mirrors! I would say, get your telescope as nearly collimated as you can using the procedures I describe in the foregoing notes and limit your use of the laser collimator to the final fine-tuning.

A better approach to using a laser collimator with a Newtonian reflector is described by Nils Olof Carlin in the January 2003 issue of *Sky & Telescope*. The first stages are carried out as previously described. His innovation is for fine-adjusting the tilt of the primary mirror. For this final stage he suggests inserting the laser into a Barlow lens and plugging this into the telescope drawtube. The bottom of the Barlow lens is closed by a disk with a small central hole in it (to allow the diverging laser light to emerge into the telescope). Instead of a central spot on the primary mirror is a small central ring (a self-sticky 'reinforcement ring' used normally for sheets of paper to go into a ring-file might do very well).

When the telescope is correctly collimated a silhouette, in laser light, of this small ring appears concentric with the hole in the disk at the bottom of the Barlow lens (if it is too far up inside the focuser for you to see it use a small mirror hand-held inside the mouth of the telescope tube). The clever thing about Carlin's innovation is that the broadness of the cone of light makes the system insensitive to secondary-mirror offset and drawtube misalignment, while still being very sensitive to primary-mirror misalignment. Hence the situation of an inaccurately offset secondary mirror is not made far worse by introducing a tilt to the primary mirror in order to bring the laser spot to target as it would in the usual way of using the device.

Laser collimators are a real boon for the final fine-tuning of Cassegrain, Schmidt–Cassegrain and Maksutov telescopes. One problem that might arise is slop in the telescope drawtube or a loose fitting of the collimator in the drawtube. Either may cause the laser beam to go a little off line, particularly so when tightening any fixing screws.

APPENDIX 2: FIELD-TESTING A TELESCOPE'S OPTICS

The majority of today's amateur astronomers do not make their own telescope optics. Nonetheless most of us are aware of the basic procedures involved and are familiar with the basic principles of the Foucault and Ronchi grating optical tests. I would, however, wager that very few telescope users realise that simple versions of these tests can be used to evaluate the overall accuracy of both figuring and alignment of the assembled telescope's optics with much greater ease than the toiling mirror maker can assess the individual components in his/her workshop.

The easiest method involves using a Ronchi grating. Many optical firms, especially those dealing in telescope optics and telescope-making materials, sell these gratings. They are very cheap to buy. If a choice is given select a grating of at least four lines per millimetre (100 lines per inch), though more lines per millimetre will provide a more sensitive test.

Simply mount a small piece of the grating over the central hole in a 'dummy eyepiece' (a collimation tool described in Appendix 1) and you have all you need to make the evaluation.

Actually I mounted my own grating in place of the eyelens of a cheap discarded eyepiece – all the lenses first being removed from it. I store the unit in a container, to keep it clean, and so it is always ready for use at a moment's notice.

Using this device could not be easier. Simply set the telescope on a fairly bright star in the normal way. Centre it accurately in the field of view. Then replace the normal eyepiece with the grating device. Peering through the grating you will see the primary mirror/objective of the telescope flooded with light from the star but crossed with bark stripes. These stripes are effectively the highly magnified image of part of the grating.

You will find that rotating the grating produces a consequent rotation of the pattern. Adjusting the focuser causes a dramatic change in the apparent magnification of the pattern and so a change in the apparent broadness of the dark bands and the number of them crossing your view of the telescope objective/primary mirror. The closer to the telescope's focal plane the grating is, the smaller the number of stripes visible.

Figure A2. Ronchi patterns for a telescope field tested on a star. (a) pattern that would be seen for good optics in accurate collimation. (b) pattern that would be obtained for a spherically overcorrected system. For instance, a pattern like this might result from the primary mirror of a reflecting telescope being ground too deep at its centre, or perhaps the change is temporary and is caused by the mirror cooling rapidly, in which case the error will slowly lessen as the mirror approaches thermal equilibrium. (c) This pattern is produced by a spherically undercorrected system (in this case if the mirror is cooler than the ambient air temperature, that fact ought to be evident because it will soon be covered in dew!) (d) This pattern is produced by the 'turned edge', a very common manufacturing fault in telescope primary mirrors. (e) This pattern results from a telescope afflicted by zonal errors. Obviously an almost infinite number of variations of this are possible. (f) This pattern illustrates astigmatism but a similar result would be generated from misalignment of the optics.

For this test I recommend adjusting the focuser until four dark stripes are seen to cross the image, the grating being just intrafocal (inside the focal plane – in other words, with the focuser rack inwards such that further inwards adjustment increases the number of stripes visible).

Assuming the grating is not faulty or dirty, the image of the stripes you see should be straight and like that shown in Figure A2(a). Grains of dust trapped in the grating will show up as a jaggedness along the edges of the stripes. Any faults in the figuring of the optics or their alignment will be immediately apparent as shown in the other views in Figure A2 and described in the caption accompanying it. Rotating the grating through 180° in several steps will allow all of the optics' radial zones to be evaluated.

The beauty of this test is that the complete optical system of the telescope is tested in field conditions – and this is surely what really counts. Also, this test is applicable to **all** telescopes. Even better, this is a *null test*. In other words, the stripes appear straight and regular when all is well. This is because the star is at infinity and the grating is then close to the principle focal plane, unlike the much more complicated situation for the mirror maker who works with the light source and the grating both close to the centre of curvature and who has to assess curved stripes.

Mounting a razor blade to half cover the hole in a dummy eyepiece enables a version of the Foucault test to be carried out. Set up as in the same way as for the Ronchi test. With your eye close to the hole and peering past the edge of the razor blade you will see the telescope objective/primary mirror flooded with light from the test star. If you move the telescope **very** slightly so that the blade begins to cut off the light you will see a black shadow sweep across the pool of light.

Adjust the focuser until moving the telescope causes the shadow's edge to become more blurred. Keep adjusting the focuser until moving the telescope causes the whole of the mirror to darken evenly. In other words, at the correct position you will not be able to decide whether the shadow sweeps across the image from the right or from the left when you move the telescope.

The razor blade is now coincident with the focal plane of the telescope. Any figuring errors or misalignment of the optics are hugely magnified and are thrown into a 'pseudo three-dimensional relief' and become very obvious at the position where the pool of light is half extinguished. If all is well with the telescope the disk of light will appear perfectly smooth and flat and an even shade of grey. However, you will also see, superimposed on the ideal image, a moving set of swirls and corrugations caused by atmospheric turbulence and the convection of air over the telescope's optical surfaces. You will have to do your best to see past these moving waves and just judge the static underlying image.

The knife-edge test performed in the field using a star as the light source has the same advantage of ease of interpretation at the Ronchi

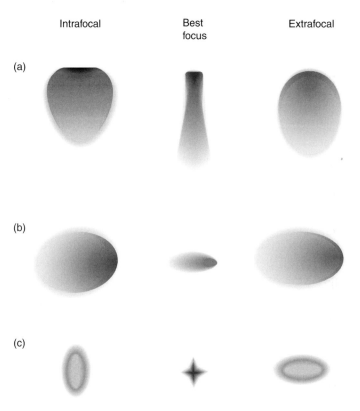

| Intrafocal | Best focus | Extrafocal |

(a)

(b)

(c)

Figure A3. Depictions of the telescopic appearances of a star at high magnification, for different focal positions, with each of the following problems: (a) tube currents; (b) misalignment of the optics (note that the orientation of the cometic image does not change either side of the best focus); (c) astigmatism.

test, over that made in the mirror maker's workshop. The Foucault test is potentially more sensitive than the Ronchi test but it is also a lot more tricky to perform. Most people will prefer to use the Ronchi test which is much more forgiving of less-than-firmly mounted telescopes.

Finally, one can inspect the focused, intrafocal and extrafocal (the eyepiece racked outwards from the best-focus position) star images when a high-powered eyepiece is plugged into the telescope. Choose a star of medium brightness as seen through the instrument, and one which is as high in the sky as possible so that the effects of atmospheric turbulence are minimised.

The precise analysis of the appearance of the intrafocal, focal and extrafocal star images is a complex business. Also, this technique is of most use for telescopes not much bigger than 200 mm aperture because in larger ones the effects of atmospheric turbulence would overwhelm the effects of all but the most serious errors on most nights. However, Figure A3 shows the star test appearances for the effects of astigmatism, misalignment and tube currents as these are the most common troubles and are by far the easiest to recognise.

APPENDIX 3: POLAR ALIGNMENT

The first thing to say is that if you are using an equatorial mount that comes with manufacturers instructions, then refer to those. For instance 'GOTO' mounts will have their own elaborate set-up procedures that should be followed for the best results. The notes I give here will be useful in the absence of instructions, particularly for the simpler models of equatorial mounting.

A good telescope mounting will have provisions for making fine adjustments to the elevation and azimuth (east–west pointing) of the polar axis. Sometimes they also have polar-alignment telescopes built into the polar axes. The eyepieces of these are fitted with a graticule of specially marked lines, and/or circles, and divisions. Using these and a chart of the area around the celestial pole (or following other supplied instructions) one can achieve a polar alignment to within a fraction of a degree accuracy.

How accurately does your telescope mount have to be aligned? The answer to that depends on what you intend doing with it. Simple visual observation is made all the easier if the polar axis is aligned within a couple of degrees of the true pole. Adjustments made in response to a squint along the polar axis may well be good enough, especially if the mount is fitted with a declination slow motion. Of course, observing is made more convenient, and hence more pleasurable, if the alignment is better than that. Photography involving exposures of a small fraction of a second is also uncritical of polar alignment. Longer exposures (rare for lunar imaging but common for deep-sky photography) demand much higher precision. In addition, if you are going to set an equatorial mount permanently into position you will surely want to bother to get the polar alignment of your telescope as true as possible. In those cases you could follow the following procedure.

Once again begin with your best go at a rough alignment (squinting along the polar axis, or using a compass and setting the polar elevation scale, if there is one, to your latitude, etc.). Then the idea is to make small corrections until the apparent north–south drift of the image of a star that is being tracked over a period of time is reduced to zero. This is done with the telescope pointing in at least two different directions.

First, set the telescope on a star that is close to the meridian (due south if you live in the northern hemisphere) and the celestial equator (i.e. with a declination close to zero degrees). Then lock the declination axis. With the telescope drive engaged monitor the apparent north–south drift of the star. Ignore any east–west drift. If you are at all unsure as to the orientation of the image, momentarily move the telescope in declination a very small amount in the direction of Polaris. The direction which the star appears to move in the eyepiece defines **south** in the field of view. Move the telescope to re-centre the star once more and continue.

If the telescope has no right-ascension (sidereal) drive then simply move the telescope every few minutes to bring the star back into the field of view. Does it come back to the **centre** of the field of view?

If the star appears to drift **southwards** over a period of time the polar axis is pointing a little **east** of the true celestial pole. An opposite error will produce an opposite direction of drift. Correct as necessary and repeat until the drift is as small as you wish to make it. I recommend you use a cross wire eyepiece, or a graticule eyepiece to help you be precise.

Remember, though, to ignore east–west drift. It is only the north–south drift of the star's image that tells you of the azimuth misalignment of the polar axis.

The other adjustment – the altitude of the polar axis – is more difficult to achieve accurately by this method. For this you should select a star that is about 6^h (that is about 90°) east of the meridian and is preferably within 20° of the celestial equator. In other words the star will be ideally due east and rather low in the sky. Centre the telescope on it and follow the same procedure as before. This time a star-image drift to the **south** indicates that the elevation of the polar axis is too **low**. As before the opposite drift indicates the opposite error. Adjust as necessary.

The fact that the star is low over the eastern horizon means it will be affected by refraction (causing it to appear higher in the sky than it really is), the effect of which will decrease rapidly as the star rises. Neither can you select a star at 6^h east of the meridian and at zero degrees declination (the theoretical ideal) as this star would be right on the horizon and virtually unobservable even if you had an uncommonly clear horizon from your observing site. The consequence is that a small error in the elevation of the polar axis will remain even if you reduce the apparent star drift to zero.

If you have the time you could set the telescope on a star about 6^h west of the meridian and repeat the procedure. This time the required corrections are opposite to before (a star-image drift to the south now means the polar axis is pointing too high). If you find any apparent error this time, after you had thought you had corrected it before, then adjust the polar axis altitude to a compromise setting.

If you are reading this in the Earth's southern hemisphere then please reverse all the foregoing directions and corrections.

Once you are sure that the polar alignment of your telescope is correct then, and only then, should you consider fine-adjusting the tracking rate of its right-ascension (sidereal) drive if there happens to be any provision for adjustment. This is because a polar alignment error will also produce east–west drifts of varying amounts dependant on where the telescope is pointing.

INDEX